NITROGEN FIXATION

Volume 2 *Rhizobium*

NITROGEN FIXATION

Volume 2 *Rhizobium*

Edited by

W. J. BROUGHTON

Max-Planck-Institut für Züchtungsforschung, Cologne

CLARENDON PRESS · OXFORD

1982

Oxford University Press, Walton Street, Oxford OX2 6DP

LONDON   GLASGOW   NEW YORK   TORONTO
DELHI   BOMBAY   CALCUTTA   MADRAS   KARACHI
KUALA LUMPUR   SINGAPORE   HONG KONG   TOKYO
NAIROBI   DAR ES SALAAM   CAPE TOWN
AND ASSOCIATE COMPANIES IN
BEIRUT   BERLIN   IBADAN   MEXICO CITY

Published in the United States by Oxford University Press, New York
© *Oxford University Press, 1982*

British Library Cataloguing in Publication Data

Nitrogen fixation.
   Vol. 2: Rhizobium
   1. Nitrogen—Fixation
   I. Broughton, W.J.
   581.1'33   QR89.7

ISBN 0-19-854552-5

Phototypesetting by Oxford Publishing Services
Printed in Great Britain
at the University Press, Oxford
by Eric Buckley,
Printer to the University

Preface

Proceedings of several conferences that have included rhizobial aspects of nitrogen fixation have recently been published. Good soil microbiology texts are also available, but not single volume has been entirely devoted to *Rhizobium*. This is an important omission since the legume–*Rhizobium* symbiosis is assuming ever increasing importance in parallel with better management of our natural resources. Furthermore, *Rhizobium*s close relationship to *Agrobacterium* (they are the only two genera in the bacterial family Rhizobizceae) is exciting molecular biologists. Along with many others, pioneers of plant-microorganism research hope that the plant transformation systems now being developed with *Agrobacterium* will also be applicable to *Rhizobium*. Thus, this volume should be a timely and extremely useful addition to the literature on nitrogen fixation.

Functionally, the book is organized so that it forms both an introduction to the field and a critical review of knowledge in the main disciplines. In Chapter 1 Gloria Lim and Joe Burton answer the question 'Which legumes are nodulated?', 'Van' Bushby discusses rhizobial distribution in the soil and factors affecting it, while Mike Trinick has provided a comprehensive account of rhizobial biology. As the biggest difference between rhizobia—that between fast- and slow-growing organisms—basically concerns varying abilities to metabolize carbohydrates, Gerry Elkan and David Kuykendall have closely examined metabolism. John Beringer, Nick Brewin, and Andy Johnston reviewed the most rapidly expanding field—the genetics of the microsymbiont, and Tsien Hsien-Chyang described the cells as seen under the microscope. Three chapters deal directly or indirectly with the outer structure of the cells—Russel Carlson with their chemistry, Jim Vincent with their immunological reactions, and Frank Dazzo and David Hubbell with the implications of various rhizobial and legume components in regulation of the interaction. Finally, Eltjo Meijer correlated observations on nodule development with known biochemical events. In short, this treatise is a thorough examination of both the freeliving *Rhizobium* and its relationship to legumes.

Preparation of this volume was largely dependent of the fifteen authors and I am exceedingly grateful to them for their expert contributions. Others whose help was essential included Fräulein Antonia Maria Schafer and Professor Fritz Lenz of the University of Bonn and Fräulein Elisabeth Schölzel and Professor Jeff Schell of the Max-Planck-Institut für Züchtungsforschung in Köln. It has been my privilege to co-ordinate the efforts of so many knowledgable and pleasant people.

Cologne   WJB

Contents

List of Contributors

J. E. Beringer,
Microbiology Department,
Rothamstead Experimental Station,
Harpenden,
Herts. AL5 2JQ,
UK.

N. J. Brewin,
Department of Genetics,
John Innes Institute,
Colney Lane,
Norwich NOR 7OF,
UK.

J. C. Burton,
The Nitrogen Co. Inc.,
3101 West Custer Avenue,
Milwaukee,
Wisconsin 53209,
USA.

H. V. A. Bushby,
CSIRO,
Division of Tropical Crops and Pastures,
St. Lucia,
Queensland 4067,
Australia.

R. W. Carlson,
Department of Chemistry,
Eastern Illinois University,
Charleston,
Illinois 61920,
USA.

F. B. Dazzo,
Department of Microbiology and Public Health,
Michigan State University,
East Lansing,
Michigan 48824,
USA.

G. H. Elkan,
Department of Microbiology,
Institute of Biological Sciences,
North Carolina State University,
Raleigh,
North Carolina,
USA.

D. H. Hubbell,
Department of Soil Science,
University of Florida,
Gainsville,
Florida 32611,
USA.

A. W. B. Johnston,
Department of Genetics,
John Innes Institute,
Colney Lane,
Norwich NOR 7OF,
UK.

L. D. Kuykendall,
Cell Culture and Nitrogen Fixation Laboratory,
United States Department of Agriculture, Research Center,
Beltsville,
Maryland 20705,
USA.

G. Lim,
Botany Department,
University of Singapore,
Bukit Timah Road,
Singapore.

E. G. M. Meijer,
Max-Planck Institut für Zuchtungsforschung,

D-5000 Köln 30,
West Germany

M. J. Trinick,
CSIRO,
Division of Land Resources Management,
Wembley,
Western Australia 6014.

H.-C. Tsien,
Department of Microbiology,
University of Minnesota,
Minneapolis,
Minnesota 55455,
USA.

J. M. Vincent,
Microbiology Department,
University of Sydney,
Sydney,
New South Wales,
Australia 2006.

1 Nodulation status of the Leguminosae

G. LIM AND J. C. BURTON

1.1 Introduction

The legumes form one of the largest families of flowering plants, ranking third in terms of world-wide occurrence, with about 600 genera and 18 000 species, although estimates of actual numbers vary considerably (from 590 to 700 genera and 12 000 to 18 000 species according to different authors; Hutchinson 1964; Airy Shaw 1966; Keng 1970; Heywood 1971; Anonymous 1979; Allen and Allen 1981).

The oldest agricultural records available indicate that leguminous crops have been cultured for centuries and that they were valued for food and soil enrichment long before their ability to work symbiotically with bacteria was understood.

Fossils of Leguminosae have been traced back to the Cretaceous or last division of the Secondary or Mesozoic era, 95–120 million years ago (Fred, Baldwin, and McCoy 1932). However, no reference was made to the presence of nodules on these ancient legumes. It has been suggested by Fred *et al.* (1932) that a re-examination of these fossils for the occurrence of nodules should be made to determine the antiquity of the *Rhizobium*–leguminous plant association.

Our knowledge of the beneficial association of rhizobia with leguminous plants in utilizing atmospheric nitrogen (N_2) is very recent when compared to the very long period that leguminous crops have been cultured and valued for food and soil enrichment. According to Fred *et al.* (1932), credit for the first published picture of nodules should be given to Fuchs who portrayed nodules on the roots of *Aphaca, Vicia faba,* and *Trigonella foenum-graecum* in the first edition of *Historia stirpium commentarii insignes* in 1542. No mention was made in this book of the nodules, which were apparently considered to be normal plant structures rather than hypertrophies induced by bacteria. It was not until 1886 that Hellriegel (1886) revealed the true origin and function of the nodule which enabled the leguminous host plant to utilize atmospheric nitrogen. Two years later the bacteria (rhizobia) were isolated from the nodule and cultured on laboratory media by a Dutch microbiologist (Beijerinck 1888).

In the tropics, leguminous plants are prevalent and constitute one of the largest groups of flora. Norris (1956) suggested that the legumes originated in the tropics; in fact tropical trees comprise a large part of the family which is said to be basically a tropical arborescent family (Tutin 1958).

In the humid tropical Malaysian peninsula and Singapore, the legumes form the fourth largest family with 66 genera and 266 species (Keng 1970), most of them woody species. According to Whitmore (1972) however, there are 70 genera and 270 species in the lowland and mountain forests of Malaya of which 53 species are trees reaching at least 1 m girth. He also considered the legumes to be amongst the loftiest trees in Malayan forests, especially the species of *Koompassia* and *Intsia*, and he regarded the family as only second to the Dipterocarpaceae in abundance among the emergent trees of the lowland rain forest. Among this emergent layer, appreciable numbers of legumes occur, mainly of the genera *Dialium, Koompassia,* and *Sindora* (Whitmore 1975). In fact, in the Malayan peninsula, legumes form the second most important timber-producing family after the Dipterocarps, *Intsia palembanica* being the most important individual species, while *Pterocarpus indicus* is said to outshine teak in natural beauty (Whitmore 1972).

In the monsoon forests of the Far East, in the Malesian region, the major dominant trees include a large number of legumes, ten species out of 26 typical species belong to the Leguminosae family, while the rest belong to various other families (Whitmore 1975). These ten species are two each of *Acacia* and *Albizia,* and one species of each of *Butea, Caesalpinia, Cassia, Dalbergia, Dichrostachys,* and *Tamarindus.*

Among the common cultivated plants in Malaysia and Singapore are legumes such as the various beans, ornamentals such as *Cassia, Caesalpinia,* and *Mimosa* spp, and park, roadside, and garden trees and shrubs such as *Acacia* spp, Flame of the forest (*Delonix regia*), raintree (*Samanea saman*), Angsana (*Pterocarpus indicus*), Bauhinias, and tamarind.

In India, the Leguminosae are considered to be the second most dominant family in order of abundance. Some genera occurring in the forests are *Ougeinia, Mastersia,* and *Wagatea.* In the delta forests of the gangetic plains, *Cynometra ramiflora* are found, and in the sandal forests of south India legumes occur including *Dalbergia latifolia, Albizia* spp, *Pterocarpus marsupium,* and *Pongamia glabra.* Other species found in various parts of India include *Indigofera* spp, *Sesbania aculatea, Aeschynomene* spp, *Alysicarpus vaginalis, Acacia* spp, and *Rhynchosia minima* (Puri 1960).

Many legumes are also cultivated for food and feed, for example, *Cicer, Crotalaria, Phaseolus, Tephrosia,* and *Vigna* spp, as well as many other grain and pulse legumes.

A large number of indigenous legume species occur in Africa and many more have been introduced and cultivated. As a result, legumes are fairly well-represented among the flora, although some areas are richer in legume species than others. In West Africa for instance, legumes are well-represented in forests with species of *Piptadeniastrum, Albizia,* and *Tetrapleura* occurring, and some rainforests are dominated by *Cynometra ananta* (Lawson 1966). *Pericopsis* is abundant in western Ashanti but rare elsewhere; *Afzelia africana* and *Parkia clappertoniana* are said to be the most common trees in the Guinea savannas (Lawson 1966); and in the subtropical regions, *Acacia* spp abound.

In the East African region, *Cynometra* spp are abundant in Tanzania and Western Uganda, and may constitute a dominant part of the lowland forests, forming 70–80 per cent of single layered canopy (Lind and Morrison 1974). Genera such as *Brachystegia, Isoberlinia,* and *Julbernardia* are dominant over wide areas of woodland in tropical Africa. In northern Nigeria, *Parkia filicoidea* is a very common tree of park savanna (Purseglove 1968).

The neotropics are also rich in legume flora with a variety of growth forms. Over 50 per cent of the trees in some forest associations in British Guiana were found to consist of legume species (Davies and Richards 1934). Jenny (1950) reported on an estimate of 50 per cent legume trees on forest soils examined in Colombia. In fact central and south America are said to be promising areas for finding species useful for improvement of soil nitrogen and for grazing (Williams 1967).

In temperate regions legumes are less predominant among the flora, and in terms of number of genera and species, fewer legumes occur there than in tropical areas. A good account of the distribution of legumes has been given by Norris (1956). Examples of legumes which are found in temperate regions are those of tribe Vicieae such as species of *Vicia,* e.g. *V. faba, Pisum* spp, and *Lathyrus* spp such as *L. odoratus* (sweet pea); members of tribe Loteae such as *Lotus* spp and *Anthyllis* spp; and Trifolieae genera such as *Trifolium, Medicago,* and *Melilotus.*

The Leguminosae is generally subdivided into four subfamilies— Mimosoideae, Caesalpinioideae, Swartzioideae, and Papilionoideae— the latter being the largest and Swartzioideae the smallest. Tutin (1958) and Allen and Allen (1961) have given excellent accounts of the descriptions, classification, size, and distribution of members of this family, and the latter reference additionally reviewed the scope of nodulation in the Leguminosae up to 1959.

Tropical legumes represent all four subfamilies, members comprising very few herbaceous species, mainly being shrubs, woody climbers, and trees, some of which are very large trees in rainforest areas. Although the humid tropics are regarded as the ancestral home of legumes, and indigenous species abound, the leguminous flora of many tropical countries contains a fairly large proportion of introduced species. These latter species were brought into a country for agricultural, economic, or ornamental purposes. Whatever the reasons, many such plants thrive well and have become established as part of the country's present-day flora.

The subfamily Mimosoideae is normally regarded as the most primitive of the four subfamilies, in terms of evolutionary development, although Hutchinson (1964) considered the Caesalpinioideae to be the most primitive. Members of the subfamily with the exception of *Acacia*, are said to occur exclusively in the tropics and most abundantly in rainforests and semi-arid subtropics (Allen and Allen 1961). Many species of *Acacia* occur in temperate and alpine regions of Australia.

The subfamily Caesalpinioideae is also regarded as being confined to the warmer parts of the world. Tropical species are largely woody shrubs, trees, and climbers. Temperate ones consist of only three genera, *Gleditsia*, *Cercis*, and *Ceratonia*.

Swartzioideae, with only 9–10 genera and about 100–150 tropical species, is native to South America and Africa, and is regarded as intermediate between Papilionoideae and Caesalpinioideae in taxonomic position (Corner 1951). All species are woody.

Papilionoideae is the only subfamily with equal distribution in tropical and temperate countries, though many more genera are tropical. The tropical species are predominantly woody, large trees, or big, woody climbers; the temperate ones are predominantly shrubs or herbs.

1.2 Rhizobiaceae

The Rhizobiaceae family of bacteria consists of only two genera: *Rhizobium* and *Agrobacterium*. In discussing the genus *Rhizobium*, it is appropriate to begin with nodulation, because the ability to incite cortical hypertrophies or nodules on leguminous roots is the one criterion currently accepted in distinguishing rhizobia from other bacteria (Buchanan and Gibbons 1974). As Vincent (1977) points out, however, the genus *Rhizobium* also includes bacteria which have lost their invasive properties, as long as there is authentic proof of clonal descent from a culture capable of inducing nodules to form on leguminous plants.

In contrast to *Rhizobium* the other genus, *Agrobacterium*, has the ability to incite hypertrophies (galls) on the roots and stems of diverse plant species. These hypertrophies may be mistaken for nodules, but they are tumorous galls consisting of unorganized tissues or structures and are usually harmful to their host. Only one species, *Agrobacterium radiobacter*, does not produce galls. The latter, because of similarity in morphology, is sometimes mistaken for fast-growing rhizobia which have lost their ability to produce nodules. However, these bacteria can be differentiated easily from rhizobia by cultural methods (Graham and Parker 1964; Vincent 1977).

Rhizobium *species*

Early studies of nodulation were concerned primarily with readily accessible leguminous plants, cultured in temperate climates mainly in the United States and Europe, as sources of food. Observations that all leguminous crops did not respond similarly to inoculation with *Rhizobium* cultures led to a grouping of plant hosts according to their nodulation responses. Leguminous plants nodulated by the same nodule bacteria constituted a group. Rhizobia isolated from nodules on a plant in the group usually produced nodules on other plants in the same group. These plant groups were called 'cross-inoculation' groups. Rhizobia able to nodulate plants in one of these groups were considered a *Rhizobium* species. Only six species of *Rhizobium* have been named, but other groups of plant genera are mutually susceptible to nodulation by a common *Rhizobium* (Table 1.1). The rhizobia without species names were identified with their parent host as cowpea, lotus, crownvetch, and other kinds of rhizobia.

TABLE 1.1
Groups of leguminous plants (genera)
nodulated by a single species or kind of Rhizobium

Rhizobium species or kind	Legume genera
1. *Rhizobium meliloti*	*Medicago, Melilotus, Trigonella* spp
2. *Rhizobium trifolii*	*Trifolium* spp
3. *Rhizobium leguminosarum*	*Lathyrus, Lens, Pisum, Vicia* spp
4. *Rhizobium phaseoli*	*Phaseolus coccineus, P. augustimus, P. vulgaris*
5. *Rhizobium japonicum*	*Glycine* spp
6. *Rhizobium lupini*	*Lupinus* and *Ornithopus* spp, *Lotus* (certain species)
7. *Rhizobium* spp (cowpea)	*Acacia, Alysicarpus, Andira, Apios, Arachis, Baptisia, Cajanus, Cassia, Canavalia, Crotalaria, Cyamopsis, Cytisus, Desmodium, Dolichos, Erythrina, Indigofera, Lespedeza, Phaseolus, Macroptilium, Pueraria, Stylosanthes, Vigna,* and others
8. *Rhizobium* spp (lotus)	*Lotus, Anthyllis, Dorycnium*
9. *Rhizobium* spp (crownvetch)	*Coronilla, Onobrychis, Petalostemum, Leucaena, Dalea* spp

The rhizobia which induce nodules on certain genera of leguminous plants are currently considered to differ from any of those listed in Table 1.1. These are *Amorpha, Amphicarpaea, Astragalus, Caragana, Cicer, Laburnum, Robinia, Sesbania, Strophostyles,* and *Wistaria.* In some genera, not all plant species are nodulated by the same cultures of rhizobia. *Vicia floridana* and *Vicia acutifolia* are examples of this (Carroll 1934); they are not nodulated by *R. leguminosarum.* Several species of *Phaseolus: P. lunatus, P. acutifolius, P. wrightii, P. ritensis, P. adenanthus,* and *P. heterophyllus* are nodulated by cowpea rhizobia whereas *P. vulgaris, P. coccineus,* and *P. augustifolia* are nodulated by *Rhizobium phaseoli.* There is a similar situation in the genus *Lupinus.* New groups will undoubtedly be formed and new additions to old groups will be made as our knowledge of the nodulating characteristics of leguminous species expands. So far, studies have involved only about 10 per cent of the total leguminous species.

The cross-inoculation concept, based on mutual susceptibility of plants within a group to nodulation by a common *Rhizobium,* has not proven acceptable because of the numerous irregularities. None the less, the grouping has been very useful in searching for good nitrogen-fixing *Rhizobium* strains. Leguminous plants can derive benefit only from rhizobia capable of nodulating their roots. Groupings based on nodulation permits focusing on good prospects when searching for highly effective *Rhizobium* strains, regardless of bacterial taxonomy.

The grouping of leguminous plants which tend to give effective nitrogen-fixing response to the same strains of rhizobia, appears to be far more useful than groupings based entirely on susceptibility to nodulation (Burton 1979). *Lotus* species are all nodulated by the same strains of rhizobia, but *Lotus* species fall into two groups when nitrogen fixation is considered (Fig. 1.1). Strains of rhizobia effective on birdsfoot trefoil, *Lotus corniculatus,* are not effective on deervetch, *Lotus pedunculatus,* and vice versa. All of the *Lotus* species studied so far can be classed either as the birdsfoot trefoil or big trefoil types.

Certain strains of cowpea rhizobia work effectively on many genera, species, and varieties of legumes. In contrast, the clover cross-inoculation group which embodies only *Trifolium* species, requires different strains of rhizobia for many of its species (Burton 1979).

Numerical taxonomy of rhizobia based on the Adasonian principle, DNA-base composition, and homology, is being studied in an attempt to clarify classification of *Rhizobiaceae.* This is discussed later in this volume (Chapter 3).

Nodules

The nodule is the focal point of reaction between the *Rhizobium* and its host plant. When a susceptible, leguminous plant and a compatible

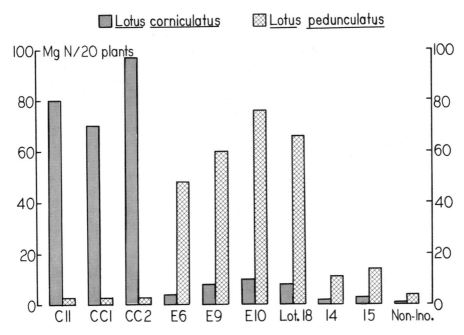

Fig. 1.1. Nitrogen fixation by two *Lotus* species. *Rhizobium* strains C11, CC1, and CC2 were isolated from and are effective on *Lotus corniculatus*. Strains E6, E9, E10, and Lot. 18 were isolated from and are effective on *Lotus pedunculatus*. All *Rhizobium* strains produced nodules on their parent host and on the other test species of *Lotus*.

strain of rhizobia are brought together under conditions favourable for growth and infection, a nodule will form. The surface chemistry of the *Rhizobium* and of the root hair, partly determines the compatibility of the two symbionts and whether or not a nodule will form. The intricate chemistry and mechanics of the rhizobia and of the root hair, are covered in Chapters 7 and 9 of this volume.

Not all leguminous plants form nodules. Up to 1981, it was thought that 90 per cent of 388 plants examined in the subfamily Mimosoideae were nodulated; 98 per cent of 2462 species in the subfamily Papilionoideae bore nodules; and only 28 per cent of 258 species examined in the subfamily Caesalpinioideae were nodulated (Allen and Allen 1961). For nodulation to occur, the host plant must be susceptible and a compatible infective strain of *Rhizobium*, capable of multiplying, must be present on the root. It is not known whether lack of nodule development in some leguminous species is due to lack of appropriate rhizobia, or to structural problems in the legume root system. The possibilities are discussed in detail by Funk (1956). Susceptibility to nodulation may be inherited in soyabeans (Williams and Lynch 1954), in peanuts (Gorbet and Burton 1979), and possibly in many other legumes.

Leguminous nodules may differ in many ways depending on the host plant, the strain of *Rhizobium*, and environmental factors. Histologically, nodules are classed as *'exogenous'* when they arise from infection of the parenchymatous cells of the root cortex exclusively. Most nodules are of this type. *'Endogenous'* nodules arise from infection and proliferation of pericylic cells. The peanut, *Arachis hypogaea*, and American jointvetch, *Aeschynomene americana*, are typical of the endogenous nodule (Allen and Allen 1954).

The legume host is apparently the dominant factor governing gross morphology of the nodule. Corby (1971) studied nodulation of more than 400 species of wild legumes indigenous to Zimbabwe, and related nodule shape to the tribal classification of the host. None the less, plant species of the same genus may have very different sizes and shapes of nodules. Effective nitrogen-fixing nodules may be elongate (clover), globose (soyabean), semi-globose and endogenous (peanut), or elongate coralloid (guar) (Fig. 1.2). Effective nodules may also appear as a collar surrounding the main root (Dart 1977). Effective nodules on four tropical legumes are shown in Fig. 1.4: *Acacia albida* and *Robinia pseudoacacia* are trees, and bear nodules perenially.

The leguminous plant with nodules on its roots induced by an effective strain of *Rhizobium*, is considered to be in symbiosis with the bacteria. The symbionts are considered to be mutually beneficial, but this relationship may shift with age. Allen and Allen (1954) suggest that in the early stages of nodule development, the bacteria are the dominant factor. As the nodule matures structurally, there is a period of balance when each symbiont is supplying and being supplied by the other. This is a period of mutual benefit; nitrogen is being supplied to the plant by the bacteria, and carbohydrates are being supplied to the bacteria for growth and nitrogen fixation. When the plant matures and begins fruiting, it becomes the dominant controlling factor and eventually initiates senescence in the nodule. The length of these stages will vary with different leguminous plants and environmental conditions (see Volume 3, Chapters 3 and 4).

This type of nodulation which results in nitrogen fixation is called 'effective' nodulation. The nodules tend to be large, concentrated on the upper root system, and usually have reddish interiors from the haemoglobin, which characterizes an effective nodule (Fig. 1.3). When nodules develop, but fix little or no nitrogen they are called 'ineffective'. These nodules are usually small, numerous, and widely scattered throughout the root system. Ineffective nodules may sometimes appear to be moderately large (Viands, Vance, Heichel, and Barnes 1979).

In describing nodule function, the terms effective and efficient are

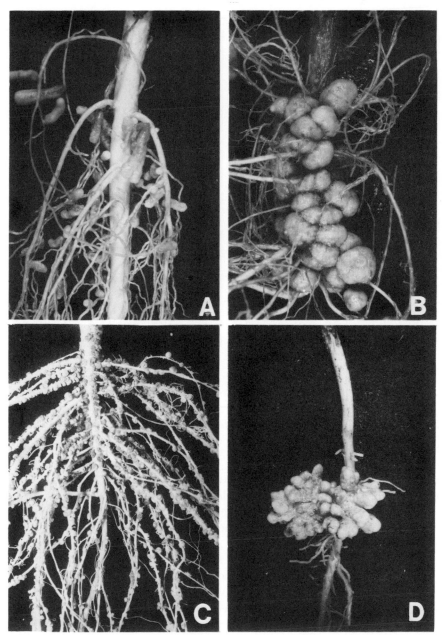

Fig. 1.2. Nitrogen-fixing nodules. (A) Elongate nodules on *Trifolium amabile*. (B) Globose nodules on *Glycine max* L. Merr. (C) Endogenous semi-globose nodules on *Arachis hypogaea*. (D) Elongate coralloid nodules on *Cyamopsis tetragonoloba*.

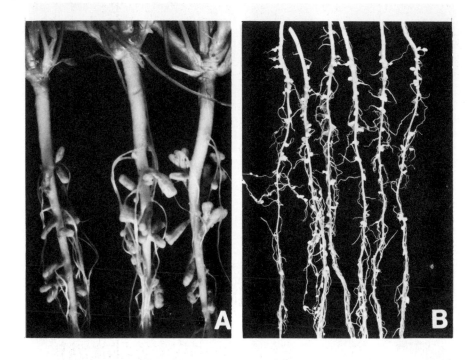

Fig. 1.3. Effective and ineffective nodules as determined by leguminous species. (A) Effective elongate nodules on white clover, *Trifolium repens*. (B) Ineffective nodules on subclover, *Trifolium subterraneum*, produced by a strain of rhizobia isolated from and effective on white clover.

often used interchangeably for nodules which fix considerable amounts of nitrogen. Likewise, the terms ineffective and inefficient are often used interchangeably for nodules which fix little or no nitrogen. In both cases, the terms are not synonymous and should not be used inter-changeably.

Vest, Weber, and Sloger (1973) suggest that the term ineffective be used to describe cortical proliferations on the root which remain small and have white or green centres, but which provide no nitrogen to their host. Vest *et al.* (1973) subdivided effective nodulation into two classes: inefficient and efficient. 'Inefficient' effective nodules may be moderately large, have green, white, or sometimes pink centres, but fix little or no nitrogen. Plants with this type of nodules remain chlorotic. In contrast, 'efficient' effective nodules are large, have pink or red centres and provide plenty of nitrogen. Plants are dark green and healthy. While there is some merit to this system, double terms cause confusion.

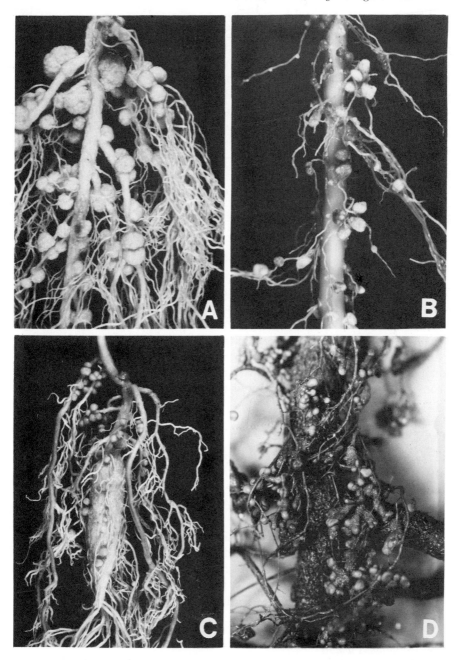

Fig. 1.4. Effective nodulation of tropical legumes of great interest. (A) Winged bean, *Psophocarpus tetragonolobus*. (B) *Acacia albida*. (C) Mexican yam bean, *Pachyrrhizus erosus*. (D) Black locust, *Robinia pseudoacacia*.

The term efficiency has recently been used in another way by Schubert and Evans (1976) and Schubert (1977). In studying the energy required by different legumes inoculated with various strains of rhizobia, it was learned that certain *Rhizobium*–legume combinations bring about hydrogen evolution and loss of energy. In studying this phenomenon, it was found that only 40–60 per cent of the electrons transferred to nitrogenase were used in fixing nitrogen; the remainder were used in producing hydrogen which was of no benefit to the host plant. Other *Rhizobium*–legume associations were able to recycle the hydrogen and utilize it in fixing nitrogen. Schubert and Evans (1976) suggest the term, 'relative efficiency', to define a leguminous *Rhizobium*–plant association's ability to utilize energy. The following formula was suggested for its derivation.

$$\text{Relative efficiency} = 1 - \frac{H_2 \text{ evolution in air}}{C_2H_2 \text{ reduction}}$$

This concept has merit because it connotes production or nitrogen-fixation with a minimum of waste. In studies involving eight leguminous species, Gibson (1978) failed to note any positive correlation between relative efficiency and nitrogen fixation.

The term inefficient is very descriptive of nodules produced by rhizobia having good nitrogen-fixing capacity, but which cannot function properly because of lack of a nutrient such as molybdenum, cobalt, or iron. Nodules under severe water or temperature stress will also function very inefficiently.

Pseudo-nodules

Hypertrophies or structures on leguminous roots induced by auxins, auxin-like substances, and kinetin may be mistaken for nodules. These bulbous swellings may encircle the root and form a collar at the root juncture (Allen and Allen 1950; Allen, Allen, and Newman 1953). These structures show no evidence of infection by micro-organisms at any stage of growth. They lack organization and tissue differentiation, and are called false or pseudo-nodules. They do not fix nitrogen or benefit plant growth. The occurrence of starch grains indicates that they may sometimes be used for storage. Chemically induced pseudo-nodules may also occur on non-leguminous plants (Nirmal, Skoog, and Allen 1959).

Non-leguminous nodules

About 170 species in 15 genera of non-leguminous plants: *Alnus, Casuarina, Ceanothus, Cercocarpus, Colletia, Comptonia, Coriaria, Discaria, Dryas, Eleagnus, Hippophae, Myrica, Purshia, Rubus,* and *Shepherdia* bear nitrogen-fixing nodules incited by Actinomycetes (Torrey 1978). These micro-organisms differ greatly from rhizobia and likewise the nodules they produce differ from legume nodules. The differences and similarities are described in detail by Allen and Allen (1954). The leguminous nodule is highly organized and has four distinct areas: a nodule cortex, a vascular system, a bacteroid area, and a meristematic region. Non-leguminous nodules appear as modified roots with little differentiation of areas.

Numerous attempts have been made over the past century to isolate the endophyte which induces nodulation of non-leguminous plants, but Callaham, Del Tredici, and Torrey made the first successful attempt in 1978. Using a yeast-extract medium, an Actinomycete or filamentous bacterium was isolated from a nodule on *Comptonia peregrina* L. This organism, when cultured in the laboratory and used as an inoculum for *C. peregrina*, produced nodules on sand-grown seedlings in repeated tests. Further, these nodules were very active in reducing acetylene to ethylene using the standard procedures used in studying legume nodules.

More recently, Berry and Torrey (1979) isolated the nodule endophyte of *Alnus rubra* Bong. This micro-organism, designated 'AR13', nodulates other *Alnus* species as well as *Myrica* and *Comptonia* species.

The only instance of nodulation of a non-legume by rhizobia has been reported by Trinick (1973). *Trema aspera*, later identified as *T. cannabina*, is nodulated by a *Rhizobium* of the cowpea type. Further studies (Ackermans, Abdulkadir, and Trinick 1978) led to the conclusion that this plant really belongs to the Ulmaceae and is in fact *Parasponia cannabina*. To date, all other non-leguminous species nodulated by rhizobia are considered to be Ulmaceae.

1.3 Nodulation status

Nodule occurrence

This chapter on the nodulation of legumes attempts to update the information, especially of tropical legumes. Many authors have commented on the limited knowledge of nodulation of tropical legumes (Banados and Fernandez 1954; Bowen 1956; Masefield 1958; Lange 1959; De Souza 1966).

Among the earlier contributors who have provided information regarding nodulation of tropical legumes are Allen and Allen (1936, 1947), Martin (1948), Allen and Baldwin (1954), Banados and Fernandez (1954), Bowen (1956), Masefield (1958) and Lange (1959). Later surveys included the list of nodulated legumes in South Africa (Grobbelaar, Van Beyma, and Synnove Saubert 1964; Grobbelar, Van

Beyma, and Todd 1967) showing the occurrence of nodules in 44 legume species, comprising several species of *Acacia, Albizia,* and *Elephantorrhiza* of the subfamily Mimosoideae and a number of species belonging to Lotoideae genera of Papilionoideae, and 246 legume species (comprising 17 of Mimosoideae, 12 of Caesalpinioideae, and 217 of Papilionoideae respectively) De Souza (1966) studied the indigenous legumes of Trinidad, and out of 79 species studied, 32 were previously unlisted; legumes distributed among several genera were found to bear nodules, including the first reported observations of nodulation in three species of *Swartzia.* Norris (1969) studied rainforest legumes in Amazonia and Guyana and recorded nodules on 30 out of 53 tree species examined. Of these 30 species from the four subfamilies, 14 belonging to nine genera were first time records, of which three species were *Swartzia* members.

More recently, other workers have made contributions to our knowledge on legume nodulation in the tropics. Dubey, Woodbury, and Rodriguez (1972) surveyed the island of Puerto Rico for indigenous, nodulated legumes and reported 13 previously unrecorded species out of the 52 species which have nodules. These 13 species belonged to various genera, including only one species (*Cassia swartzii*) from Caesalpinioideae, while the other 12 were papilionaceous members. In a qualitative study of nodulating ability, Grobbelaar and Clarke (1972) examined 215 legume species indigenous to Southern Africa, most of which were papilionaceous; 207 species had not been examined previously for nodules. In a later paper, Grobbelaar and Clarke (1974) added further to the growing number of legume species examined; out of the 213 species studied, 76 had not been examined previously. Twenty-one species were observed to be consistently without nodules, 11 of which were caesalpinioid members, while the other 10 were papilionaceous. Out of these 10 species, six had been reported previously by other workers to have nodules. Allen and Allen (1976) give a profile on nodulation in the genus *Cassia.* Broughton and John (1979) listed some Malaysian legume species from which rhizobia isolates were obtained, as well as legume species that form effective nodules with rhizobia. In a survey of the rhizobia population of legumes in Thailand, Kampee and Lohtong (1979) also gave a list of 52 legume species which have nodules. These 52 species consisted of 36 herbaceous plants, nine trees, five shrubs and two woody climbers. Lim and Ng (1977) reported on the nodulation of legumes in Singapore, and found 33 species out of 35 to have nodules, the nodulated species belonging largely to Mimosoideae and Papilionoideae subfamilies, with only one species (*Delonix regia*) of the Caesalpinioideae, and constituting the first reported observation of nodules in this species. They also reported, for the first time, the occurrence of nodules on *Adenanthera pavonina*

(Mimosoideae). In another report on legumes in Singapore Lim (1977), examined 68 species and noted nodulation in 31 species, distributed amongst the different subfamilies. This report also contained the first recorded observation of nodules in *Calliandra inaequilatera* (Mimosoideae). The nodulating ability of legumes in subtropical Pakistan was studied by Athar and Mahmood (1978) who reported 52 species of Papilionoideae to have nodules, five of which were new records. Further work by Athar and Mahmood (personal communication, 1979) on 39 legume species showed that all except three species of *Cassia* had nodules. Their study also confirmed the occurrence of nodules in *A. pavonina* and *D. regia*.

Thus over the last 20 years, investigators from several countries have continued to survey and examine tropical legumes for nodule occurrence. However, the data collected so far is still not sufficiently comprehensive, and a large number of species in all four subfamilies, especially the rare ones, have yet to be collected and studied for their nodulating habit.

Nodulation in legumes associated with crops and pastures

It has long been recognized that nodulated legumes contribute to soil fertility, and early reports exist on the use of legume species as green manures in connection with the growing of tea, cacao, rubber, coconuts, and rice. In Sri Lanka, for example, the species *Crotalaria, Phaseolus vulgaris, Albizia falcata (= A. moluccana), Erythrina,* and *Mimosa pudica* were used as 'nitrogen collectors'. Nodules were noted on these plants and described, and a list of legumes possessing nodules was given by Wright (1903, 1905). The practice of using green manures extended to other tropical countries, such as India, Malaysia, Indonesia, Fiji, and Mauritius.

Obviously by experimenting, planters selected and used those species that confer definite benefits to their crops. The species of legumes used were indigenous or introduced species.

In Malaysia, Corbet (1935) noted that the legume *Mimosa pudica,* a common weed at roadsides and neglected corners of gardens, bears nodules. He also observed that species of legumes used as cover crops on rubber plantations, such as *Crotalaria anagyroides, Dolichos lablab, Centrosema pubescens,* and *C. plumieri* possess nodules. Beeley (1938), also working in Malaysia, noted that legume cover-plants were used on old rubber estates for regeneration of the soil and further stated that agriculturally, the value of leguminous plants is three-fold; (i) as food and fodder crops, (ii) as fertilizers of soil, and (iii) as cover plants. *Centrosema pubescens, Crotalaria, Desmodium, Pueraria, Tephrosia,* and *Vigna* spp were reported to be used as cover plants.

Some of the legumes used in India as green manures are as follows:

Tephrosia purpurea, Crotalaria juncea, and *Sesbania aculeata* in rice fields in some southern provinces, *Dolichos lablab* in rotation with rice, and *T. candida* and *C. juncea* in the central provinces (Anon 1936). Other examples are *Albizia chinensis* and *A. odoratissima* used as shade plants for tea and coffee, and *Centrosema plumieri,* used as a cover crop (Anon 1936). Many other species are cultivated as pulse crops for food, for example, *Phaseolus* spp, *Vigna unguiculata* (cowpea), *Cicer arietinum* and *Cajanus cajan;* for oil seeds, such as *Arachis hypogaea,* and for other useful products, such as hemp, *Crotalaria juncea* for fibre, and *Indigofera tinctoria* for dye.

In the African tropics, legumes have been utilized as cover crops, as green manures, and in crop rotation. Some of the species used are *Glycine max, Vigna sinensis, Phaseolus aureus,* and *Crotalaria juncea.* Others grown as food or feed include *Vigna* spp, *Cajanus cajan, Lens esculenta, Voandezeia subterranea, Arachis hypogaea,* and *Phaseolus* spp (Cobley 1976).

In tropical America, legumes have been used as green manure, cover crops and in crop rotation. In Mexico *Canavalia ensiformis, Crotalaria juncea,* and *Stizolobium deeringianum* have been used as green manure, with maize as a subsequent crop. Other species used as forage crops in the neotropics are *Centrosema pubescens, Glycine javanica, Leucaena glauca, Lotononis bainesii, Phaseolus atropurpureus, Pueraria phaseoloides,* and *Stylosanthes gracilis.*

It can be seen, therefore, that much of the early knowledge on nodulation of tropical legumes was confined to species of the few genera which are useful in agriculture, and species that are grown for special purposes. A gap thus exists in our knowledge which, in part, may be attributed to the inaccessibility of specimens, many of which are to be found in areas which are not easily reached by collectors. Some genera and species are confined to certain parts of the world. Moreover, woody tree genera particularly the larger ones, are by no means easy to examine for the presence or absence of nodules. An additional factor contributing to the problems in obtaining a more comprehensive body of information is the difficulty in obtaining , or the unavailability of seeds of a number of tropical species, since seeds may often be required for confirmation of nodulating ability. Thus a complete survey of nodulation among tropical legumes is not always feasible and may not be achieved easily.

Nodulating ability

The nodulating ability of legume species varies tremendously. This ability has no correlation with whether the species is indigenous or otherwise (Schofield 1941; Martin 1948; Davies 1951; Bowen 1956; Bonnier 1957; Masefield 1958; Corby 1967, 1974; Lim 1977; Lim and Ng 1977). In fact some introduced non-native species develop and form abundant root nodules, while some indigenous species do not bear nodules. There are, however, exceptions where introduced species require the presence of the appropriate rhizobia from the country of host origin in order to form nodules, as shown by Norris (1958a) for *Lotononis bainesii* and Lie (1980) for *Pisum sativum*.

The variation in nodulating ability of legumes has led to a number of conflicting reports. Some species form root nodules at the seedling or young tree stage, but lack nodules as mature trees. This applies particularly to the large, woody species, e.g., *Delonix regia*, which bears nodules only in the seedling state (Lim and Ng 1977). Therefore, data presented by different investigators may appear contradictory, depending on the age of the plant from which the specimens were collected. It is quite probable then, that a number of species reported to be lacking in nodules may well have had nodules if they had been examined at the seedling state.

Alternatively, as pointed out by Norris (1956), who cited Hannon (1949), cyclical nodulation is common in tropical legumes and the 'wet' season might be the time of maximum nodule occurrence.

Furthermore, physical factors such as light intensity, high soil temperature, aeration of the soil, soil moisture content, and soil pH may affect nodulation (see Lie 1980). Bonnier (1957) noted that more nodules occur on various legumes in the light forests of the Congo than in the denser forests. He also noted that nodules develop particularly on *Gilbertiodendron* growing on well-aerated soils. In Malaysia, Masefield (1955, 1957) has shown that moisture content of the soil had the greatest effect on nodulation. Earlier, Phillips (1939) also observed that with soyabeans in Trinidad, the higher the water content of the soil the greater the nodulation that occurred.

Besides such physical factors, chemical factors may also play some part in influencing nodulation. Some of these factors are the presence of phosphates, which are said to increase nodulation (Bumpus 1957), lime, which increases nodulation on *Pueraria phaseoloides* (Samuels and Landrau 1952), and organic matter which results in heavier nodulation in *Pueraria phaseoloides (= P. javanica)* in the Congo (Bonnier 1957). Some chemical substances inhibit nodule formation, such as that demonstrated by Bowen (1961). However, Norris (1958b, 1965) in reviewing the studies of various workers on the effect of lime on nodulation of tropical legumes, concluded that the picture for tropical legumes is different from that for temperate legumes; tropical legumes in fact gave a negative response to liming.

Nutritional factors may also affect nodulation, especially the carbohydrate : nitrogen ratio (Hallsworth 1958). Perhaps in tropical forests where the nitrogen is not depleted, nodulation occurs infrequently (Bonnier and Seeger 1958).

Microbiological factors such as presence of other micro-organisms, bacteria, fungi (Lim 1961), bacteriophages and so on may influence nodulation. The number of nodule bacteria present in the soil or root zone of the plant also affect nodulation (Purchase and Nutman 1957).

Undoubtedly, another factor that has to be considered is the genetic make-up of the host plant, which determines the susceptibility or otherwise of the plant species and variety to infection of its roots by the appropriate rhizobia. Genetic lines of clover resistant to infection are well-known. Is there a parallel in tropical legumes? Observations indicate that within a species some plants are very much less susceptible than others to infection by rhizobia and do not nodulate, whereas a neighbouring plant in the same plot may bear abundant nodules. Such genetic and field differences were noted by Masefield (1958).

The presence or absence of the appropriate rhizobia is a factor of prime importance in determining nodulation. The cowpea type rhizobia which nodulates many tropical legumes is culturally slow growing, symbiotically promiscuous, and represents the ancestral type of rhizobia common to tropical legumes (Norris 1956). As a general rule therefore, many legumes in the tropics nodulate without the benefit of inoculation.

Recent studies (Lim and Ng 1977) showed that many isolates from tropical legumes in Malaysia and Singapore were fast growing, and there were as many slow-growing 'cowpea' types as fast-growing types.

Variation in ineffectiveness of *Rhizobium* strains also exists. Lack of nodulation can also be due to root secretions of some legume species (Elkan 1961). Bonnier (1954) found that the roots of some legumes inhibit nodulation by some strains of *Rhizobium*, but not by others. Root diffusates from weeds and non-leguminous plants are also said to reduce nodulation (Rice 1971). Competition between strains of rhizobia, such as between virulent infective strains and non-infective ones, was shown by Winarno and Lie (1979) to inhibit nodule formation.

Thus numerous factors can explain the absence of nodules on a plant species. Can one guarantee that a species is non-nodulating? Perhaps within the context of certain habitats, localities, or situations, it is correct to suggest that a species is non-nodulating, but this may not hold for other conditions.

Variation in nodulation in subfamilies

Many workers have observed variation in nodulation between the subfamilies of Leguminosae. All have reported that of the four sub-families, Caesalpinioideae members tend to lack nodules. Leonard (1925) noted this and argued that there must be a fundamental factor intrinsic in these plants which inhibits nodule formation.

Allen and Allen (1936) listed plants of the Caesalpinioideae which lacked nodules. In a later paper (Allen and Allen 1947) they again commented on this and showed that 11.8 per cent of Mimosoideae, 3.3 per cent of Caesalpinioideae, and 84.9 per cent of Papilionoideae had nodules. They also tabulated nodule distribution within tribes, genera, and species, noting that in the subfamily Mimosoideae, most of the nodulated species examined belong to only three genera, *Acacia*, *Albizia*, and *Mimosa*. Most of these species are indigenous to tropical parts of the world, such as tropical Australia, New Guinea, Philippines, Africa, and South America.

In their study on *Rhizobium*–legume relationships, Allen and Baldwin (1954) stated that only 10–12 per cent of legume species have been examined for nodules, and of these, about 88 per cent are known to bear nodules. The majority of these nodulated species, about 86 per cent are members of the Papilionoideae. Of 97 species of Caesalpinioideae examined by Allen and Baldwin, about only 33 per cent possessed nodules. Members of the *Cassia*, *Bauhinia*, *Caesalpinia*, *Saraca*, *Gleditsia*, and *Gymnocladus* are the best known genera where non-nodulated species occur. More recently, Allen and Allen (1961) provided another numerical estimation which again indicated low incidence of nodulation among the Caesalpinioideae.

In surveying the legumes of the Philippines for nodulation, Banados and Fernandez (1954) found that a large proportion of members of the Caesalpinioideae did not bear nodules. Similarly, De Souza (1966) also found among indigenous Trinidad legumes, that of 18 species of Caesalpinioideae examined, only three had nodules. Grobbelaar *et al.* (1964), in their survey of indigenous and introduced legumes in Pretoria, South Africa, found that most of the species listed as not having nodules belonged to the Caesalpinioideae. In a further study Grobbelaar *et al.* (1967) found that only eight species did not nodulate, all of which belonged to the Caesalpinioideae.

Subba Rao (1972) suggests that 65 per cent of Caesalpinioideae do not possess nodules, compared to 10 per cent of Mimosoideae and 6 per cent of Papilionoideae. Grobbelaar and Clarke (1972) also noted that members of the Caesalpinioideae especially lack nodulating ability, and commented in a later paper (Grobbelaar and Clarke 1974) that lack of ability to nodulate is frequently high among species of the Caesalpinioideae.

More recently, Lim (1977) observed that of 27 species of Caesal-pinioideae examined, only four had nodules. She also reported a low percentage of nodulation among the Caesalpinioideae, only 17 per cent compared to 60 per cent in Mimosoideae and 89 per cent in Papilio-noideae (1976). A global survey of nodule occurrence on *Cassia*, the largest genus of Caesalpinioideae was reported by Allen and Allen (1976), who estimated that currently about 70 per cent of Caesal-pinioideae members lack nodules. Their comprehensive and excellent account listing the nodulating and non-nodulating species of *Cassia*, as well as species of conflicting reports, provides the most up to date information on this group of legumes.

Additional studies have also shown that nodulation among the Caesalpinioideae is much less common than that of members of the other subfamilies. Corby (1971) studied nodule shapes and root colour in the legumes and found that coloured roots were more common in Caesalpinioideae than in either Mimosoideae or Papilionoideae. He also noted that non-nodulating species tend to be more common amongst those with coloured roots than amongst those with near-white roots.

Current estimates would therefore place the nodulation status of legumes in the region of 25–30 per cent for Caesalpinioideae, 60–70 per cent for Mimosoideae and 90–95 per cent for Papilionoideae. The nodulation status of Swartzioideae cannot be estimated accurately until more genera and species have been examined.

As the observations of many workers agree that the Caesalpinioideae are largely non-nodulating, it may be inferred that they are less capable of forming root nodules than the other subfamilies. Why this should be so has intrigued many investigators. Comparatively few reports provide an explanation for this difference. Some earlier workers such as Nobbe, Schmid, Hiltner, and Hotter (1891) and McDougal (1921) believed that physical character of the root hairs in some Caesalpinioideae members inhibit nodule formation. Other intrinsic factors however may be involved, such as the differences in the root or root hair characteristics and/or, anatomical, physiological, genetical, biochemical features between non-nodulating and nodulating members of the Caesalpinioideae.

Swartzioideae

The first reported account of root nodules in this subfamily was made by De Souza (1966) in his study of the indigenous legumes of Trinidad. He examined three species of *Swartzia*, *S. simplex*, *S. pinnata*, and *S. trinitensia*, and found nodules on all of them. Norris (1969) in his observations of rainforest legume species in Amazonia and Guyana, also reported the occurrence of nodules on three other species of *Swartzia*, *S. benthamina*, *S. leiocalycina* and *S. oblanceolata*. Corby (1974) recorded nodules on *S. madagascariensis*.

In this small subfamily only seven species have been found to have nodules. All seven species belong to the genus *Swartzia*. Other genera of this subfamily have not been extensively examined for nodules and therefore the nodulation status of this subfamily cannot be definitely stated. However, since its taxonomic position is intermediate between the Caesalpinioideae and the Papilionoideae (Corner 1951; Tutin 1958), and since all members are woody species, its nodulation status may be somewhere in between that of Caesalpinioideae and Mimosoideae. Corby (1974) on the other hand, argued that *Swartzia*, because of its ability to form nodules, should be placed in the Papilionoideae. Since the nodulating ability of other genera has not been thoroughly studied [apart from two instances of absence of nodules reported in *Holocalyx balansae* (de Rothschild 1968) and in *Cordyla africana* (Corby 1974; Grobbelaar and Clarke 1975)], his suggestion should be considered only when the nodulating habit of other genera and species is known.

Caesalpinioideae

As mentioned earlier, this subfamily is known to contain many more non-nodulating members than other subfamilies. There are 152 genera and nearly 2800 species in Caesalpinioideae (Hutchinson 1964), most of them tropical and sub-tropical, and most numerous in tropical America. Surprisingly, few genera have been examined for the presence of nodules. Information on nodule occurrence in *Brachystegia*, *Isoberlinia*, and *Julbernardia* from tropical Africa would be particularly helpful. Similarly, information is lacking regarding the nodulating ability of many legumes from the Malay Peninsula including woody climbers of such genera as *Mezoneuron* and *Pterolobium*, and trees of *Amherstia*, *Crudia*, *Cynometra*, *Dialium*, *Hymenaea*, *Intsia*, *Koompassia*, *Leucostegane*, *Peltophorum*, *Sindora*, *Sympetalandra*, and *Tamarindus*. Many of these genera are represented by one or just a few species and are not readily available for collection or survey. *Crudia* trees, for instance, are rare, and although 13 species are said to occur in Malaysia, only five have ever been seen at a time (Whitmore 1972). Similarly, there are 10 species of *Dialium*, but only two have been seen. Most occur in forests and are located in remote or isolated parts of the Malayan peninsula.

Species of even familiar genera such as *Bauhinia* and *Caesalpinia* have not been completely surveyed for nodules in this part of the world. It is fairly certain that many species will never be investigated because the forests are being rapidly and indiscriminately felled. Species and even whole genera will therefore disappear soon. Those like *Crudia* and *Ormosia* (Papilionoideae) which are excessively rare, according to Whitmore (1972), are on the way to extinction. That a similar situation may exist in other tropical areas of the world is not improbable. Thus it is highly likely that exotic tropical genera, especially those single or few species genera growing in remote forests or lands, may never be examined for nodule occurrence and the nodulating profile of Caesalpinioideae cannot be fully determined in view of these missing examples.

According to Allen and Allen (1961), 40 genera and 115 species have been examined. Of these, only 26 species have nodules, 82 species lack nodules, and conflicting data surrounds seven species, all of which are species of *Cassia*. Some of the genera which have been examined and found to have nodules, albeit only represented by single nodulated species, are ten in number, while five other genera contain both nodulated and non-nodulated species (Allen and Allen 1961). Absence of nodules in 20 genera was also noted by Allen and Allen (1961), who stated that the nodulation status of species in more than 50 genera is unknown.

Since their publication in 1961, more genera and species have been studied. The additions are De Souza's (1966) study of indigenous Trinidad legumes, and Norris's (1969) observations, in which one species of *Crudia*, several of *Eperua*, two of *Tachigalia* and one of *Macrolobium* were found to be non-nodulating. Grobbelaar *et al.* (1967) found that of 12 species of Caesalpinioideae examined, eight did not have nodules, these being four species of *Cassia*, and one each of *Bauhinia*, *Gleditsia*, *Pterolobium*, and *Peltophorum*. The four nodulated species were four other *Cassia* species.

In studying 215 species of legumes of Southern Africa, Grobbelaar and Clarke (1972) listed only four species of Caesalpinioideae, of which only *Cassia capensis* had nodules, while the others, *Colophospermum mopane*, *Schotia capitata*, and *S. latifolia* lacked nodules. They further examined 213 species and found 11 out of 12 species of Caesalpinioideae to be consistently without nodules (Grobbelaar and Clarke 1974), these being one species each of *Bauhinia*, *Ceratonia*, *Cercis*, and *Parkinsonia*, five of *Cassia*, and two of *Caesalpinia*. In another study of 184 legume species a year later (Grobbelaar and Clarke 1975), six out of ten species of Caesalpinioideae were found to be consistently without nodules, these being one species each of *Burkea* and *Schotia* and two species each of *Bauhinia* and *Cassia*.

Corby (1974) provided a list of Caesalpinioideae members whose nodulating ability was studied. Species with no nodules included *Afzelia quazensis*, *Baikiaea plurijuga*, four species of *Bauhinia*, seven of *Brachystegia*, *Burkea africana*, *Caesalpinia decapetala*, 12 species of *Cassia*, *Colophospermum mopane*, *Cordyla african*, *Cryptosepalum maraviense*, *Dialium engleranum*, two species of *Guibourtia*, *Hoffmanseggia burchelii*, *Julbernardia globiflora*, *Peltophorum africanum*, *Piliostigma thonningii*, two species of *Schotia*, *Tamarindus indica*, and *Tylosema fassoglensis*. Those which possessed nodules were 14 species of *Cassia*, two of *Erythrophleum* and *Swartzia madagascariensis*.

As mentioned earlier, an exhaustive investigation of the genus *Cassia* and a compilation of a global survey were presented by Allen and Allen (1976). They estimated Caesalpinioideae members lacking nodules at 70 per cent. They also put forward the suggestion that nodulation of *Cassia* is a relatively recent physiological adaptation. Data presented by them also suggest that nodulating ability is a useful criterion for defining phylogenetic relationships.

Further additions to species examined are listed among the legumes studied by Lim (1977), where one species of *Baikiaea, B. insignis,* and one of *Intsia, I. bijuga* among several other species belonging to the more common genera such as *Bauhinia, Brownea, Caesalpinia, Cassia,* and *Peltophorum* were found to lack nodules. She reported sparse nodules on *Cassia tora,* though many previous accounts recorded this species as non-nodulating (Allen and Allen 1976). Another member of Caesalpinioideae previously reported to be non-nodulating but now regarded as having nodules is *Delonix regia* (Lim and Ng 1977). Athar and Mahmood (personal communication, 1979) confirmed the presence of nodules on *D. regia.* They also reported that of nine Caesalpinioideae members examined, *Cassia fistula, C. holosericea,* and *C. siamea* did not have nodules.

Genera whose nodulating habits are not known still number over 100 (total number of genera based on Hutchinson 1964), many of which are single species genera found in tropical America and tropical Africa. Thus about two-thirds of the genera in Caesalpinioideae have yet to be examined for nodulation. All are tropical genera, while the nodulating habit of the three temperate genera *Cercis, Ceratonia,* and *Gleditsia* are known (Allen and Allen, 1961). Some of these tropical genera are *Apuleia, Batesia, Campsiandra, Cenostigma, Discorynia, Melanoxylon, Paloue, Poeppigia, Pterogyne,* and *Recordoxylon* found in tropical America, *Bussea, Didelotia, Distemonanthus, Gilletiodendron, Hylodendron, Loesenera, Pellegriniodendron, Polystemonanthus, Scorodophloeus, Talbotiella, Tessmannia,* and *Zenkerella,* occurring in tropical Africa, *Kalappia* found in Celebes, and *Schizoscyphus* found in New Guinea.

Mimosoideae

This subfamily is considered by some to be the most primitive of all the subfamilies, although Hutchinson (1964) thought otherwise and considered the Caesalpinioideae to be more primitive. Its distribution is mainly tropical and subtropical; nearly all are woody, the majority being trees, shrubs, or climbers. Unlike the Caesalpinioideae, its members often bear nodules and the nodulation status of the subfamily is thus much higher than that of the Caesalpinioideae. According to Allen and Allen (1961) the nodulation data available cover 146 species distributed among 21 genera. This means that out of a total of 56 genera and 2800 species (Hutchinson 1964), slightly less than half the genera have been examined, and a large number of species have not been investigated. The explanation for this may lie in the fact that many of them are tropical trees found in the more remote areas, where physical access may be difficult.

Allen and Allen (1961) mentioned that species of *Adenanthera* and *Pentaclethra* lacked nodules and that an examination of 100 plants of *A. pavonina* showed absence of nodules. Another species, *A. intermedia,* was also reported to lack nodules in the Philippines (Banados and Fernandez 1954), though related species of *Parkia* were said to have abundant nodules (Allen and Allen 1961). Lim and Ng (1977), however, observed nodulation in *A. pavonina*. Confirmation of nodulation in this species was given by Athar and Mahmood (personal communication, 1979) in Pakistan.

Among the various investigators who have recorded nodulation in Mimosoideae, Grobbelaar *et al.* (1964) found healthy nodules in several species of *Acacia*, *Albizia*, and *Elephantorrhiza*. De Souza (1966), in examining indigenous legumes in Trinidad, found nodules on several species of *Acacia*, *Calliandra*, *Entada*, *Inga*, *Leucaena*, *Mimosa*, *Pentaclethra*, *Pithecellobium*, and *Zygia*. Norris (1969) found nodules on species of *Inga*, *Pentaclethra*, and *Pithecellobium*, while some species of *Acacia*, *Mimosa*, and *Parkia* were without nodules. Thus both De Souza and Norris found nodules on *Pentaclethra macroloba*, though Allen and Allen (1961) suggested that species of *Pentaclethra* lacked nodules, and another species, *P. macrophyllum*, did not have nodules in the Congo (Bonnier 1957).

Grobbelaar *et al.* (1967) examined 246 species, and found all 17 Mimosoideae species examined to have nodules. Among them were 11 species of *Acacia*, two of *Mimosa*, and one each of *Albizia*, *Desmanthus*, *Dichrostachys*, and *Elephantorrhiza*.

Another list published by Grobbelaar and Clarke (1972) showed six species of *Acacia* indigenous to South Africa, South West Africa, Botswana, Lesotho, and Swaziland as having nodules. Their studies two years later(Grobbelaar and Clarke 1974) listed 38 Mimosoideae members as having nodules. These were *Leucaena*, *Dichrostachys*, and *Prosopis*, and *Mimosa* species. A further list (Grobbelaar and Clarke 1975) showed only two species of Mimosoideae out of 24 to be without nodules; these two species being *Acacia ataxacantha* and *Entada spicata*.

Among the Rhodesian legumes studied, Corby (1974) found 62 out of 69 species of Mimosoideae to have nodules. Those with nodules included a large number of *Acacia* species, (three without nodules), 16 species of *Albizia*, four of *Dichrostachys*, four of *Elephantorrhiza*, three of *Entada*, one of *Mimosa*, and one of *Neptunia*. Besides the three *Acacia* species, those thought to be non-nondulating were two species of *Newtonia*, *Amblygoncarpus andongensis*, and *Xylia torreana*.

Lim (1977) noted the occurrence of nodules in *Calliandra inaequilatera*, *Mimosa sepiara*, *Parkia javanica*, *P. speciosa*, and *Samanea saman*. Her observations on *C. inaequilatera* is the first report on nodulation in this species. Earlier Banados and Fernandez (1954) reported lack of nodules in the species. She also listed several species of Mimosoideae lacking nodules, being distributed in several genera—*Acacia*, *Albizia*, *Adenanthera*, *Calliandra*, *Piptadenia*, and *Pithecellobium*. This agrees with the conclusion of Allen and Allen (1961) that both nodulated and non-nodulated members are found in the genera *Acacia*, *Calliandra*, and *Mimosa*. This feature may be extended to include more genera such as *Adenanthera*, *Pentaclethra*, *Pithecellobium*, and *Parkia*.

There remain a number of genera (again mainly tropical) that have not been examined for nodules. Information on these will be a useful addition to our current knowledge. Such genera number over 30 and among them are some occurring in tropical America such as *Mimozyganthus* of tribe Mimozygantheae, *Desmanthus*, and *Dinizia* of tribe Mimoseae, *Parapiptadenia* and *Stryphnodendron* of tribe Adenanthereae and *Affonsea* and *Cedrelinga* of tribe Ingeae. Some are found in tropical Africa, such as *Aubrevillea* of tribe Mimoseae, *Calpocalyx*, *Piptadeniastrum*, and *Fillaeopsis* of tribe Adenanthereae, and a few such as *Delaportea* of the tribe Adenanthereae found in the Indo-china region, *Serianthes* of tribe Ingeae which is scattered throughout Malesia and New Guinea, and *Wallaceodendron* also of tribe Ingeae found in the Celebes. Many such genera occur in remote habitats, some of which, under the present political circumstances, are hazardous and often impossible to reach. It may be a long time, if ever, before additional data on the nodulating habits of these genera is forthcoming. Nevertheless, despite this short-coming, an estimate of the nodulation status of Mimosoideae based on the number of species and genera so far examined and reported in literature would be about 60–70 per cent nodulated.

Papilionoideae

This subfamily, the largest, is heterogenous with members equally distributed in tropical and temperate zones. The more primitive, woody genera are found mostly in the tropics while the herbaceous, more advanced ones occur in temperate regions. About 482 genera and 12 000 species comprising herbs, shrubs, trees, and climbers are distributed throughout the world (Hutchinson 1964). Many species are important pulse crops in the tropics and are widely cultivated; others are grown as cover crops, green manures, and fodders in many tropical, countries, and still others provide important oil seeds, fibres, or various miscellaneous products (Purseglove 1968; Cobley 1976). In temperate regions, all the important pasture legumes are members of this sub-family, and many cultivated vegetables and food crops also belong to the Papilionoideae. Many of the members that have been examined were reported to be nodulated and data show that of 1024 species examined, 959 species distributed among 175 genera, had nodules (Allen and Allen 1961).

In the numerous, more recent publications by many investigators, the nodulated legume species observed largely belong to this subfamily. Grobbelaar *et al.* (1964), in examining indigenous and introduced legumes in South Africa, listed a large number of papilionaceous members as having nodules compared to fewer members of the other subfamilies. For instance, 37 species of Papilionoideae distributed among 16 genera had nodules, out of the 44 legume species that had nodules. The few non-nodulating species of Papilionoideae were one each of *Sesbania, Sophora,* and *Virgilia.* All 40 species of Papilionoideae belonging to 26 genera examined by De Souza (1966) in Trinidad had nodules. In a study of rainforest legumes, Norris (1969) reported nodulation in 13 out of 18 species of Papilionoideae examined, the nodulated species being distributed over 11 genera. Those without nodules were one species of *Alexa, Andira, Diplotropis, Dipteryx,* and *Myroxylon.* Among the species observed to have nodules, new records were for *Clathrotropis macrocarpa, Hymenolobium* sp, *Ormosia coccinea, O. coutinhoi,* and *Sweetia praeclara.*

Grobbelaar *et al.* (1967) examined 247 species of legumes, of which the majority, 217 species were papilionaceous, and all 217 species belonging to 69 genera had nodules. Further new records of nodulating Papilionoideae members were found in Puerto Rico (Dubey *et al.* 1972), where 12 indigenous species distributed among 11 genera were noted.

Among the 215 species of legumes indigenous to Southern Africa examined by Grobbelaar and Clarke (1972), 205 were papilionaceous species distributed among 26 genera, and they all had nodules. Later (Grobbelaar and Clarke 1974) they examined 213 legume species of which 163 were Papilionoideae members, and noted that only ten did not have nodules. Of these ten species, six had been previously found by workers elsewhere to bear nodules. The four species without nodules were *Castanospermum australe, Cladastris lutea, Ononis viscosa,* and *Sesbania punicea.*

An extensive list of 410 nodulating papilionaceous species is given by Corby (1974). Only two species, *Baphia massaiensis* and *Calpurnia aurea,* were found not to form nodules.

In another study Grobbelaar and Clarke (1975) found that out of 149 species of Papilionoideae examined, only two were consistently without nodules, being *Calpurnia aurea* and *Virgilia divaricata.*

Other additions to the list of species examined are given by Lim and Ng (1977) where again the majority of species found to bear nodules were papilionaceous ones. In another survey, Lim (1977) found non-nodulating Papilionoideae species in five genera, some of which also contained nodulated species.

In sub-tropical Pakistan, a study of the nodulating habit of legumes showed 52 species of Papilionoideae to have nodules (Athar and Mahmood 1978), distributed among more than 20 genera, with new records for five species belonging to four different genera, *Crotalaria, Tephrosia, Trigonella,* and *Sesbania.*

Thus of all the subfamilies, Papilionoideae is the one whose members have been examined in greatest numbers and which contains the largest number of nodulated species. This in part is due to the larger number of genera and species contained in the subfamily, but also to the fact that so many members of this subfamily are widely cultivated. As such they are familiar to most people, and are easily accessible. Notwithstanding this, the nodulation status of Papilionoideae is definitely highest of all the subfamilies.

The non-nodulating species in the subfamily belong to widely different genera. *Chaetocalyx*, an example of a non-nodulating papilionaceous legume (Diatloff and Diatloff 1977) is reported to lack nodules despite attempts to induce nodules by artificial inoculation and various treatments on two species. A search for nodules on the roots of *C. latsiliqua* in Costa Rica, Panama, and Nicaragua was also unsuccessful, though the genus is indigenous.

More genera of Papilionoideae need to be collected for nodulation studies, especially those not reported so far in literature. For instance, genera such as *Fordia* said to be rare in Malaysia, though commoner in Borneo occurring there in lowland forests; *Inocarpus* found in swampy forests in southern Malaysia and on Pacific islands, *Ormosia* rare or uncommon in Malaysia and on the way to extinction in this region, and *Pericopsis* also very rare in Malaysia in coastal forests. The following genera, belonging to various tribes, also need to be studied:

Tribe Cadieae	—*Cadia* from tropical Africa, *Riedeliella* from Brazil and Paraguay;
Tribe Sophoreae	—*Neoharmsia* from Madagascar, *Vexillifera*, *Uleanthus*, *Dussia*, and *Uribea* from the American tropics, *Angylocalyx*, *Camoensia*, *Platycelyphium*, *Leucomphalos*, and *Bowringia* from tropical Africa, the latter genus also found in tropical Asia;
Tribe Lotononideae	—*Melliniella* and *Robynsiophyton* from tropical Africa;
Tribe Crotalarieae	—*Heylandia* from Ceylon and India;
Tribe Robinieae	—*Afgekia* from Thailand, *Coroya* from Indochina, *Poitaea*, *Notodon*, *Corynella*, and *Sabinea* from West Indies, and *Sauvallella*, *Bembicidium*, and *Cajalbania* from Cuba;
Tribe Barbierieae	—*Barbieria* from tropical America;
Tribe Millettieae	—*Dewevrea*, *Platysepalum*, and *Schefflerodendron* from tropical Africa, *Taralea* and *Poecilanthe* from tropical America, *Burkillia* the Malay

	Peninsula, and *Antheroporum* from Thailand and Indo-china;
Tribe Lonchocarpeae	—*Behaimia* from Cuba, *Muellera* from tropical America, *Ostryoderris* and *Ostryocarpus* from tropical Africa, and *Kunstleria* from the Malay Peninsula and Philippines;
Tribe Pterocarpeae	—*Vatairea, Centrolobium, Vataeriopsis,* and *Cyclolobium* from tropical America;
Tribe Tephrosieae	—*Caulocarpus* from tropical Africa and *Papilionopsis* from New Guinea;
Tribe Pseudarthrieae	—*Pycnospora* from tropical Africa, Asia, and South-East Asia;
Tribe Indigofereae	—*Rhynchotropis* from tropical Africa;
Tribe Cajaneae	—*Atylosia* from tropical Asia, *Dunbaria* from Malaysia and New Guinea, *Cylista* from Indo-china and Mauritius, and *Carrisoa* from tropical Africa;
Tribe Diocleae	—*Macropsychanthus* from Philippines and New Guinea;
Tribe Galactieae	—*Spatholobus* and *Nogra* from tropical Asia, *Cruddasia* from Thailand and Burma, and *Mastersia* from Indo-Malaysian region;
Tribe Erythrineae	—*Rhodopis* from West Indies;
Tribe Phaseoleae	—*Condylostylis* from tropical America, *Physostigma, Haydonia,* and *Spathionema* from tropical Africa, *Peekelia* from New Guinea, and *Endomallus* from Indo-china;
Tribe Glycineae	—*Platycyamus* from tropical America, *Dumasia* and *Teyleria* from the Malesian region, *Shuteria* from tropical Asia, *Diphyllarium* from Indo-china, and *Paraglycine* and *Amphicarpaea* from tropical Africa and Asia;
Tribe Loteae	—*Gamwellia* from tropical Africa;
Tribe Aeschynomeneae	—*Pictetia, Brya,* and *Poiretia* from West Indies and tropical America, *Nissolia* from tropical America, *Geissaspis* from tropical India and Burma, and *Bakerophyton* and *Humularia* from tropical Africa;
Tribe Desmodieae	—*Nephromeria* from tropical America, *Dendrolobium* from tropical Asia, *Monarthrocarpus* and *Hanslia* from Indo-Malesian region and New Guinea, *Mecopus* from the Malay Archipelago and Indo-china, *Dicerma* from Burma, and *Kerstingiella* and *Uraria* from tropical Africa;

Tribe Stylosanthe   —*Pachecoa* from tropical America;
Tribe Lespezeae   —*Eleiotis* from Ceylon and India, *Phylacium* from the Malay Archipelago, Philippines, and India, and *Neocollettia* from Burma and Indonesia.

These genera are among those which will help to provide additional information on the nodulation status of the Papilionoideae.

Although very many members of the Papilionoideae (every tribe represented) have been examined for nodules, about half the tropical genera, some with one or few species in each genus, still need to be collected and their nodulating habits determined. The majority of these genera are in the tropical regions of America and Africa. On the basis of published literature therefore, the current estimate on the nodulating status of Papilionoideae would be around 80–90 per cent nodulation.

Future trends of research

There is still much scope for those interested in the study of nodulation of tropical legumes. It is highly likely that more instances of nodulating species both at generic and specific levels await discovery, particularly in the less frequented parts of the world. The combined co-operative efforts of persons trained in different disciplines will be required if a systematic survey to collect the rarer genera and species is to be successful. A forester or taxonomist together with a microbiologist as a team, for instance, would be more likely to accomplish a thorough collection and study of the nodulating habits of the desired genera and species growing in a particular region or country.

In studying legume specimens in the future, concentration ought to be on those genera and species which are in imminent danger of extinction, on the one and few species genera, and on those that are considered rare and whose collection and observation will not be repeatable. These genera and species are mentioned above.

Another area of research would be to elucidate the cause or causes underlying the very different nodulating abilities of members of the Caesalpinioideae. This line of investigation may also provide insights into the relationship between the subfamilies and their phylogeny.

1.4 Nitrogen and food from the Leguminosae

On a global basis, Larsen (1979) suggests that 140 million tons of nitrogen are fixed biologically each year, and that about four-fifths of this (112×10^6 metric tonnes) is fixed by nodulated, leguminous plants. According to Burris (1978), the world has a factory production capacity of about 60×10^6 metric tonnes (t) of anhydrous ammonium a year, but only about 50×10^6 t are produced. More than 40×10^6 t of this is used as fertilizer. The biological process appears to fix three times the amount produced by chemical fixation or about 150×10^6 t annually. In order to increase world food supplies, it will be necessary to rapidly accelerate our studies of leguminous plants and develop more effective ways of effectively utilizing nodulated leguminous plants (Dawson 1970). This is particularly true now with the rapidly decreasing supply of fossil fuels and the rising cost of nitrogen fertilizers.

With about 18 000 species of leguminous plants, it is surprising to note the small number 100–150 species, which are actually cultivated. The nine grain legumes used to produce our vegetable food protein are listed in Table 1.2, with areas and production attributed to each during 1971 (Camacho 1977).

TABLE 1.2
World production of food leguminous crops in 1971[a]

Leguminous plant	Area planted (ha)	Production (t)
Drybeans, *Phaseolus vulgaris*	22 868 000	10 218 000
Soyabeans, *Glycine max*	21 086 000	36 181 000
Peanut, *Arachis hypogaea*	16 689 000	15 779 000
Chickpea, *Cicer arietinum*	10 150 000	6 670 000
Dry peas, *Pisum sativum*	8 999 000	10 360 000
Horsebean, *Vicia faba*	4 673 000	5 143 000
Cowpea, *Vigna unguiculata*	3 088 000	1 144 000
Pigeon pea, *Cajanus cajan*	2 914 000	1 975 000
Lentils, *Lens culinaris*	1 698 000	1 075 000

[a] Camacho (1977).

The latest data on soyabean indicate that the production of this crop has more than doubled since 1971. Around 48×10^6 ha were planted to this crop in 1979 (Soybean Blue Book ASA 1979). However, the increases in areas planted to other grain legumes have not paralleled that of soyabeans; in fact, hectares planted to some of the grain legumes appear to be declining (Hardy 1974).

Rhizobium *Germplasm*

In studying the prospects of increasing world food supplies through more extensive utilization of the *Rhizobium*–leguminous plant association, both symbionts must be considered. With 18 000 or so species of leguminous plants, the potential for *Rhizobium* germplasm appears enormous. Yet fewer than 10 per cent of these leguminous species have been examined from a relatively small proportion of these.

There are many collections of *Rhizobium* cultures. Many of these may be registered in the World Data Center for micro-organisms located at the University of Queensland in St. Lucia, Queensland, Australia. However, specific information on parent hosts, strain characteristics, and nitrogen-fixing potential is scarce or lacking. Information on the spectrum of leguminous plants represented is needed also.

The particular collection with which one of the authors (J.C.B.) has been involved, contains 1748 strains isolated from 300 leguminous species in 85 different genera. From this collection, effective rhizobial inocula for 385 leguminous species in 87 different genera have been formulated.

Global outlook

Future generations may be greatly dependent upon biological nitrogen fixation for survival. The leguminous plant–*Rhizobium* system certainly offers the most promise in providing food. Much attention will have to be focused on making the leguminous host plant more efficient in providing the nodule bacteria with energy. Another possibility is to discover new leguminous species with high production capabilities. More effective strains of rhizobia will have to be discovered or developed, and these 'super' strains will have to be aggressive and more acceptable to their particular host than those currently in use. Then perhaps it will be possible to bring about increased nitrogen fixation even in soils which already harbour numerous, highly infective strains of the microsymbiont. It is imperative that techniques for using laboratory grown cultures of bacteria are developed.

The task ahead is difficult and will require the combined efforts of many disciplines of biological research. The symbiosis must involve scientists as well as bacteria and plants.

References

Ackermans, A. D. L., Abdulkadir, S., and Trinick, M. J. (1978). *Pl. Soil* **49**, 711–15.

Airy Shaw, H. K. (1966). *A dictionary of the flowering plants and ferns*. (ed. J. C. Willis) 7th edn. Cambridge University Press.

Allen, E. K. and Allen, O. N. (1950). *Bacteriol. Rev.* **14**, 273–330.

—— —— (1958). *Handbuch der Pflanzenphysiologie* (ed. W. Ruhland), Vol. 8, p. 48. Springer-Verlag, Berlin.

—— —— (1961). In *Recent advances in Botany*, pp. 585–8. 9th Int. Bot. Congr., University of Toronto Press.

—— —— (1976). In *Symbiotic nitrogen fixation in plants* (ed. P. S. Nutman) pp. 113–22. Cambridge University Press.

—— —— and Klebesadel, L. J. (1963). *Proc. 14th Alaskan Sci. Conf.*, Anchorage, Alaska (ed. G. Dahlgren) pp. 54–63.

—— —— and Newman, A. S. (1953). *Am. J. Bot.* **40**, 429–35.

Allen, O. N. and Allen, E. K. (1936). *Soil Sci.* **42**, 87–91.

—— —— (1947). *Soil Sci. Proc.* **12,** 203–8.

—— —— (1954). *Abnormal and pathological plant growth,* Brookhaven Symposia in Biology No. 6, pp. 209–34.

—— —— (1981). *The Leguminosae: a sourcebook of uses and nodulation.* University of Wisconsin Press, Madison.

—— and Baldwin, I. L. (1954). *Soil Sci.* **78,** 415–27.

Anon. (1936). *Int. Inst. Agric.,* Rome.

Anon. (1979). *Tropical legumes: resources for the future.* Ad Hoc. Panel Report, pp. 332. National Academy of Sciences, Washington, DC.

Athar, M. and Mahmood, A. (1978). *Pak. J. Bot.* **10,** 95–9.

Banados, L. L. and Fernandez, W. L. (1954). *Philipp. Agric.* **37,** 529–33.

Beeley, F. (1938). *J. Rubber Res. Inst. Malaya* **8,** 149–62.

Beijerinck, M. W. (1888). *Bot Ztg.* **46,** 726–55, 741–50, 757–71, 781–90, 797–804.

Berry, A. and Torrey, J. G. (1979). In *Symbiotic nitrogen fixation in the management of temperate forests.* (eds. J. C. Jordan, C. T. Wheeler, and D. A. Perry) Forest Research Laboratory, Corvallis, Oregon.

Bonnier, C. (1954). *C. r., Soc. Biol.* **148,** 1894–6.

—— (1957). Symbiose *Rhizobium*–Legumineuses en region equatoriale. INEAC Ser. Sci. No. 72, Brussels.

—— and Seeger, J. (1958). Symbiose *Rhizobium*–Legumineuses en region equatoriale (deuxieme communication). Publ. Inst. nat. agron. Congo Belge Ser. Sci. **76,** pp. 66.

Bowen, G. D. (1956). *Queensl. J. agric. Sci.* **13,** 47–60.

—— (1961). *Pl. Soil* **15,** 155–65.

Broughton, W. J. and John, C. K. (1979), In *Soil Microbiology and Plant Nutrition* (ed. W. J. Broughton, C. K. John, J. C. Rajarao, and B. Lim) pp. 113–36. University of Malaya Press, Kuala Lumpur.

Buchanan, R. E. and Gibbons, N. E. (1974). *Bergey's manual of determinative bacteriology,* 8th edn. Williams and Wilkins, Baltimore.

Bumpus, E. D. (1957). *E. Afr. agric. J.* **23,** 91–9.

Burris, R. H. (1978). *Devel. Ind.* Microbiol. **19,** 1–13.

Burton, J. C. (1979). In *Microbial technology—microbial processes,* Vol. 1, 2nd edn. (ed. H. J. Peppler and D. Perlman) pp. 29–55. Academic Press, New York.

Callaham, D., Del Tredici, P., and Torrey, J. G. (1978). *Science N.Y.* **199,** 899–902.

Camacho, L. H. (1977). In *FAO plant production and protection,* paper 9. *Food legume crops improvement and production,* pp. 66–72. FAO, Rome.

Carroll, W. R. (1934). *Soil Sci.* **37,** 117–35.

Cobley, L. S. (revised by W. M. Steele) (1976). *An introduction to the botany of tropical crops.* Longmans, London.

Corbet, A. S. (1935). *Biological processes in tropical soils with special reference to Malaysia.* Heffer, Cambridge.

Corby, H. D. L. (1967). *Proc. Grassl. Soc. S. Afr.* **2,** 75–81.

—— (1971). *Pl. Soil* Special Volume, 305–14.

—— (1974). *J. Nat. Herb. Salisbury,* **9,** 301–29.

Corner, E. J. H. (1951). *Phytomorphol.* **1,** 117–50.

Dart, P. (1977). In *A treatise on dinitrogen fixation,* Section III, *Biology* (ed. R. W. F. Hardy and W. S. Silver) pp. 367–472, John Wiley, New York.

Davies, J. G. (1951). *J. Austr. Inst. Agric. Sci.* **17,** 54–66.

Davis, T. A. W. and Richards, P. W. (1934). *J. Ecol.* **22,** 106–55.

Dawson, R. C. (1970). *Pl. Soil* **32,** 655–73.

De Souza, D. I. A. (1966). *Trop. Agric.* **43,** 265–7.

Diatloff, A. and Diatloff, G. (1977). *Trop. Agric.* **54,** 143–7.

Dubey, H. D., Woodbury, R., and Rodriguez, R. L. (1972). *Bot. Gaz.* **133,** 35–8.

Elkan, G. H. (1961). *Can. J. Microbiol.* **7,** 851–6.

Fred, E. B., Baldwin, I. L., and McCoy, E. (1932). *University of Wisconsin studies in science*, No. 5. *Root nodule bacteria and leguminous plants.* University of Wisconsin Press, Madison.

Funk, H. B. (1956). Ph.D. Thesis University of Wisconsin. *Summaries of Doctoral Dissertations* **16,** 4–5.

Gibson, A. H. (1978). *Proc. Chinese Austr. Symp. Biological Nitrogen Fixation,* pp. 77–84.

Gorbet, D. W. and Burton, J. C. (1979). *Crop. Sci.* **19,** 727–8.

Graham, P. H. and Parker, C. A. (1964). *Pl. Soil* **20,** 383–95.

Grobbelaar, N. and Clarke, B. (1972). *J. South Afr. Bot.* **38,** 241–7.

—— —— (1974). *Agroplantae* **6,** 59–64.

—— —— (1975). *J. South Afr. Bot.* **41,** 29–36.

—— Van Beyma, M. C., and Synnove Saubert, M. (1964). *South Afr. J. agric. Sci.* **7,** 265–70.

—— —— and Todd, C. M. (1967). Publ. Univ. Pretoria, New Ser., No. 38, 1–9.

Hallsworth, E. G. (1958). In *Nutrition of the legumes* (ed. E. G. Hallsworth) Butterworths, London.

Hannon, N. (1949). Cited in Norris (1956).

Hardy, R. W. F. (1974). In *Proc. 1st Int. Symp. Nitrogen Fixation,* Vol. 2. (eds W. E. Newton and C. J. Nyman) pp. 693–717, Washington State University Press, Pullman.

Hellriegel, H. (1886). *Z. Ver. Rübenzucker—Ind. Dt. Reichs* **36,** 863–77.

Heywood, V. H. (1971). In *Chemotaxonomy of the Leguminosae,* (eds J. B. Harborne, D. Boulter, and B. L. Turner) Academic Press, London.

Hutchinson, J. (1964). *The genera of flowering plants.* Clarendon Press, Oxford.

Jenny, H. (1950). *Soil Sci.* **69,** 63–9.

Kampee, T. and Lohtong, N. (1979). *Symp. Somiplan, Kuala Lumpur.*

Keng, H. (1970). *J. Trop. Geogr.* **31,** 43–56.

Lange, R. T. (1959). *Antonie van Leeuwenhoek J. Microbiol. Serol.* **25,** 272–6.

Larsen, J. A. (1979). *Univ. Ind. Res. Newslett.* **13,** 5–7.

Lawson, G. W. (1966). *Plant life in West Africa.* Oxford University Press, London.

Leonard, L. T. (1925). *Soil Sci.* **20,** 165–7.

Lie, T. A. (1980). In *Nitrogen fixation.* I. *Ecology* (ed. W. J. Broughton), Oxford University Press.

Lim, G. (1961). Microbiological factors influencing infection of clover by nodule bacteria. Ph.D. thesis, University of London.

—— (1977). *Trop. Agric.* **54,** 135–41.

—— (1979). *J. Soil Microbiology and Plant Nutrition* (ed. W. J. Broughton, C. K. John, J. C. Rajarao, and B. Lim), pp. 159–75. University Malaya Press, Kuala Lumpur.

—— and Ng, H. L. (1977). *Pl. Soil* **45,** 1–11.

Lind, E. M. and Morrison, M. E. S. (1974). *East African vegetation.* Longman, London.

Martin, W. P. (1948). USDA Reg. Bull. No. 107, Plant study ser. No. 4, 1–10.

Masefield, G. B. (1955). *Emp. J. Agric.* **23,** 17–33.

—— (1957). *Emp. J. exp. Agric.* **25,** 139–50.

—— (1958). In *Nutrition of the legumes* (ed. E. G. Hallsworth) Butterworths, London.

McDougal, W. B. (1921). *Am. J. Bot.* **8,** 171–5.

Nirmal, A., Skoog, F., and Allen, O. N. (1959). *Am. J. Bot.* **46,** 610–13.

Nobbe, F., Schmid, E., Hiltner, L., and Hotter, E. (1891). Cited in Allen and Allen (1936).

Norris, D. O. (1956). *J. exp. Agric.* **24,** 247–70.

—— (1958a). *Austr. J. agric. Res.* **9,** 629–32.

—— (1958b). In *Nutrition of the legumes* (ed. E. G. Hallsworth) Butterworths, London.

—— (1965). *Proc. 9th Int. Grassland Cong.,* Brazil, **2,** 1087–92.

—— (1969). *Trop. Agric.* **46,** 145–51.

Phillips. T. A. (1939). Cited in Masefield (1958).

Pierce, C. (1979). *Univ. Ind. Res. Newslett.* **13,** 7.

Purchase, H. F. and Nutman, P. S. (1957). *Ann. Bot.* **21,** 439–54.

Puri, G. S. (1960). *Indian forest ecology,* Vol. 1. Primlani, India.

Purseglove, J. W. (1968). *Tropical crops, Dicotyledons* I. Longmans Green, London.

Rice, E. L. (1971). *Am. J. Bot.* **58,** 368–71.

Rothschild, de, D. I. (1968). Cited in Corby (1974).

Samuels, G. and Landrau, P. (1952). *Proc. Soil Sci. Soc. Am.* **16,** 154–5.

Schofield, J. L. (1941). *Queensl. Agric. J.* **56,** 378–88.

Schubert, K. R. (1977). In *Recent developments in nitrogen fixation* (eds W. E. Newton and J. R. Postgate) pp. 469–85. Academic Press, New York.

—— and Evans, H. J. (1976). *Proc. natn. Acad. Sci.* U.S.A. **73,** 1027–11.

Skoog, N. A. and Allen, O. N. (1959). *Am. J. Bot.* **46,** 610–13.

Subba Rao, N. S. (1972). *Current Sci.* **41,** 1–9.

Torrey, J. G. (1978). *Bio Science* **28,** 586–92.

Trinick, M. J. (1973). *Nature, Lond.* **244,** 459–60.

Tutin, T. G. (1958). In *Nutrition of the legumes* (ed. E. G. Hallsworth) Butterworths, London.

Vest, G., Weber, D. F., and Sloger, C. (1973). In *Soybeans, improvement, production and uses.* (ed. B. E. Caldwell) ASA Monograph 16, pp. 353–90. Am. Soc. Agronomy, Madison.

Viands, D. R., Vance, C. P., Heichel, G. N., and Barnes, D. K. (1979). *Crop Sci.* **19,** 905–8.

Vincent, J. M. (1974). In *The biology of nitrogen fixation* (ed. A. Quispel) pp. 265–341. North Holland, Amsterdam.

—— (1977). In *A treatise on dinitrogen fixation.* Section III, *Biology* (eds R. W. F. Hardy and W. S. Silver) pp. 277–366. John Wiley, New York.

Whitmore, T. C. (1972). In *Tree flora of Malaya* (ed. T. C. Whitmore) Vol. 1. Longmans, Singapore.

—— (1975). *Tropical rain forests of the Far East.* Clarendon Press, Oxford.

Williams, L. F. and Lynch, D. L. (1954). *Agron. J.* **46,** 28–9.

Williams, W. A. (1967). *Trop. Agric.* **44,** 103–15.

Winarno, R. and Lie, T. A. (1979). *Pl. Soil* **51,** 135–42.

Wright, H. (1903). *Circ. Agric. J. R. Bot. Gardens Ceylon* **2,** 77–81.

—— (1905). *Circ. Agric. J. R. Bot. Gardens Ceylon* **3,** 181–98.

2 Ecology

H. V. A. BUSHBY

2.1 Introduction

Rhizobial ecology is a subject that is spoken of frequently, but about which little is known. Perhaps this is not surprising as it encompasses such topics as seed inoculation, soil micro-organism–*Rhizobium* interactions, soil–*Rhizobium* associations, and of course rhizosphere effects upon the latter two topics. Until recently, the lack of techniques to quantify or qualify populations of *Rhizobium* in soils and rhizosphere has been a major handicap. However, the use of antibiotic markers or fluorescent antibodies provides a means of more accurately monitoring population dynamics under natural conditions, and should provide a better insight into the mechanics of rhizobial ecology in these environments. The use of scanning and transmission electron microscopy will aid in the description of soil–micro-organism–root spatial relations. Ultimately, the reasons for attempting to understand the processes and events governing the history of rhizobial inoculum are economic, as it is hoped that by application of such knowledge, techniques or procedures will be developed which will lead to increased crop yields or pasture production.

In this review, the aim has been to indicate some recent findings from soil science that might be relevant to rhizobial ecology. To this end, a brief section has been devoted to those aspects of clay mineralogy which form the basis of a generalized differentiation between tropical and temperate soils, or more correctly between highly weathered and less weathered soils. Where possible, results of ecological studies with *Rhizobium* are related to these principles, but generally it becomes obvious that very little work has been done with rhizobia in highly weathered soils. Much of our knowledge of ecology of *Rhizobium* comes from studies in the less weathered soils of the temperate regions of the earth.

2.2 Aspects of soil science relevant to ecology of bacteria in soils

There are fundamental differences between the highly weathered soils of the tropics and most soils from temperate regions of the world. The basis of these differences resides in the soil mineral fraction, or more specifically, in the origin of the surface charge on the soil minerals. The main characteristics of these minerals have been described by van Raij and Peech (1972), Keng and Uehara (1973) and Uehara (1977), but a brief description is given here.

All soil minerals can be placed into one of two groups, the constant surface charge (CSC) and the constant surface potential (CSP) groups. Minerals in the former group have a permanent and constant charge due to isomorphous substitution of an ion of lower valence in positions within the particle normally occupied by an ion of higher valence, whereas the charge on CSP minerals arises from sorption and desorption of hydrogen ions. Hence the latter have a charge which can vary in sign and magnitude with pH, ionic strength of the solution, and type of ions in solution. These are the pH-dependent charge minerals or the so called low activity clays, whereas the CSC minerals are commonly called high activity clays. Soils with both high and low activity clays are widely distributed, but as a generalization, the proportion of low activity clays increases as latitude decreases (Uehara 1977).

The relation between the most common soil minerals has been schematically outlined in Table 2.1 (for a more complete outline, see Uehara 1978).

TABLE 2.1

Relationships between the most common soil particles

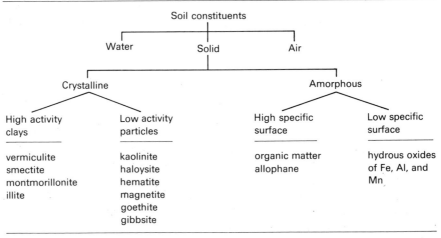

From Uehara (1978).

Examples of pH-dependent clays are the oxides and hydrous oxides of iron (haematite, goethite, and magnetite), aluminium (gibbsite), manganese (pyrolusite), and although not a clay, organic matter has a pH-dependent charge. Kaolinite is a common pH-dependent charged clay. As indicated in Table 2.1, the hydrous oxides of iron, aluminium, and manganese can exist both as discrete crystalline particles or as amorphous gel-like materials, which according to Jones and Uehara (1973), thinly coat and connect the layer-silicate particles. This commonly held belief, that the surfaces of clays are often coated by such amorphous hydrous oxides has been challenged by Greenland and Mott (1978). These authors suggest, on the basis of electron micrographs, that iron oxides in tropical soils do not always coat kaolinite particles but exist in many such soils as small, discrete particles. If surface coatings were a common phenomenon, potentially dramatic changes in the surface charge characteristics of the coated particles would be expected. However, in several soils including a wide variety of red tropical soils, removal of iron oxides did not substantially alter the negative charge of the soil (Deshpande, Greenland, and Quirk 1964; Tweneboah, Greenland, and Oades 1967). The fact that the cation exchange capacities of soils can usually be related to clay mineralogy (high in soils containing montmorillonite, low in soils containing kaolinite), suggests that substantial alteration of clay surfaces by iron oxides is the exception rather than the rule (Greenland and Mott 1978). Clarification of these points is important for soil microbiologists as microbial activity can be markedly altered depending on the presence or absence of cutans (Gray, Baxby, Hill, and Goodfellow 1968; Marshall 1975).

As stated above, the charge on CSP minerals is dependent on adsorption of H^+ or OH^- from the soil solution. It is apparent, therefore, that these minerals may have a net zero charge when an equal quantity of H^+ and OH^- ions are adsorbed. The pH at which this occurs is known as the point of zero charge, and representative values for a range of minerals are presented in Table 2.2.

Specific ion adsorption may occur at sesquioxide surfaces. With the adsorption of anions, the positive charge decreases, as does the point of zero charge (Hingston, Posner, and Quirk 1972), thus phosphate adsorption, a common phenomenon in many tropical soils, descreases the point of zero charge of iron oxides (Mekaru and Uehara 1972; Breeuwsma and Lyklema 1973). The opposite trend occurs with adsorption of cations.

Because the charge characteristics of highly weathered tropical soils are dependent upon pH, and ionic strength and composition of the soil solution, their surface charge cannot be determined without knowledge of their solution strength and composition. This is not the case with less

TABLE 2.2

pH values of the points of zero charge (PZC) for some soil minerals

Mineral	Formula	PZC	Source
Quartz	SiO_2	1–3	Parks (1965)
Manganese dioxide	MnO_2	about 4.0	Parks (1965)
Magnetite	Fe_3O_4	6.5	Parks (1965)
Haematite	Fe_2O_3	8.3–9.0	Parks (1965)
Goethite	FeOOH	5.9–7.2	Parks (1965)
Amorphous iron hydroxide	$Fe(OH)_3$	8.5	Stumm and Morgan (1970)
Amorphous aluminium hydroxide	$Al(OH)_3$	7.5–8.0	Parks (1965)
Gibbsite	$Al(OH)_3$	7.8–9.5	Hingston *et al.* (1972)
Montmorillonite		< 2.5	Parks (1965)
Kaolinite		3.0	Tschapek, Tcheichvilli, and Wasouski (1974)
Kaolinite		7.3 (edge)	Rand and Melton (1975)

From Bell and Gillman (1978).

weathered soils from temperate regions whose surface charge is independent of soil-solution characteristics. Gillman and Bell (1978) have studied the soil solution composition of highly weathered soils from northern Queensland and found that ionic strengths for Ca^{2+}, Mg^{2+}, and K^+ are usually at least ten times lower than those recorded for temperate soils. Bell and Gillman (1978) estimate that a realistic total ionic strength for the soils studied was 0.005, with values of less than 0.002 being common in the majority of horizons. This is far lower than values measured for temperate soils, e.g. for 16 soils Larsen and Widdowson (1968) obtained values from 0.004 to 0.113, with an average of 0.043. These differences between temperate and tropical soils could have major effects on ecology of *Rhizobium* as will be discussed in Section 3.

Consideration of these results is imperative when the surface charges of weathered soils are determined, as gross overestimates are obtained by conventional methods (Gillman and Bell 1976). It is important that soil microbiologists realize this, as solution composition and strength, together with surface charge of particles can influence adsorption between bacteria and clays. Factors such as these may well determine the suitability of a particular *Rhizobium* strain to a soil and hence its saprophytic competence in the absence of its host legume.

In surface soils, organic matter adds to the complexity can frequently be tightly bound to sesquioxides and layer silicates (Greenland, Oades, and Sherwin 1968; Bell and Gillman 1978) so as to prevent its microbial degration (Greenland 1965; Theng 1979). Some uncharged polysaccharides attach to montmorillonite by precipitation rather than adsorption (Parfitt and Greenland 1970). In contrast, it is currently thought that breakdown products of organic matter, humic and fulvic acids, attach to clay particles by complexing with iron, aluminium, or calcium at the clay surface, rather than forming surface coatings or argillans (Greenland 1971), i.e. association of these acids with clays is via exchangeable cations. Under conditions found in soils, humic and fulvic acids do not adsorb in the interlayers of expanding lattice clays. Different mechanisms of attachment of humic and fulvic acids to sesquioxides appear to exist as they can be difficult to separate (Greenland and Mott 1978). The different mechanisms of organic matter attachment to soil particles must have effects upon microbial activity, but as yet this is unknown. It is interesting, however, that skins of organic matter on clays may not be as common as once thought, but may be common on hydrous oxides of iron and aluminium. If it is true, that humic and fulvic acids attach to clays via exchangeable cations, this should make them available to soil micro-organisms, which may contrast with the availability of these materials attached to sesquioxides.

2.3 Description of soil—*Rhizobium* relations

Sorptive interactions between soil particles and micro-organisms

Marshall (1971, 1976), states that sorptive interactions between micro-organisms and particulate matter are of prime importance in the ecology of microbes in many environments including aquatic, enteric, and soil. Marshall (1971, 1976) and Stotzky (1972) have reviewed the literature dealing with the mechanisms of interaction, their effects upon certain activities of micro-organisms, and methods of studying these phenomena.

There are three main classes of interaction between bacteria and the solid constituents of soil which depend on the relative sizes of bacteria and soil particles involved. They are adsorption of micro-organisms on to large soil particles, adsorption between bacteria and soil colloids of approximately the same dimensions, and adsorption of very small particles on to bacteria (Zvyagintsev 1962).

The general observation for adsorption of bacteria to large soil particles is that organisms are unevenly distributed over the surface of particles (Zvyagintsev 1962; Gray *et al.* 1968; Marshall 1971). This has led to the concept of microhabitats in which an organism can exist and proliferate with minimal interaction with adjacent species in neighbouring microhabitats, which may be only a few millimetres away. Generally, the only bridge interconnecting these habitats will be water. In view of the uneven distribution of charge on clays (Gillman and Bell

1976) and its dependence for sign and magnitude on pH, ion type, and ionic strength of the soil solution (see Section 2), it is not surprising that non-random distribution of bacteria has been the usual observation. The implication of this is that the relative organization of clays, cutans, and organic matter should be investigated when undertaking sorption studies of micro-organisms to large particles. Marshall (1971) indicated that much of this information can be obtained by examination of thin sections of soil with a polarizing microscope.

Mutual adsorption of colloidal particles describes interactions between bacteria and soil particles of similar dimensions. Such mutual adsorption has been demonstrated by Peele (1936) and Hubbel and Chapman (1946), and observed with the electron microscope by Jackson, Mackie, and Pennington (1946), Marshall (1971) and Fehrmann and Weaver (1978).

One of the basic questions concerning the mechanisms of mutual adsorption and adsorption of bacteria to larger particles asks if electro-static forces are involved, and if so, how does adsorption occur between particles of like net charge? Under laboratory conditions Santoro and Stotzky (1967, 1968) and Stotzky and Bystricky (1969) observed that bacteria adsorbed to montmorillonite only when the cells were positively charged (i.e. the pH of the ambient solution was less than the isoelectric point) and the clay remained negatively charged. Sorptive interactions also occurred when polyvalent cations (Al^{3+} and Fe^{3+}) reduced or completely reversed the surface charge on bacteria. Under natural conditions, however, both mutual adsorption and adsorption of bacteria to large soil particles occur at pH values above the isoelectric point and in the absence of large quantities of trivalent cations. The isoelectric point of bacteria can be altered markedly by the composition of the soil water. In mono- or divalent solutions at low ionic strengths (3.10^{-4} M) bacteria had isoelectric points ranging from pH 2.5 to 3.5, but in the presence of Fe^{3+} or Al^{3+} at the same concentration, the pH was 7.0 (Stotzky 1972). The latter pH value is the same as that for the PZC of iron and aluminium hydroxides (Table 2.2) and probably indicates precipitation (rather than adsorption) and envelopment of the bacteria by the hydrous oxides. Under all the conditions used by Stotzky (1972), the montmorillonitic clays were negatively charged. Stotzky's (1972) explanation may not be valid however, if variable charge clays are employed where the surface charge also varies with pH, ionic strength, and composition of the ambient environment. Peele (1936) prepared a positively charged fraction from soil and found very strong adsorption of bacteria. Zvyagintsev (1962) noted that cells of *Serratia marcescens* attached to soil particles by their sides but on some, attachment of bacteria was by their pole. This suggested a variation in the surface charge-density of the bacteria. Surfaces not negatively charged may have regions which are positively charged, at which sorption of negatively charged bacteria might occur.

Marshall, Stout, and Mitchell (1971) differentiated between an initial rapid but reversible sorption and a time-dependent, irreversible phase of mutual adsorption, suggesting two different mechanisms. Reversible sorption decreased to zero as the electrolyte concentration decreased or as the thickness of the electrical double-layer round bacteria and surface increased. The electrolyte concentration at which bacteria were repelled depended on the valency of the cation. The reversible phase was explained as a balance between attractive van der Waals forces and repulsion due to electrical double-layer energies. When repulsive forces were reduced to a minimum, they were still sufficiently great to prevent bacteria from actually touching the surface. Bacteria attached by this mechanism were easily removed by agitation (Marshall *et al.* 1971). Irreversible adsorption implies firm adhesion and involves bridging between the bacterium and the test surface by a polymer produced by the bacterium. Such a mechanism has been proposed for the orientation of bacteria at a range of solid interfaces (Marshall and Cruickshank 1973), and can be seen in many electron micrographs (Marshall *et al.* 1975).

Sorption of small particles by larger bacterial cells is described by Lahav (1962) and Marshall (1967, 1968*a*, *b*, 1969*a*, *b*). Both authors used sodium-saturated clays of the constant surface-charge type (montmorillonite), although Marshall (1969*a*, *b*) did include non-expanding illite. Lahav described an increase in the electrophoretic mobility (EPM) of *Bacillus subtilis* with increasing montmorillonite concentrations, which indicates a progressive envelopment of bacteria by the smaller clay particles. Marshall (1969*a*, *b*) demonstrated a similar phenomenon with rhizobia which had differing EPM values in the absence of clay, but the same mobility at a sodium–montmorillonite concentration of 100 μg/ml^{-1}. When illite was used, the EPM of those rhizobia with greater mobilities than the clay decreased with increments in concentration of sodium–illite, whereas those strains with lower mobilities increased as they were mixed with increasing concentrations of the clay. At 120 μg sodium–illite ml^{-1} all rhizobia had the same EPM (Marshall 1969*a*). There was no obvious relation between the surface charge-density of the rhizobia and the amount of clay adsorbed, however the amount of clay adsorbed per unit area of cell surface was related to the nature of the surface ionogenic groups of the bacteria (Marshall 1968*b*). The surfaces of some rhizobia were of a simple carboxyl-type, while others contained a predominance of acidic (carboxyl) groups along with basic (amino) groups. Rhizobia with an exclusively carboxyl-type surface sorbed about twice the amount of sodium–montmorillonite and soduim–illite as those with the more complex carboxyl-amino groups. An explanation for this lies in the orientation and packing of the clay particle round the bacterial cell. Because of the plate-like structure of montmorillonite, Lahav (1962) suggested three possible orientations at bacterial surfaces: (1) face-to-face sorption with the broad particle surface facing the bacterial surface, (2) edge-to-face sorption with the edge of the particle facing the bacterium, (3) mixed sorption with platelets sorbed in both ways. Marshall (1969*b*) suggested a predominantly edge-to-face sorption of clay platelets to rhizobia with simple carboxyl surfaces, giving a closely packed array of platelets round the cells. In contrast, those rhizobia with carboxyl and amino groups probably have some face-to-face associations due to attraction between the positively-charged amino groups and the negatively-charged surface of the flat clay platelet surfaces. This less ordered, more loosely packed arrangement of clay would explain the lower amount of clay adsorbed to these strains of *Rhizobium* (Marshall 1969*b*). Sorption of the two clays studied by Marshall appears to be independent of the surface charge properties of the clay, as the same amount of illite and montmorillonite was adsorbed per cell. Electrical polarity (Bushby unpublished data) and polarity indicated by uneven attachment of antibody (Bohlool and Schmidt 1976; Tsien and Schmidt 1977) may also be factors which influence the arrangement of clay platelets around the cell walls of rhizobia.

The work reported on the sorption of small clay particles to larger bacterial cells has been carried out with clays which were homoionic to monovalent cations. Lahav (1962) reported briefly that montmorillonite homoionic to calcium reduced the EPM of *Bacillus subtilis*. All the clays have been of the high activity, constant surface-charge type, nothing has been done with soil particles from the low activity group such as kaolinite and the various sesquioxides (Table 2.1) which often predominate in tropical soils. It is certain that interactions between bacteria and constant surface potential colloids will differ from those described above, especially with pH variations of the ambient solution. It would seem, for example, that the rapidity with which rhizobia might be enveloped with clay in soils would be dependent upon the mobility of the colloidal mineral fraction. In this context, the amount of water-dispersible clay (WDC) (Gillman 1974), might be relevant. The amount of WDC contained by a soil is usually governed by the presence of an excess of surface negative or positive charge on the clay surface. The greater the excess, the more easily the clay is dispersed. Samples at or near their point of zero charge (Table 2.2) have little or no WDC. It is probable therefore, that in less weathered soils, with significant quantities of highly charged montmorillonitic clays, bacteria would be rapidly enveloped in the manner described by Marshall (1969*b*). In many tropical soils, however, with minerals of low electrostatic repulsion due to their high content of low activity clays, the amount of WDC can be negligible. Under these circumstances, surface coatings of bacteria by clays might take longer to form or even be a rarity. Such differences in the state of bacteria between different soils could have profound effects upon ecology of *Rhizobium*.

Adsorption between bacteria and soil particles is a complex phenomenon. Other factors, such as the coating of particles by inorganic and organic molecules, can alter the wetability and charge of the surface, as well as the nutrient status and pH surrounding particles. The occurrence of pili (or fimbriae) on bacterial surfaces may be of ecological significance. The role of these structures has been reviewed by Ottow (1975) and their occurrence on rhizobia discussed by Tsien (see Chapter 6). Fimbriae alter the net surface charge of bacteria (Brinton, Buzzell, and Lautter 1954) and appear to be especially abundant on bacteria in, or freshly isolated from, their natural environments. The enhanced adhesive properties attributable to fimbriae are considered by Ottow to be of great importance in the ecology of pathogenic bacteria.. Perhaps the same applies to rhizobia?

Effects of sorptive interactions on behaviour of micro-organisms in soil

The complicated and frequently contradicting literature dealing with the effects of sorption on the metabolic activities of micro-organisms has been reviewed by Marshall (1971) and Stotzky (1972), but there are very few publications on the effect of sorptive interactions on microbial ecology in soil. Some, Giltner and Longworthy (1916), and Marshall, Mulcahy, and Chowdhury (1963) have suggested that soil texture is an important characteristic in determining the ability of certain rhizobia to survive desiccation and heat. Marshall (1964) demonstrated that under certain conditions fast-growing rhizobia were more susceptible to desiccation and dry heat than slow-growing strains. The survival of *R. trifolii* under these extremes seemed to be related to soil texture, in that poor survival was obtained in grey and yellow sands, while good survival occurred in red sands and soils of heavier texture. Under similar conditions, however, slow-growing species of *Rhizobium* were less susceptible, especially in the light grey and yellow sands. Addition of montmorillonite or illite but not kaolinite to the grey sand protected *R. trifolii* from both desiccation and heat. Soils amended with montmorillonite did not increase survival of the slow-growing strains (Bushby and Marshall 1977*a*, *b*). Haematite but not goethite afforded some protection to *R. trifolii* (Marshall 1964), but in similar experiments, neither mineral improved survival of desiccated rhizobia (Bushby unpublished). It is interesting that protection is afforded only by the constant surface-charge minerals (high activity clays) and not by the constant surface potential particles. This may have practical application in soils dominated by the CSP minerals that are subjected to seasonal cycles of desiccation and heating.

Weaver and Frederick (1974) reported that the percentage of clay and organic matter in the soil affected the competitive ability of *R. japonicum* strain 110, but not strain 138. Similarly Ham, Frederick, and Anderson correlated the occurrence of certain strains in nodules with soil texture, as well as soil pH, organic matter, and nitrogen, phosphorus, and potassium status of the soil.

Clays may influence survival of micro-organisms in soils by modifying the host–parasite or host–predator interactions. Roper and Marshall (1978*a*) demonstrated direct protection of *Escherichia coli* from phage attack by adsorbed particles of montmorillonite which had a spherical diameter of less than 0.6 μm. If the particle size distribution exceeded 0.6 μm, *E. coli* was not protected. The authors suggested that in the former case, a complete envelope round the bacterium afforded protection, but the larger particles (> 0.6 μm diameter) may not form a continuous envelope, enabling phage to contact the host surface and cause lysis. Other studies by Roper and Marshall (1978*b*) emphasized the subtlety of interactions between clay particles, bacteria, and predatory micro-organisms. The interaction between *Bdellovibrio* and *E. coli* was partially inhibited by the presence of colloidal montmorillonite. However, if the clay envelope round *E. coli* was thin enough, because of its high mobility, *Bdellovibrio* could penetrate the barrier. Colloidal clay had little effect upon predation of *E. coli* by the myxobacterium, *Polyangium*, and had no effect on the activity of the amoeba *Vexillifera*. Crude clay however, physically separated predator and prey, which completely inhibited the interaction between *E. coli* and *Polyangium* and reduced the rate of engulfment of *E. coli* by *Vexillifera*.

The above interactions demonstrate that constant surface charge minerals do have modifying effects on bacterial survival in soils. Again, it has become obvious that little work has been done with the constant surface potential minerals.

Interactions between organic matter and soil micro-organisms

Qualitative or quantitative descriptions of associations between rhizobia and organic matter are rare. Attempts to relate colonization of various soils in the U.S.A. to particular serogroups of *R. japonicum*, have resulted in several correlations between some soil characteristics and the success of a particular serogroup. Amongst the features of soils which appear to be important is soil organic matter (Bezdicek 1972; Ham *et al.* 1971). The distribution of *R. japonicum* 123 in Maryland (Bezdicek 1972) and Iowa soils (Ham *et al.* 1971) was negatively correlated with organic nitrogen whereas serogroup 110 was positively correlated with soil organic matter and nitrogen (both authors) and serogroup 94 was similarly related to soil nitrogen status in some Maryland soils. Serogroup 110 was further positively correlated with several organic nitrogen fractions and extractable ammonium and nitrate which suggested that strains within the group could persist in soils with high nitrogen availability and nodulate soybean under these conditions.

Dart, Roughley, and Chandler (1969) examined the association between *R. trifolii* and peat. Rhizobia were found on surfaces and in crevices of particles but were rarely in groups of more than ten cells. Individual cells within the groups were several microns apart. Although the peat culture was over a month old, the rhizobia were structurally similar to log phase, broth cultured cells, except for electron-dense regions between the cell-wall membrane and the plasma membrane. The rhizobia were enmeshed in a fibrillar matrix of unknown composition and origin, but was probably extracellular gum. There was no particular orientation of the bacteria relative to peat surfaces. An increase in the population of 100-fold occurred from the time the peat was inoculated to when the observations were made. The fact that the number of cells increased without the formation of large colonies, suggests that the interaction between bacteria and peat is reversible and that the cells were able to migrate throughout the peat via water films covering the surfaces of the peat particles. The reversible nature of the interaction has been clearly demonstrated by Weaver (1979) for two different sources of peat.

The mechanism by which organic matter enhances bacterial survival in natural environments is largely unknown, but alteration of prey–predator relationships may be involved (Roper and Marshall 1974). These authors demonstrated protection of *Escherichia coli* from phage attack by natural organic matter–clay mixtures. Although much of the protection was attributed to adsorption of bacteriophage by the clay fraction, organic matter was considered to be important, possibly through mechanisms similar to that of clay, i.e. by attraction and fixation of phage to the highly reactive surfaces of organic matter. Alternatively, its activity may be via effects upon growth. Both the humic and fulvic acid fractions of soils appreciably improve growth of *R. trifolii* when added as the sodium salt to culture media (Bhardwaj and Guar 1972), although the method of action is obscure as ^{14}C-labelled humic acid was not taken up by *R. trifolii* (Bhardwaj and Guar 1971). Enhanced survival of pH-susceptible *Leucaena* strains in soil was attributable to the buffering capacity of a neutral peat (Bushby unpublished data).

It is assumed that the presence of organic matter should be beneficial to survival of rhizobia but this may not always be the case. Increases in soil organic matter, due to superphosphate application for growth of *Trifolium* spp, have been accompanied by a tendency for soil pH to fall

(Donald and Williams 1954; Barrow 1964; Lee 1980). The magnitude of the pH decrease can be large. For example, in sandy soils of Western Australia the pH dropped from 5.0 to about 4.2 over a period of 40 years (Barrow 1964). Williams and Donald (1957) demonstrated that the cation-exchange capacity of these soils increased due to the increased level of organic matter, but occurred without a sufficient increase in exchangeable cations. Thus, the exchange complex became less saturated, the pH fell, and lime is required to raise it again. The mechanism of organic matter-induced pH changes under pasture has been discussed by Lee (1980), who pointed out that manganese and aluminium toxicity problems may occur if the pH falls below 5.5 and 5.0 respectively (see Section 2.4, p. 50). It is possible therefore, that organic matter accumulation, accompanied by a decline in soil pH, may lead to poorer persistence of some strains of *Rhizobium*.

Rhizosphere effects

The word rhizosphere was introduced by Hiltner (1904, cited by Rovira 1978) to define the regions in soil which are affected by living plant roots. It is spatially poorly delineated, but includes the soil immediately adjacent to the root surface where there is the greatest chemical and microbial activity, and can extend into the soil away from the root for a distance of about 20 mm. This region is characterized by declining microbial and chemical activity as distances from the root increases.

The non-random nature of the distribution of microbial cells on roots has been demonstrated to occur at both the level of the clustering of microcolonies and over larger regions of the root by Newman and Bowen (1974). Reasons for this are related to the non-random distribution of microbes throughout the soil which, is determined predominantly by the distribution of clay mineral and organic matter fractions (see Sections 2 and 3). Other reasons for the non-random distribution of microbes on the root surfaces are: the greater exudation of organic compounds at the junctions between epidermal cells and, the pockets of sloughed-off cells which remain in the rhizosphere as the root tip advances (Bowen and Rovira 1976).

Transmission- and scanning-electron microscopy (TEM and SEM, respectively) have demonstrated that the rhizosphere of *Trifolium subterraneum* has bacteria which vary greatly in cell morphology and cell-wall structure, whilst the rhizosphere of *Paspalum dilatatum* consists of a thick layer of intact organic matter, with microcolonies of bacteria scattered throughout. The *Triticum vulgare* rhizosphere is confined to the root surface or invaded epidermal cells (Foster and Rovira 1976; Rovira 1978). Scanning electron micrographs of the roots of *T. subterraneum* show alternating regions of bare root and mucilage, with the mucilage confined to the root surfaces. Bacteria were identifiable, especially on the bare root surface, and were frequently enclosed by their own polysaccharide or enveloped by clay (Rovira 1978). In the rhizosphere of a *P. dilatatum* however, the mucilage was not confined to the root and spread out to envelope sand and clay particles, making differentiation of bacteria quite difficult.

Using a combination of SEM and electron microbeam analysis (EMA), Tan and Nopamornbodi (1979) presented pictorial distributions of a range of elements in the soil, rhizosphere, and plant tissue of cotton, pine, and peanut. The potassium distribution pattern was similar in all cases, being high in the plant tissue, low in concentration immediately outside the root, and then rising to higher concentrations farther away in the bulk soil. Calcium concentrations in peanut were greater in the nodule than in the root or soil, and within a nodule much of the calcium was concentrated round the peripheral cells. Calcium was also concentrated in the rhizoplane. The distribution of magnesium within the peanut–rhizosphere–bulk soil complex showed concentration within the root and nodules, but not preferential partioning between these organs. In agreement with Rovira (1978), the SEM presented by Tan and Nopamornbodi (1979) showed the presence of a mucilage layer enveloping root surfaces. The authors postulated that this, plus the embedded soil particles, were active in chelating or complexing elements from the nutrient pool around the root surface. If, as Rovira (1978) has indicated, the mucilage of legumes is discontinuous, the consequent uneven mineral accumulation could add to or help explain the non-random nature of bacterial distribution on root surfaces.

The ionic balance of the soil solution can be a factor in the accumulation of ions in the rhizosphere. For instance, nitrate fed *G. max* showed an accumulation of HCO_3^- and a consequent increase in the pH of the rhizosphere (Riley and Barber 1969). The opposite might be expected in nodulated plants due to excretion of H^+ ions to maintain electrical neutrality, as atmospheric nitrogen is fixed and organic acids accumulated (Israel and Jackson 1978). The practical effects of this in the form of reduced soil pH under legume-based pastures have been discussed in the previous section.

The large distances that can occur between individual bacteria and between microcolonies on root surfaces, together with their frequent envelopment by either clay- or plant-produced mucilage suggests that interaction between rhizosphere microbes may be much less than originally envisaged. No doubt there are exceptions, such as at the points of exit of lateral roots where enhanced proliferation of bacterial activity might be expected to such an extent that competition, predation, and antagonism might occur between micro-organisms. As these regions are frequently the sites of nodule formation within the tropical legumes (Dart 1974; Date personal communication), interactions between rhizobia and other organisms could determine the ability of rhizobia to form a sufficiently large population for nodule formation.

In a series of laboratory experiments van Egeraat (1975*a, b, c, d, e,* 1976) determined the site of exudation of nutrients from *Pisum sativum* roots, and studied their effects as modified by environmental conditions, upon root-nodule bacteria. Exudates from young pea roots were localized to the growing tips of the tap and lateral roots, as well as at the points of lateral root emergence. The ratio of unidentified amino acids to homoserine decreased with distance from growing points, and relatively large quantities of homoserine were released from damaged plant root-cells during lateral root emergence. *Rhizobium leguminosarum,* *R. trifolii, R. phaseoli,* and *R. meliloti* were tested for their ability to grow in either homoserine or glutamic acid as their sole source of carbon, nitrogen, and energy. Satisfactory growth of *R. trifolii, R. phaseoli,* and *R. meliloti* occurred only on glutamic acid, and was entirely absent or greatly reduced on homoserine. For the pea strains of *R. leguminosarum,* abundant and equal growth occurred on homoserine and glutamic acid. Van Egeraat (1975*d*) suggested this as a mechanism by which *R. leguminosarum* might be differentially stimulated within the rhizosphere in the presence of a mixture of rhizobial strains belonging to different cross-inoculation groups. An interesting phenomenon noted was that *R. leguminosarum* tended to be stimulated in the regions of lateral emergence on pea roots, but was evenly distributed over the root surface of *Trifolium, Medicago, Lupinus,* and *Vigna radiata* (van Egeraat, 1975*c*).

Dazzo, Napoli, and Hubbell (1976) studied the root hairs of clover in the presence of non-infective *R. trifolii, R. meliloti,* and infective *R. trifolii* strains. Adsorption of the later strains were four to five times greater than non-infective rhizobia of the same or different species. These phenomena (Dazzo *et al.* 1976; van Egeraat 1975*c, d*), indicating preferential stimulation of growth or attachment of nodulating strains of rhizobia to specific sites on the host, have been studied by Currier and Strobel (1976) and Vittal and Patil (1978), by monitoring chemotactic responses of rhizobia to root exudates. Currier and Strobel were not able to demonstrate any specificity in rhizobial chemotactic-responses as root exudates of non-legumes attracted the bacteria. Vittal and Patil (1978) measured positive chemotactic-responses of *Rhizobium* sp towards chickpea (*Cicer arietinum*) root extracts. The exudate was separated into anionic, cationic, and neutral fractions for determination of their chemotactic characteristics. The response was greatest to the cationic, intermediate to the neutral, and least to the anionic fractions. Histidine from the cationic fraction strongly attracted rhizobia. Although these results might seem to compliment those of van Egeraat

(1975*c*, *d*) in that specific attraction is implied, they do suffer from insufficient comparisons both between rhizobia and non-rhizobial bacteria and other plants including non-legumes. If such comparisons were made, it might well be found that chemotactic specificities were of negligible importance in the nodulation process. Non-specific attraction of a range of *Rhizobium* sp to a glycoprotein extracted from *Lotus corniculatus*, further demonstrated the low order of importance of chemotaxis in the nodulation process (Currier and Strobel 1977).

From the economic standpoint, one of the most important aspects of ecology of *Rhizobium* is interstrain competition within the rhizosphere for nodule formation. Due to the complex nature of the problem, very few unequivocal statements can be made. For instance, Robinson (1969*a*) found that in mixed culture, most nodules formed on *Trifolium* were effective, even when the ratio of effective to ineffective strains was as low as 1:100. Thus the legume exercised a strong selective preference for the effective strain within a suite of strains within the soil. However, this is not always the case as Ireland and Vincent (1968) found that naturally occurring less effective strains prevented the formation of nodules by effective strains introduced as seed inoculum. An extreme example has been noted by Diatloff and Brockwell (1976) in the nodulation of soyabean cultivar Hardee by *R. japonicum* strains CB1809 and CB1795. Strain CB1809 was unable to form nodules on Hardee, but suppressed nodule formation by CB1795 to such an extent that plant growth was no better than uninoculated controls, even though CB1795 alone was able to abundantly and effectively nodulate the cultivar. Similarly, Nicol and Thornton (1941), Vincent and Waters (1953), van Rensburg and Strijdom (1969), Franco and Vincent (1976) and Johnston and Beringer (1976) observed that the ineffective strain formed the majority nodules.

Robinson (1969*a*) suggested that the host plant can distinguish between effective and ineffective strains at a very early stage in the infection process. It is difficult, however, to imagine how such a defect is recognized early in infection, such that the ineffective strain is rendered less competitive than the effective one. Mytton and de-Felice (1977) concluded that the ability to infect and effectiveness were not necessarily linked and the plant genotype could modify the relative competitiveness of strains.

Recently, Marques Pinto, Yao, and Vincent (1974), Labandera and Vincent (1975) and Franco and Vincent (1976) attempted to quantify many facets in the nodulation process. Using paired strains, their measurements included the strain ratio in the inoculum, strain representation on the root surface, and strain representation in nodules. Thus under certain conditions the competitive ability of a strain can be quantified. Date and Brockwell (1978) have summarized the results of

Vincent and co-workers as follows: that inoculum strain ratios are not always reflected in initial root-surface ratios; that strains differ in their ability to colonize root surfaces; a dependence of nodulation success on root-surface representation when the host influences strains equally and an independence when the host has an overriding influence on determining which strains form nodules; a relation between symbiotic effectiveness and competitive ability; and a marked influence of environment, especially temperature, on interstrain competition.

Except for the work of Vincent and co-workers, many conclusions are based on the percentage of nodules due to the inoculum strains, which may not be an accurate representation of rhizosphere events. The question has been posed, does the pictorial evidence provided by Rovira (1978) suggest that intercolony competition under natural conditions is minimal? At some stage it will be necessary to reconcile the idea that competition between rhizosphere micro-organisms is intense, with the evidence that little direct contact seems to exist between microbes over much of the root surface (Rovira 1978).

2.4 Inoculation of legumes and persistence of inoculum strains in soils

The literature on the need for inoculation, preparation of inoculants, methods of inoculation, and estimating the results of inoculation have been collated by numerous authors (Vincent 1965; Norris and Date 1976; Roughley 1976; Date 1975, 1977; Date and Roughley 1977). There is less information available concerning the factors that affect persistence of inoculum and colonization of soil and rhizosphere.

Persistence of inocula has been measured indirectly as the proportion of nodules formed by the inoculum strain, employing such techniques as serological identification (Means, Johnson, and Date 1964; Date 1974), antibiotic markers (Brockwell, Schwinghamer, and Gault 1977), black nodule character in *Centrosema pubescens* (Cloonan 1963) the pink strain from nodules of *Lotononis bainesii* (Diatloff 1977), and induction of chlorosis in the tops of *G. max* by *R. japonicum* serogroup 94 (Weber and Miller 1972). Most quantitative measurements have been carried out using the plant infection method (Vincent 1970) where native rhizobia, able to nodulate the test host are either non-existent or present in very low numbers (Tazimura, Watanabe, and Shi 1966; Robinson 1967; Cloonan and Vincent 1967). Only recently have fluorescent antibody techniques (Reyes and Schmidt 1979) and antibiotically marked strains (Bushby 1980*a*, *b*) been used for direct estimation of soil and rhizosphere numbers. Reyes and Schmidt (1979) demonstrated modest rhizosphere stimulation of *R. japonicum* strain 123 by the soyabean cultivar Chipperwa and slight stimulation by *Zea mays* rhizosphere. High populations occurred only at crop maturity and harvest, and were associated with disintegration of the root system and nodules. This high population level declined rapidly after harvest. These results are in agreement with those of Diatloff (1969) who reported that when nodules were intact, rhizosphere populations of rhizobia on soyabeans were comparable to populations in the rhizospheres of non-legumes. The increased soil population occurred through the release of rhizobia from decaying nodules. The situation with black gram *Vigna mungo* cv. Regur) differed in that the level of the inoculum (*Rhizobium* sp CB756) decreased during the early stages of root development, but subsequently increased. This increase continued for 184 days at which stage roots and nodules were decaying. The overall generation time for the rhizosphere population was 17 days, however, this was subdivided and generation times for various phases of the growth curve calculated (Table 2.3). The growth rates at all times were remarkedly slow. The soil population increased only after nodule decay (Bushby 1980*b*). This early decrease and later increase in rhizosphere populations of CB756 was similar to that observed by Cloonan and Vincent (1967) for the same·strain of *Rhizobium* on *Lablab purpureus*. This phenomenon is variable and may depend upon soil type. The population of rhizobia in the rhizosphere of *V. mungo* grown in a sandy podzolic increased, and there was no sign of the early decrease that occurred in the red earth with the same strain of *Rhizobium* (Bushby unpublished data).

The method of inoculation can prevent the initial drop in the number of seed-applied rhizobia. When inoculated in the usual way (Norris and Date 1976), strain NGR8 for *Leucaena leucocephala* cv. Cunningham was not detectable in the rhizosphere or soil one week after sowing into a red podzolic soil. Strain CB81 persisted for a much longer period, but also

TABLE 2.3
Generation time for Rhizobium *sp CB756 in the rhizosphere of field grown* Vigna mungo *cv. Regur*

Time period after sowing (days)	Treatment	Generation time (days)
12–19	Unpelleted	4.2
12–19	Pelleted	1.4
19–35	Pelleted plus unpelleted combined	9.0
19–54	Pelleted plus unpelleted combined	17

From Bushby (1980*b*)

declined rapidly in numbers after sowing. If the inoculum consisted of a plug of peat (0.02 g peat per seed), the micro-environment was sufficiently altered to allow multiplication and persistence such that peat plugs retrieved from the soil contained $10^7 - 10^8$ rhizobia g^{-1} of peat 12 months later. Such a procedure reduced the time to nodulation from about 50 days with the former method of inoculation to less than 14 days. The method of inoculating seed with granules of peat is being widely used in USA for soyabean inoculation.

Inadequate nodule formation has frequently been attributed to inhibited multiplication or death of poorly adapted strains of *Rhizobium*. However, this might not always be the case. Nodulation of *V. mungo* in the absence of lime was reduced (Bushby 1981*b*), but this was not due to any effect upon the rhizosphere population of inoculum, but rather to the lime treatment raising the soil pH from 5.0 to 5.5, a more favourable condition for nodule formation.

The understanding and solution of the 'second year mortality' problem with *Trifolium subterraneum* in some Western Australian soils, is a classic in the study in rhizobial ecology (Chatel and Parker 1973*a, b*; Chatel and Greenwood 1973*a, b, c*; and Chatel, Shipton, and Parker 1973). The problem was due primarily to low numbers of the inoculum rhizobia (*R. trifolii* strain TAI) in the soil at the end of the first growing season, and was aggravated by high soil temperatures during the summer months. Marked differences in both root and soil colonization were noted between strains of *R. trifolii* during the first year, and these differences were reflected in the second year nodulation results. Differences were also noted between species of rhizobia, with *R. lupini* being superior in soil and rhizosphere colonization than any strain of *R. trifolii*, to the extent that no problems were encountered with lupin (*Lupinus* spp) or serradella (*Ornithopus sativus*). Both these legumes are nodulated by *R. lupini*. In addition to differences in their ability to persist in soil and colonize host-legume rhizospheres, strains of *R. trifolii* differed in ability to colonize non-host rhizospheres (Chatel and Greenwood 1973*a*). This is a significant factor in the persistence of rhizobia and especially relevant to pasture legumes that depend on the formation of new nodules for their nitrogen supply. Chatel and Greenwood (1973*c*) also examined the distribution of inoculum strains among the various components of the soil by partitioning soil cores from under senesced swards into soil, legume roots, nodules, and debris, and counting the number of rhizobia associated with each. Differences between species were noted, with *R. lupini* being more evenly distributed over all fractions than strains of *R. trifolii*. There was a greater decline of *R. trifolii* than *R. lupini* during the summer/autumn period. Similarly, numbers of *R. trifolii* TAI declined more than strain WU95. As the 'pool' of rhizobia available for nodule formation, colonization of the soil by rhizobia was considered more important than localization within the surrounds of decayed nodules, since roots of new season plants would have a greater chance of encountering suitable inoculum rhizobia. A better understanding of this area of rhizobial ecology is essential, especially when legumes are being introduced into new environments.

Successful colonization of the rhizosphere and formation of nodules by rhizobia introduced as inoculum with the host-legume seed implies adaptation to the soil environment and establishment within the soil microbial community. However, this has not happened in many Australian pasture soils. Characteristically, if sown into soils containing low numbers of indigenous rhizobia capable of nodulating the host, the inoculum strain will form a large proportion of the nodules in the first year, but the proportion declines at varying rates in subsequent years as serologically-unrelated strains invade the plots (Dudman and Brockwell 1968; Brockwell, Bryant, and Gault 1972; Chatel *et al.* 1973; Gibson, Date, Ireland, and Brockwell 1976; Roughley, Blowes, and Herridge 1976; Date and Brockwell 1978). However, there are exceptions. Under northern Australian conditions, *Lotononis bainesii* (Diatloff 1977) and Kenya white clover (*Trifolium semipilosum* cv. Safari) (Jones and Date 1975) are nodulated only by their respective specific-inoculum rhizobia. If the legume is nodulated by a wide range of rhizobia the proportion of strains due to the inoculum, even in the first year, can be negligible, e.g. field sowings of *Macroptilium atropurpureum* cv. Siratro (Date and Brockwell 1978), and green (*Vigna radiata* cv. Berken) and black (*V. mungo* cv. Regur) grams (Bushby unpublished results). In many such cases, the native strains that form the major proportion of the nodules are effective nitrogen fixing organisms and there is no decline in plant yield (Date and Brockwell 1978). It has been suggested that under such circumstances, inoculation has few, if any, long term benefits (Roughley *et al.* 1976). It is these situations, as well as those soils which contain high populations of ineffective to moderately effective indigenous rhizobia, that provide the greatest resistance to the introduction and persistence of inoculum strains.

There are two major stages of plant development during which our knowledge of the fate of inocula is deficient. The first is the period following sowing and just prior to nodule appearance, and the second is the time following release of rhizobia from old nodules. These periods are in need of intensive investigation.

Factors affecting colonization of soil and rhizosphere

pH and related factors
Extremes of pH frequently reduce nodule formation by reducing

colonization of soil and the legume rhizosphere by rhizobia. The effect of both low and high pH is dealt with in this section.

Growth of *Rhizobium*

Effects of high hydrogen ion concentrations on *Rhizobium* are well documented. Pure culture studies (Fred, Baldwin, and McCoy 1932; Graham and Parker 1964) have shown that the lowest pH at which rhizobia will grow varies from about 4.0 to 6.0, but with the general grouping of the slow-growing *R. japonicum, R. lupini,* and the cowpea-type rhizobia being less sensitive than faster-growing strains such as *R. meliloti.* Within a group (i.e. fast- or slow-growing) quite large strain-to-strain variability exists (Graham and Parker 1964; Munns, 1965; Rerkasem 1977). Recently, Date and Halliday (1979) found that some strains isolated from *Stylosanthes* sp growing in acid soil (pH 4.5) grew best on agar at pH 4.5, and two strains out of 50 were obligate in their requirement for low pH conditions for growth. Their paper has reopened the question of the ecological role of acid or alkali production by rhizobia. Norris (1965) postulated that slow-growing rhizobia were, by virtue of alkali production on yeast–mannitol agar, more likely to be adapted to acid soils than the fast-growing, acid producing rhizobia. It was assumed that alkali production also occurred in the field and was sufficient to alter the micro-environment in favour of survival of these strains under acid conditions. Therefore, the demonstration that alkali production by slow-growing rhizobia was dependent on the carbohydrate in the culture medium, together with the obligate requirement of low pH for growth of some strains isolated from acid soils, seriously questions the validity of Norris's hypothesis.

Soils of high acidity frequently have low levels of phosphorus, calcium, and molybdenum and high concentrations of aluminium and manganese. Until recently, there had been little work on the effects of these low-pH associated factors on rhizobia. Rerkasem (1977) and Carvalho, Bushby, and Edwards (1981) have demonstrated the ability of some strains to survive high aluminium concentrations at low pH values. The ameliorative effect of calcium on aluminium toxicity in *Rhizobium*–legume symbiosis has been noted (Sartain and Kamprath 1975; Andrew 1976), but studies by Rerkasem (1977) and Keyser and Munns (1979*b*) report the interaction effects of calcium and aluminium on *Rhizobium per se.* Rerkasem noted that calcium prevented the decline in viability of a fast-growing strain at pH 4.3, but had no effect on aluminium toxicity. A slow-growing strain that was unaffected by either low pH or by aluminium was not affected by calcium supply (10 mM). Calcium in the range 50–100 μM had little protective effect against aluminium or acidity for three Vigna strains (Keyser and Munns 1979*b*).

Tolerance of rhizobia in broth cultures to low pH (4.5), low phosphate (5–10 μM), and high aluminium (50 μM) has been studied in detail for ten 'cowpea type' rhizobia, and a further 52 'cowpea' and 13 *R. japonicum* strains screened against these stresses (Keyser and Munns 1979*a*). Low phosphorus reduced the total yield of all strains and decreased the growth rate of some, whereas acidity affected most strains by either increasing the lag time and/or decreasing the growth rate. It stopped growth of about 30 per cent of the strains tested. The authors concluded that tolerance of low pH did not necessarily imply tolerance to aluminium, as this element altered or completely stopped the growth of nearly all acid-tolerant strains. Excessive levels of manganese (200 μM) and low calcium concentrations (50 μM) in broth cultures (pH 4.6) slowed the growth of some of the 23 'cowpea type' rhizobia and ten strains of *R. japonicum* tested (Keyser and Munns 1979*b*). The effect of the manganous ion is similar to that described by Wilson and Reisenauer (1970). Continued culturing of *R. trifolii* on high manganese media led to the reversible production of ineffectiveness in nitrogen fixation (Holding and Lowe 1971), a result supported by Masterson (1968) who found an increased proportion of ineffective isolates of *R. trifolii* when isolated from high manganese soils, and a reversal of this trend when field soils were limed (Jones 1966). The results of Keyser and Munns (1979*b*), differ somewhat from those of Date and Halliday (1979), who found that the ability of a strain, isolated from an acid soil to grow at low pH was not altered by low phosphorus (10 μM), high aluminium (50 μM), low calcium (50 μM), or high manganese (200 μM). Date and Halliday (personal communication) have extended these results to include a number of other strains, and concluded that aluminium at low pH did not alter the reaction over and above a strain's reaction to low pH.

Carvalho *et al.* (1981) concluded that aluminium-induced reduction of nodule formation in *Stylosanthes* spp could not be attributed to an effect on the viability of *Rhizobium*. In an attempt to determine the mechanism by which nodule formation was affected, a microelectrophoretic method (Marshall 1967) was used to determine adsorption of aluminium (0–400 μM) at pH 4.5, to 16 species of rhizobia (Bushby unpublished data). Unexpectedly, very little interaction was observed between aluminium and negatively-charged slow growing rhizobia. This is similar to data obtained for *Bacillus subtilis* (Beveridge and Murray 1976), i.e. aluminium did not adsorb to this Gram-positive bacterium. The results for some, but not all, fast-growing rhizobia were inconclusive, as flocculation occurred, frequently making accurate readings difficult. Rerkasem (1977) noted that the phenomenon of flocculation requires further study as it may be important in the nodulation process via an effect on the number of rhizobia available for plant root infection. For the slow-growers, however, it is unlikely that aluminium interferes in nodule formation by adsorption to the rhizobial surface and consequent interruption of the recognition events which take place between the host and strain. This does not preclude adsorption of aluminium to root surfaces or the uptake of aluminium by rhizobia, with consequent effects upon nodule formation.

Inhibition of soil and rhizosphere colonization by low pH and calcium has been reported for many fast-growing temperate rhizobia, e.g. *R. meliloti* (Robson and Loneragan 1970a, b), *R. trifolii* (Rovira 1961; Jones 1966). Similar results are reported from solution culture studies (Munns 1968; Lie 1969; Lowther and Loneragan 1968). Invariably, substantial increases in rhizobial growth occurred with increases in pH or calcium concentration. However, there is a distinct lack of information on rhizosphere populations for tropical legumes nodulated by slow-growing rhizobia. Working with four slow-growing rhizobia for *Macroptilium atropurpureum* cv. Siratro and *Stylosanthes hamata* cv. Verano in solution culture at pH values 4, 5, and 6, Bushby (unpublished data) concluded that pH did not affect rhizosphere numbers over a 10 day period. Between 10 and 30 days after inoculation, however, all strains disappeared at pH 4.0, the rate of disappearance varying with strain. Survival was not significantly different on the two host root systems.

The information available concerning the effects of alkaline conditions *per se* on rhizobia is scant. Graham and Parker (1964) reported growth of a range of strains in broths that initially ranged from pH 8.0 to 10.0. Only one slow-growing strain (of 31) would grow at pH 9.0, but almost all strains of *R. meliloti* grew at pH 9.5. Other fast-growing strains were between these two extremes, and no growth of any strain occurring at pH 10.0. At pH values likely to be encountered in alkaline soils (pH 8.0–8.5), all the fast-growing strains and 90 per cent of the slow-growing ones were able to grow at pH 8.0, whereas at pH 8.5 81 per cent of fast- and 65 per cent of slow-growing rhizobia showed growth. The main problem with these and other results (Damirgi, Frederick, and Anderson 1967) is that variation in initial broth pH is likely to have occurred during growth.

Nodulation process

The effects of soil acidity with the related factors of toxic concentrations of aluminium and manganese, and inadequate supply of phosphorus, calcium, and molybdenum on nodule formation and growth of the host has been studied for a wide range of legumes, and the literature has been adequately reviewed by Munns (1977, 1978) and Carvalho (1978). Each of the four 'steps' in the formation of an efficient nitrogen-fixing system on legumes has been implicated as being the sensitive 'link in the chain' which contributes most to reduced nodule formation. These steps are:

multiplication of rhizobia in the soil and rhizosphere; infection of the root; nodule formation; and maintenance of an efficient nitrogen-fixing system. The first step has been discussed in the previous section. Literature referring to the infection and nodule formation steps is summarized in Table 2.4a, and that for effects on nitrogen fixation in Table 2.4b.

Calcium and pH interact on all steps of the nodulation process, such that generally, a lower pH can be tolerated if calcium is high and vice versa.

Recently, Carvalho (1978) provided direct evidence, from solution culture transfer experiments, of an aluminium-induced reduction in infection and/or nodule initiation in *Stylosanthes hamata* and *S. scabra*. Restricted nodulation occurred prior to any significant reduction in host-plant growth and nitrogen fixation in nodulated plants was unaffected by the aluminium concentrations used (up to 100 μM). Thus nodulation was more sensitive than host-plant growth to high solution aluminium concentrations. Aluminium delayed the appearance of first nodules in both species and reduced the total number of nodules. Nodulation of *Stylosanthes* sp in the presence of aluminium is enhanced by additions of calcium. The mechanism of this interaction is not known (Carvalho 1978).

There is a very wide host-dependent range of minimum pH values at which nodulation can take place (Table 2.4a and b). For example, all plants of *Macroptilium lathyroides* were nodulated at pH 4.0, but this did not occur until pH 6.0 for *Neonotonia wightii*. This suggests that, for CB756 at least, inhibition of nodule formation is not due to the inability of the strain to multiply, but rather, there is a later acid-sensitive step which varies with the host. Munns, Keyser, Fogle, Hohenberg, Righetti, Lauter, Zaroug, Clarkin, and Whiteacre (1979) made similar observations for two cultivars of *Vigna radiata* and three cultivars of *V. unguiculata* (Keyser, Munns, and Hohenberg 1979).

Munns and Fox (1976, 1977) and Munns, Fox, and Koch (1977) altered the pH of an oxisol in which 18 legumes were grown, from 4.7 to 7.1 (Table 2.4b). There was no distinct pattern in responses between temperate and tropical legumes, as might be expected from the work of Andrew and Norris (1961) and Norris (1965, 1967). Excessive lime application reduced growth of many legumes but some were unaffected, notably *A. hypogea*, *G. max*, and *V. unguiculata* which are also very tolerant to low pH and calcium supply (Munns 1978).

The possibility of low pH altering the competitive ability of paired strains of *R. meliloti* on *Medicago truncatula* was investigated by Couper (Couper and Vincent, personal communication). Acidity (pH 5.2) favoured one strain (SU51) in preference to the other (SU126). Distribution of some strains of rhizobia, e.g. *R. japonicum* 135, is determined

TABLE 2.4a

Summary of literature referring to the effect of acidity on the root-infection and nodule formation steps of the legume–Rhizobium symbiosis

Step	Host	Strain	Factors	Reference
Infection step	*Medicago sativa*	*R. meliloti*	pH (interaction with Ca)	Munns (1968, 1970)
	Trifolium subterraneum	*R. trifolii*	pH (interaction with Ca)	Lowther and Loneragan (1970)
	Pisum sativum	*R. leguminosarum*	pH (interaction with Ca)	Lie (1969)
	Stylosanthes hamata	*Rhizobium* sp CB756	Al	Carvalho (1978)
	Stylosanthes scabra			
Nodule formation	*Medicago sativa*	*R. meliloti*	pH	Rice, Penney, and Nyborg (1977)
	Trifolium pratense	*R. trifolii*	pH no effect from 4.5 to 7.2	Rice et al. (1977)
	Medicago sativa	*R. meliloti*	pH, Al (interaction with Ca)	Rice (1975)
	Glycine max	*R. japonicum*	Al (interaction with Ca)	Sartain and Kamprath (1975, 1977)
	Vigna radiata	several sp of *Rhizobium*	pH	Munns et al. (1979)
	Vigna mungo	*Rhizobium* sp CB756	pH	Bushby (1980b)
	Macroptilium lathyroides	*Rhizobium* sp CB756	pH (<4.0)[a] (no interaction with Ca)	Andrew (1976)
	Lotononis bainesii	*Rhizobium* sp CB376	pH (<4.0) (no interaction with Ca)	Andrew (1976)
	Stylosanthes humilis	*Rhizobium* sp CB756	pH (<4.5) (negative interaction with Ca)	Andrew (1976)
	Desmodium uncinatum	*Rhizobium* sp CB627	pH (<4.5) (no interaction with Ca)	Andrew (1976)
	Neonotonia wightii (formerly *Glycine wightii*)	*Rhizobium* sp CB756	pH (6.0) (interaction with Ca)	Andrew (1976)
	Medicago sativa	*R. meliloti* U45 and SU47	pH (6.0) (interaction with Ca)	Andrew (1976)
	Medicago truncatula	*R. meliloti* U45 and SU47	pH (6.0) (interaction with Ca)	Andrew (1976)
	Medicago scutellata	*R. meliloti* U45 and SU47	pH (5.0) (interaction with Ca)	Andrew (1976)
	Trifolium repens	*R. trifolii* UNZ29	pH (5.0) (interaction with Ca)	Andrew (1976)
	Trifolium semipilosum	*R. trifolii* CB778	pH (4.5) (interaction with Ca)	Andrew (1976)
	Trifolium rueppellianum	*R. trifolii* CB780	pH (4.5) (interaction with Ca)	Andrew (1976)
	Neonotonia wightii	*Rhizobium* sp	pH	Philpotts (1975)
	18 legumes	Several rhizobia	pH (interaction with Ca)	Munns et al. (1977)

[a] Values in parentheses are minimum pHs at which nodulation, measured as percentage of plants nodulated, was equal to 100 per cent.

TABLE 2.4b

Summary of literature referring to the effect of acidity on nitrogen fixation and/or yield of the legume–Rhizobium symbiosis

Step	Host	Strains	Factor	Reference
Nitrogen fixation or yield	Medicago sativa	R. meliloti	pH and Al (interaction with Ca)	Rice (1975)
	Neonotonia wightii	3 Rhizobium sp	Mn	Rice et al. (1977)
	6 Stylosanthes sp	3 Rhizobium sp	Al (interaction with Ca)	Philpotts (1975)
	Vigna radiata	40 Rhizobium sp	pH	Carvalho (1978)
	Glycine max	R. japonicum	Al (interaction with Ca)	Munns et al. (1979)
				Sartain and Kamprath (1975, 1977)
	Macroptilium lathyroides	Rhizobium sp CB756	pH (4.5)[a] (no interaction with Ca)	Andrew (1976)
	Lotononis bainesii	Rhizobium sp CB376	pH (5.0) (interaction with Ca)	Andrew (1976)
	Stylosanthes humilis	Rhizobium sp CB756	pH (5.0) (interaction with Ca)	Andrew (1976)
	Desmodium uncinatum	Rhizobium sp CB627	pH (6.0) (no interaction with Ca)	Andrew (1976)
	Neonotonia wightii	Rhizobium sp CB756	pH (5.0) (no interaction with Ca)	Andrew (1976)
	3 Medicago sp	R. meliloti U45 and SU47	pH (6.0) (all interact with Ca)	Andrew (1976)
	Trifolium repens	R. trifolii UNZ29	pH (6.0) (interacts with Ca)	Andrew (1976)
	Trifolium semipilosum	R. trifolii CB778	pH (5.0) (no interaction with Ca)	Andrew (1976)
	Trifolium rueppellianum	R. trifolii CB780	pH (5.0) (no interaction with Ca)	Andrew (1976)
	Vigna unguiculata	Rhizobium sp	pH (5.6) (interacts with Ca)	Munns and Fox (1976, 1977)
	Arachis hypogea	Rhizobium sp	pH (5.6) (no interaction with Ca)	Munns and Fox (1976, 1977)
	2 Stylosanthes sp	Rhizobium sp	pH (4.8) (negative interaction with Ca)	Munns and Fox (1976, 1977)
	Desmodium intortum	Rhizobium sp.	pH (5.6) (interacts with Ca)	Munns and Fox (1976, 1977)
	13 legumes	Rhizobium sp.	pH (6.0–7.0) (variable interaction with Ca)	Munns and Fox (1976, 1977)

[a] Minimum pH at which maximum nitrogen-fixation occurs or maximum yield is obtained.

primarily by pH, occurring only in soils of pH > 7.5 (Damirgi *et al.*
1967; Ham *et al.* 1971). However, a soil of high pH (7.2–8.3) was not
considered to be of primary importance in determining the distribution
of several plasmid-defined subgroups of serogroup 135 (Gross, Vidaver,
and Klucas 1979). Strains of serogroup 135 have performed poorly in
high pH (7.4 and 8.3) soils in Australia, but this may be due to host
incompatibility. Alkalinity (pH > 8.0) was considered to be one of the
major factors involved in nodulation failure of soyabean in Queensland
(Diatloff 1970).

Highly alkaline soils also tend to be high in sodium chloride,
bicarbonate, and borate, and frequently differentiation between these
factors is not made. This is especially important for high pH and
sodium chloride effects. Wilson (1931) showed *R. trifolii* and
R. leguminosarum to be sensitive to salinity, especially that caused by
more than 0.4 per cent sodium bicarbonate, while potassium, sodium,
sulphate, and chlorine seem to be less harmful; sodium chloride was
found to be less harmful than sodium sulphate, potassium sulphate, or
potassium chloride. Sodium chloride (120 mM) strongly inhibited
nodulation of soyabean, the depression in growth being greater in
symbiotic than nitrogen fed plants. In contrast, *Medicago sativa* was
unaffected, and both nitrate fed and symbiotic plants were equally
affected by salt (Bernstein and Ogata 1966). Salinity (148 mM) also
inhibits nitrogen fixation as well as nodule formation in *Neonotonia
wightii* (Wilson 1970). Nodules formed prior to the stress were not
damaged and rapidly regained their nitrogen-fixing ability when the
salt was removed. This may have been due to limited salt accumulation
by nodules, which indicated that nitrogen fixation in salt-stressed
plants may be host controlled. Hutton (1971) ranked eight legumes in
order of their response to 60 mM sodium chloride. Lucerne was most
resistant and *Neonotonia wightii* and *Desmodium* the most susceptible.
Nodulation was not prevented, but nitrogen fixation was reduced.
Three of the legumes were nodulated by the same *Rhizobium* strain
(CB756), but were ranked differently, which supports Wilson's (1970)
suggestion that salt effects on the host rather than on the *Rhizobium*
strain are the limiting factors. This was given further support by Wilson
and Norris (1970) who demonstrated reduced nodule-formation due to
salt stress (148 mM) of *Neonotonia wightii*, but developed nodules were
quite resistant to the stress. Growth of *Rhizobium* strain CB756 was
greatly retarded on media containing 80 mM sodium chloride, and
completely inhibited at 160 mM sodium chloride. The high salt
tolerance of *Medicago sativa* is in keeping with the ability of *R. meliloti* to
grow in 340 mM sodium chloride.

Problems due to salt toxicity of *R. trifolii* in peat cultures were
encountered by Steinborn and Roughley (1974), and in a subsequent

investigation, the toxic factor in sodium chloride was ascribed to the chloride ion. Unexpectedly, calcium chloride was more toxic than sodium chloride in broth and peat cultures at equivalent concentrations (Steinborn and Roughley 1975). As expected, *R. meliloti* was more tolerant of salt than *R. trifolii* and could be adapted to grow in media containing 600 mM sodium chloride. It was not possible to accurately determine the tolerances to salts in peat or soil by simple broth studies (Steinborn and Roughley ·1975). Upchurch and Elkan (1977) found that potassium chloride was more inhibitory at equivalent concentrations than sodium chloride (45 mM) to four strains of *R. japonicum*. Small colony (non-gum producing) isolates from these four strains were found to be more sensitive to salt than the large, gum-producing colonies. They concluded that the following factors were correlated: increased salt sensitivity, high symbiotic nitrogen-fixing ability, and small, non slimy colony morphology.

Soil fertility

Parker, Trinick, and Chatel (1977) have raised the question of wisdom of culturing rhizobia on media which are nutritionally luxurious, with the consequent possibility of selection of strains with reduced saprophytic competence. The results of Munns *et al.* (1979) and Date and Halliday (1979) independently support this idea, especially the fact that the success of some isolations from nodules of *Stylosanthes capitata* require media that are nutritionally stringent and have a reduced pH (Date and Halliday 1979).

Is there evidence in the literature suggesting that the soil fertility level may determine the qualitative and/or quantitative nature of the rhizobial status of a soil?

Hagedorn (1978) reported that the size of the soil populations of *R. trifolii* was related to such factors as soil texture and organic matter content, while effectiveness was positively correlated with soil base saturation level, content of exchangeable bases, and phosphorus level. Effectiveness was not associated with the size of the rhizobial population. Similar findings were reported for native islolates able to nodulate *Trifolium repens*, in that effectiveness was associated with soil base status (Holding and King 1963; Masterson 1968) but Gibson, Currow, Bergersen, Brockwell, and Robinson (1975) suggested that these conclusions may have been influenced by the use of an inappropriate host. In contrast to these findings, Gibson *et al.* (1975) could not relate effectiveness of isolates of *R. trifolii* to soil pH or management practices, no other soil characteristics were available for consideration. Hambdi, Yousef, Al-Tai, Al-Azawi, and Al-Baquari (1978) did not find any relation between soil population levels of *R. trifolii* or *R. meliloti* and pH, conductivity, lime content, or geographical location of a range of

Iraq soils. The authors did not determine the effectiveness of their isolates in relation to soil characteristics. Results of the studies mentioned are inconclusive, and may suggest that soil fertility is of minor significance in *Rhizobium* ecology. However, phosphorus and calcium can inhibit growth of broth cultures of rhizobia at levels which are higher than those frequently found in soil (Gillman and Bell 1978), as Keyser and Munns (1979*a*) have demonstrated.

Temperature

The effect of temperature on nitrogen fixation has been reviewed by Lie (1974); Dart, Day, Islam, and Döbereiner (1976), and Lie (1981).

The ecological significance of temperature on *Rhizobium* can be dramatic, in that not only does it affect survival of inoculum on seed and the level of rhizobia saprophytically resident in many soils, but the competitive ability of strains can be completely altered by changes in the temperature of legume growth.

Rhizobia are more susceptible to elevated temperatures when moist than when dried (Bowen and Kennedy 1959; Vincent *et al.* 1962; Wilkins 1967). Survival of rhizobia on seed of tropical legumes, *Trifolium* spp, *P. sativum* and *M. sativa* sown into moist soil was severely reduced at 40 °C. The degree of reduction was dependent on the size of the initial population, the time of exposure, and the strain of *Rhizobium* (Bowen and Kennedy 1959). Similarly, survival of cowpea-type *Rhizobium* sp CB756 inoculated on to seed of *Neonotonia wightii* was poorer when stored at 50 °C and partially desiccated at 75 per cent relative humidity than when desiccation was more complete at 40 per cent relative humidity. Strain differences were noted (Philpotts 1977*b*). Although there are strain differences, the upper limit for growth is in the vicinity of 40 °C (Bowen and Kennedy, 1959; Ishizawa 1953) and soil rhizobial population levels may be greatly reduced when soil temperatures exceed 40–50 °C for several hours daily.

Marshall (1964) demonstrated that montmorillonite protected desiccated *R. trifolii* and *R. meliloti* against elevated temperatures (50 and 70 °C) but the results do not reveal any effect of this clay on the slow-growing *R. japonicum* or *R. lupini*. Incubation of *R. lupini* in soil amended with clay prior to desiccation and heating increased survival over non-incubated treatments, but the montmorillonite was still without effect. The mechanism of protection against high temperatures of fast- but not slow-growing strains is unknown and, as has been pointed out by Brockwell and Phillips (1970), is the subject of controversy.

Elevated temperatures may delay nodule initiation (Gibson 1967; Roughley and Dart 1970), nodule development (Kumarasinghe and Nutman 1979), and interfere with nodule structure (Roughley 1970) and nodule functioning in temperate legumes. Few critical studies are available for tropical legumes, but as Date (1977) indicated, the optimum temperature for nodule formation and nitrogen fixation is generally close to 30 °C.

An interesting aspect of the effect of temperature on nodule formation is the dramatic effect that temperature changes have on the competitive ability of *Rhizobium* strains. Caldwell and Weber (1970) observed that the proportion of specific serogroup of *R. japonicum* in the nodules of *G. max* changed as the planting date was altered. Subsequently, Weber and Miller (1972) found that at temperatures ranging from 10 °C to 30 °C the proportion of nodules formed by a particular serogroup increased (groups 94 and the composite groups C1, C2, and C3) or decreased (groups 110 and 125). The distribution of serogroups in nodules (Weber and Miller 1972) was very similar to that found in the date-of-planting field studies (Caldwell and Weber 1970). It was assumed that temperature influenced the population of serogroups in soil by differential effects upon growth (Weber and Miller 1972). The results of intensive investigations on the competitive abilities of the paired strains SU51 and SU126 on *Medicago truncatula* and *M. sativa* (Couper and Vincent, personal communication) may reveal something of effects of temperature upon strain competition. Under optimum conditions for growth (15–16 °C for *M. truncatula* and 25–26 °C for *M. sativa*) SU126 was more competitive than SU51 in nodule formation on *M. truncatula,* but the reverse was true on *M. sativa*. However, by altering the root and shoot temperatures, 30–100 times reversal of nodulating competitiveness was observed. Hardarson and Jones (1979) demonstrated a similar phenomenon for *R. trifolii* on clover plants. Strain 75 str. was more competitive at 12 °C than strain 33 spc. The reverse applied at 25 °C. Interestingly, this trend was unrelated to the geographic origin of the *Rhizobium*, as strain 33 spc came from Iceland. Roughley *et al.* (1976) sampled *Trifolium subterraneum* cv. Woogenellup at various times up to 22 weeks after sowing, to determine if the proportion of nodules due to the inoculum strain varied during the growing season. The site supported a native population of *R. trifolii* of $1500 \ g^{-1}$. No substantial changes were noted in the percentage of nodules formed by either single- or mixed-strain inocula. Although temperature data are not provided (nor are planting dates), it is probable that during the five month period of sampling, temperatures differed significantly. These results are in contrast to those of Weber and Miller (1972) for soyabean, and the laboratory results for *Medicago* spp obtained by Vincent and Couper. Roughley *et al.* (1976) commented that their results may not apply to all soils, particularly if higher populations of native rhizobia were present and in soils which restricted the movement of rhizobia. However it is also possible that the strain populations were not differentially stimulated as Weber and Miller

(1972) postulated for strains of *R. japonicum*. Other mechanisms may be involved. Onset and rate of infection-thread formation, and success of nodule formation are temperature sensitive (Kumarasinghe and Nutman 1979).

Water availability

Soil micro-environments inhabited by micro-organisms are essentially aquatic, in that colonies are surrounded by a film of water which may exhibit a range of water activities (Stotzky 1972). The importance of desiccation as a factor in determining survival of rhizobia in soil is uncertain. In sandy soils, low in organic matter and clays, rhizobia die when desiccated (Marshall 1964; Chatel and Parker 1973*b*; Bushby and Marshall 1977*a*, *b*). However, there is very little evidence to suggest that in heavier soils, desiccation *per se* is very important under field conditions. Brockwell and Whalley (1970) demonstrated that *R. meliloti* associated with pelleted, inoculated *Medicago* seed persisted in high numbers for up to 200 days when in storage or when mixed with dry soil. Desiccation of red basaltic soils of Uruguay for 188 days, did not increase the loss of viability of *R. trifolii* over and above that observed in undried soil (Date 1965).

Differences in resistance to desiccation between fast- and slow-growing rhizobia were attributed to differential water contents at various relative humidities (Bushby and Marshall 1977*b*). Pena-Cabriales and Alexander (1979) were not able to confirm this difference. There are several possible reasons for this, the first being the differences in the soils used. Bushby and Marshall (1977*b*) reported that only the fast-growing rhizobia were protected from the effects of desiccation by clay, the viability of slow-growing strains being unaffected. Hence, clay minerals in the silt loam used by Pena-Cabriales and Alexander (1979) may have enhanced survival of the fast-growing strains such that the differences between the two groups were not manifest. Another explanation may be the desiccation rate. When dried rapidly, fast-growing rhizobia are more susceptible to desiccation than slow-growing strains, but the reverse is true when they are dried slowly.

The biphasic nature of survival curves of desiccated rhizobia has been demonstrated many times (Date personal communication; Vincent, Thompson, and Donovan 1962; Pena-Cabriales and Alexander 1979). The initial rapid decline occurs during water loss by the soil and a slower rate after all free moisture has been removed. Thus there are at least two mechanisms of death, one due to dehydration of vegetative cells and the second, subsequent to this. It is possible that the slow death-rate of desiccated rhizobia associated with the second mechanism is due to oxygen toxicity at low relative vapour pressures (cf. Webb 1965; Cox and Heckly 1973). There was negligible death of freeze-dried rhizobia stored over silica gel in an atmosphere of nitrogen, but significant loss of viability under air (Table 2.5). Variability between strains in their susceptibility to air is demonstrated (also in Table 2.5), and this may account for some of the differences observed between rhizobia after long-term storage. Preliminary experiments suggest that for most rhizobia, oxygen becomes toxic when stored at relative humidities lower than 10 per cent. Perhaps the inclusion of antioxidants in pellets may give protection to rhizobia inoculated on to seed.

TABLE 2.5

Effect of oxygen on the viability of desiccated rhizobia stored over silica gel

Strain	Host	Initial viable count[a]	Log viable count after 9 weeks storage under:	
			air	nitrogen
Slow-growing				
CB756	Broad spectrum	8.2	4.8	7.8
CB376	*Lotononis bainesii*	8.2	7.0	7.6
CB1552	*Stylosanthes guianensis*	8.2	6.0	7.2
CB627	*Desmodium*	8.2	6.0	7.6
Fast-growing				
CB782	*Trifolium semipilosum*	8.0	5.6	6.6
TA1	*Trifolium repens*	8.0	6.8	7.4
TA101	*Pisum sativum*	8.0	6.4	6.8
NGR8	*Leucaena leucocephala*	7.6	4.8	5.8

[a] Log viable numbers per millilitre after freeze drying.

Rhizobia are more susceptible to partial than complete desiccation (Vincent *et al.* 1962; Philpotts, 1977*b*; Davidson and Reuszer 1978; Bushby unpublished data).

The mechanisms of protection are unknown (Vincent *et al.* 1962; Bushby and Marshall 1977*b*). Particulate matter such as peat and certain clays can give protection. Dart *et al.* (1969) suggested that the fine matrix of fibrous material surrounding cells of *R. trifolii* in peat culture may be protective, and Marshall (1964) and Bushby and Marshall (1977*b*) proposed that the mechanism of clay protection might be via altered water relations. The CSP minerals, goethite, haematite, and kaolinite are not protective but the high activity CSC clays do protect rhizobia from desiccation (Section 3, see p. 39).

Worrall and Roughley (1976) studied the effects of moisture stress on infection and nodule formation of subterraneum clover. A 2 per cent reduction in soil water content resulted in a ten fold increase in water potential which significantly reduced infection thread formation and completely inhibited nodulation, although the rhizosphere population of *R. trifolii* was unaffected. Deformation of root hairs was common at low moisture levels and this appeared to be the reason for reduced infection thread and nodule formation. Removal of the water stress resulted in rapid recovery of infection thread formation and nodules were formed at rates equivalent to those in non-stressed plants. The authors commented that the short and swollen root hairs on the stressed plants were similar to those observed on *M. sativa* grown under highly saline conditions (Lakshmi-Kumari, Singh, and Subba Rao 1974). Effects of severe water stress on nodule structure and function has been summarized by Sprent (1976). Generally, moderate stress slowed acetylene reduction and respiratory activity, these effects being reversible. More severe stress stopped nitrogen fixation and induced structural changes within nodules. These changes were not reversible. Shedding of these nodules probably occurs with the need to form new nodules before nitrogen fixation can resume.

Little attention has been paid to the significance of waterlogging in the ecology of *Rhizobium*. The possibility of differences between rhizobia in their response to flooding is suggested by the work of De Polli, Franco, and Döbereiner (1973) who reported reduced nodulation of *Centrosema pubescens* but not soyabean when grown in previously flooded soils. *Rhizobium trifolii* and *R. japonicum* lost viability in flooded soils over a six week period, the decline in *R. trifolii* populations being greater than that of *R. japonicum* (Osa-Afiana and Alexander 1979). Predation by *Bdellovibrio* or protozoa was not involved in the decline in populations. Paterno (1979) compared survival of two soyabean and two *Vigna radiata* rhizobia under different cropping systems and inoculation regimes. Nodulation of soyabean was increased by sowing inoculated seed after rice, but the proportion of nodules due to the inoculum strain remained about the same. Thus, both soyabean strains survived the flooding period, but numbers were probably reduced. The situation was similar with *V. radiata* rhizobia, except that native rhizobia successfully competed with the inoculum strain both before and after the rice crop. Consequently, inoculation following rice increased the proportion of nodules attributable to the inoculum strain. Strain differences in response to flooding were marginal. Experience in Western Australia suggested that flooding did not seriously affect the nodulation of *Trifolium* spp (Parker *et al.* 1977).

Recent investigations with *Escherichia coli* indicate that cell–soil particle interactions occur at electrolyte concentrations sufficient to reduce the repulsive forces between like-charged bodies below a certain value (Roper and Marshall 1979). By decreasing the ionic strength of the suspending medium, a point is reached where colloidal material and bacteria repel each other, and under such circumstances, *E. coli* remains in suspension and does not flocculate out. The importance of this phenomenon in rhizobial ecology in flooded rice paddy fields and other waterlogged conditions is unknown. It is possible that survival in flooded soils is being underestimated by sampling only the sediment, as nothing is known of possible desorption of rhizobia from soil particles and their movement into the overlying water. If this occurs it could provide an explanation for survival of this obligate aerobe under seemingly anaerobic conditions. Osa-Afiana and Alexander (1979) have measured the number of *R. trifolii* and *R. japonicum* in the water phase in the laboratory and their results show a very low proportion of the total population resident in the water phase. However, the artificiality of the system could have resulted in very much higher ionic strengths of the water than occur under natural conditions. This would reduce numbers due to flocculation and co-precipitation with soil colloidal particles. An increase in the ionic strength may explain the time-dependent decrease of both rhizobia within the water phase.

It is commonly accepted that flooding reduces nitrogen fixation in nodulated legumes through a reduction in oxygen supply to nodules (Sprent 1976; Minchin, Summerfield, Eaglesham, and Stewart 1978). Some recent studies by D. E. Byth (University of Queensland), R. J. Lawn (CSIRO Division of Tropical Crops and Pastures) and co-workers, however, indicate that several legumes, including soyabean, have substantial abilities to adjust to water-logged soils, particularly when the soil saturation occurs during early seedling growth, and the water table is maintained at a constant level in the soil, e.g. at about 15 cm below the soil surface. Initial studies indicate that nodule mass was increased in high water-table culture by up to 35 times that of control plants, and in some cases accounted for 5 per cent of total plant dry matter (Hunter, de Jabrun, and Byth 1980). Subsequent studies in both the glasshouse (K. Nathanson *et al.* unpublished) and field (A. Garside *et al.* unpublished) have shown that both nodulation and plant growth may be enhanced by high water-table culture. After application of the water table, seedlings initially turned yellow–green, but subsequently greened up after 10–15 days. Thereafter, growth rates far exceeded those of well watered control plants so that ultimate seed yields were highest in the near saturated soil culture. Nodules developed in the wet soil culture are characterized by profuse development of aerenchyma tissue in the cortex, and have acetylene reduction activities similar to those of normal field-grown plants. The agronomic and physiological significance of these unexpected results are being further investigated (R. J. Lawn, personal communication).

Direct measurement of lateral movement of rhizobia in soil has indicated that moisture content, rather than soil pore-neck size distribution, was more likely to restrict movement of rhizobia via an effect on the continuity of water pathways. In each of the soils tested (coarse sand, fine sand, and silt loam), lateral movement stopped when water pathways became discontinuous (Hambdi 1971). Griffin and Quail (1968) recorded similar results for *Pseudomonas aeruginosa*; movement was restricted at field capacity. Brockwell and Whalley (1970) suggested that some of the variability in nodulation of *Medicago* species inoculated seed sown into dry soil, may have been due to the rapid disappearance of continuous water pathways after germinating rains. Movement can be related also to soil type, pH, and bacterial size (Bitton, Lahav, and Henis 1974). Bacteria–soil interactions (see Section 3) have major effects upon the spread of rhizobia through soils. Worrall and Roughley (1976) did not detect any movement of *R. trifolii* over a 15 day period, either through the soil or along the root system of *Trifolium subterraneum* roots, even though the soil was held at field capacity. However, movement was observed due to washing through the soil during watering. Paterno (1979) reported limited vertical movement in a soil held at field capacity. Experience in north-eastern Australia (H. V. A. Bushby and R. A. Date) suggests limited movement from inoculated seed in field experiments, even when plots have been awash with water due to torrential rain. Movement of *Rhizobium* sp along the root systems of tropical legumes has been measured by Date (personal communication) and Bushby (1980*b*). Lateral movement of 7 cm and vertical movement of 19 cm from the inoculated seed was reported over 58 days for *Vigna mungo* grown in a red earth (Bushby 1981*b*). In a sandy podzolic, maximum horizontal movement of 75 cm for both CB756 and CB627 on *Macroptilium atropurpureum* and on *Desmodium intortum* respectively, were measured over 68 weeks (Date, personal communication).

Biological factors

The general topic of biological factors in ecology has been reviewed by Stotzky (1972) and Gray *et al.* (1968) and specific aspects concerning *Rhizobium*, by Trinick (1970), Alexander (1978), Chowdhury (1977) and Parker *et al.* (1977). Generally, the role of biotic factors in *Rhizobium* ecology have been considered within the categories suggested by Malcolm (1966). These include antagonistic interactions such as competition, amensalism, predation, parasitism, and lysis, beneficial associations such as commensalism, proto-co-operation and symbiosis, and neutral associations.

Competition for limited nutritional resources is frequently put forward as a major reason for the disappearance of micro-organisms inoculated into soil. Garrett (1965) suggested that the most frequent cause of death of micro-organisms in soil is starvation and support for this is evident as competitive effects can be reduced when a complete nutrient solution is added to soil (Malcolm 1966).

The predatory bacterium *Bdellovibrio* and many protozoa are known to attack rhizobia (Brockwell 1974; Alexander 1978), and in some situations could place an upper limit on soil populations. However in nearly all cases, the role of these predators under natural conditions is uncertain, especially when variation of microbe–predator interactions are introduced by such soil particles as clay (Roper and Marshall 1978*a*, *b*). Bacteriophages have also been implicated as agents in control of *Rhizobium* populations in soil (Chowdhury 1977; Parker *et al.* 1977) but in all cases unambiguous statements on their role under field conditions are not possible. Recently, Evans *et al.* (1979*a*, *b*) reported on the effects of rhizoplane phage populations on nodulation and effectiveness of *R. trifolii* on *Trifolium glomeratum* grown in seedling tubes. Phage reduced the rhizoplane population of *R. trifolii* and induced growth of variant substrains which were less susceptible to bacteriophage and usually ineffective in nitrogen fixation (Evans, Barnet, and Vincent 1979*a*). Growth temperature was an important factor, as at higher temperatures the frequency of strain variants increased and the number of nodules due to the parent strain decreased. Under similar growth conditions, phage altered the competitive nodule-forming ability of a susceptible strain in a mixture of two strains (Evans, Barnet, and Vincent 1979*b*). In the presence of phage, resistant or partially-resistant rhizobia, which were normally less competitive than the phage-susceptible parent strain, were able to form a proportion of nodules which was representative of the inter-strain ratio of root populations. The authors underlined the need for caution in extrapolating their results to field conditions. Phage–*Rhizobium* interactions observed under artificial conditions could be altered by clay or organic matter in soils, so as to completely remove antagonistic effects (Roper and Marshall 1974, 1978*a*).

2.5 Concluding remarks

Isolation of rhizobia from nodules is rarely accompanied by even the most rudimentary description of the soil from which the host plant came. This wastes time and resources, as much effort is required in determining whether a particular strain is likely to be suitable for certain soils. As indicated in Section 4, (p. 50) soil acidity is a very important factor in nodule formation and nitrogen fixation. Therefore, the time taken to measure the soil pH from which a nodule was taken could be of much value. Date, Burt, and Williams (1979) found that accessions of *Stylosanthes* from alkaline soils formed effective associations only with rhizobia isolated from similar conditions. Accessions from wetter, more acidic, soils formed effective symbioses with a wider range of rhizobia. The authors made the obvious, but frequently forgotten statement that a 'legume and its associated *Rhizobium* must be adapted to the same soil conditions for successful nodulation and growth'.

Under undisturbed conditions, the effective symbiotic relationship between legume and *Rhizobium* represents an integration for that soil of many, if not all, of the factors discussed in this chapter. Therefore, when we impose new legumes, rhizobia, or any management and fertilizer practice on the environment, new equilibria must be attained. This implies that what were once adapted strains of rhizobia, may become 'aliens' (as described by Alexander 1971), and therefore not suited to the new environment. Under these circumstances, what were the dominant strains might disappear. Hence the phenomenon of strain disappearance probably occurs whenever man alters the environment by farming. So perhaps one method of tackling the question of the bases of biological and physical rejection of aliens would be to follow changes of the natural *Rhizobium* community induced by known inputs imposed by man.

In this review, the possibility that rhizobia from relatively unweathered soils are unsuited to highly weathered soils, has been suggested. If this is true, for whatever reasons, it is not surprising that in many instances the inoculated strain disappears, especially if required to live saprophytically for an extended period. If advances are to be made in the study of *Rhizobium* ecology, questions such as these need to be answered.

References

Alexander, M. (1971). *Microbial ecology.* John Wiley, New York.

—— (1978). In *Biological nitrogen fixation in farming systems of the tropics* (ed. A. Ayanaba and P. J. Dart) pp. 99–114. John Wiley, New York.

Andrew, C. S. (1976). *Aust. J. agric. Res.* **27,** 611–23.

—— and Norris, D. N. (1961). *Aust. J. agric. Res.* **12,** 40–55.

Barrow, N. J. (1964). *Aust. J. exp. Agric. Anim. Husb.* **9,** 437–44.

Bell, L. C. and Gillman, G. P. (1978). In *Mineral nutrition of legumes in tropical and subtropical soils* (ed. C. S. Andrew and E. J. Kamprath) pp. 37–58. C.S.I.R.O., Melbourne.

Bernstein, L. and Ogata, G. (1966). *Agron. J.* **58,** 201–3.

Beveridge, T. J. and Murray, R. G. E. (1976). *J. Bacteriol.* **127,** 1502–18.

Bezdicek (1972). *Soil Sci. Soc. Am. J.* **36,** 305–7.

Bhardwaj, K. K. R. and Guar, A. C. (1971). *Zentral. Bakteriol. Parasit. Infektionskraukheiten Hygiene,* **126,** 649–99.

—— —— (1972). *Ind. J. Microbiol.* **12,** 19–21.

Bitton, G., Lahav, N., and Henis, Y. (1974). *Pl. Soil* **40,** 373–80.

Bohlool, B. B. and Schmidt, E. L. (1976). *J. Bacteriol.* **125,** 1188–94.

Bowen, G. D. and Kennedy, M. M. (1959). *Qld. J. agric. Sci.* **16,** 177–97.

—— and Rovira, A. D. (1976). *A. Rev. Phytopathol.* **14,** 121–44.

Breeuwsma, A. and Lyklema, J. (1973). *J. Colloid Interface Sci.* **43,** 437–48.

Brinton, C. C. and Lauffer, M. A. (1959). In *Electrophoresis: theory, methods and applications*

(ed. M. Bier) pp. 427–92. Academic Press, New York.
—— Buzzell, A., and Lauffer, M. A. (1954). *Biochem. Biophys.* **15,** 533–42.
Brockwell, J. (1974). *Proc. Ind. natn. Sci. Acad.* **B40,** 687–99.
—— and Phillips, L. J. (1970). *Aust. J. exp. Agric. Anim. Husb.* **10,** 739–44.
—— and Whalley, R. D. B. (1970). *Aust. J. exp. agric. Anim. Husb.* **10,** 445–59.
—— Bryant, W. G., and Gault, R. R. (1972). *Aust. J. exp. Agric. Anim. Husb.* **12,** 407–13.
—— Schwinghamer, E. A., and Gault, R. R. (1977). *Soil Biol. Biochem.* **9,** 19–24.
Bushby, H. V. A. (1981*a*). *Soil Biol. Biochem.* **13,** 237–39.
—— (1981*b*). *Soil Biol. Biochem.* **13,** 241–5.
—— and Marshall, K. C. (1977*a*). *Soil Biol. Biochem.* **9,** 143–7.
—— —— (1977*b*). *J. gen. Microbiol.* **99,** 19–27.
Caldwell, B. E. and Weber, D. F. (1970). *Agron. J.* **62,** 12–14.
Carvalho, M. M. (1978). Ph.D. Thesis, University of Queensland.
—— Bushby, H. V. A., and Edwards, D. G. (1981). *Soil. Biol. Biochem.* In press.
Chatel, D. L. and Greenwood, R. M. (1973*a*). *Soil Biol. Biochem.* **5,** 433–40.
—— —— (1973*b*). *Soil Biol. Biochem,* **5,** 799–808.
—— —— (1973*c*). *Soil Biol. Biochem.* **5,** 809–13.
—— and Parker, C. A. (1973*a*). *Soil Biol. Biochem.* **5,** 425–32.
—— —— (1973*b*). *Soil Biol. Biochem.* **5,** 415–23.
—— Shipton, W. A., and Parker, C. A. (1973). *Soil Biol. Biochem.* **5,** 815–24.
Chowdhury, M. S. (1977). In *Exploiting the legume*–Rhizobium *symbiosis in tropical agriculture* (ed. J. M. Vincent, A. S. Whitney, and J. Bose) pp. 385–412. College Trop. Agric., Misc. Publ. 145, University of Hawaii.
Cloonan, M. J. (1963). *Aust. J. Sci.* **26,** 121.
—— and Vincent, J. M. (1967). *Aust. J. exp. Agric. Anim. Husb.* **7,** 181–9.
Cox, C. S. and Heckly, R. J. (1973). *Can. J. Microbiol.* **19,** 189–94.
Currier, W. W. and Strobel, G. A. (1976). *Plant Physiol.* **57,** 820–23.
—— —— (1977). *Science N.Y.* **196,** 434–6.
Damirgi, S. M., Frederick, L. R., and Anderson, I. C. (1967). *Agron. J.* **59,** 10–12.
Dart, P. J. (1974). In *The biology of nitrogen fixation* (ed. A. Quispel) pp. 381–429. North-Holland, Amsterdam.
—— Roughley, R. J., and Chandler, M. R. (1969). *J. appl. Bacteriol.* **32,** 352–7.
—— Day, J., Islam, R., and Döbereiner, J. (1976). In *Symbiotic nitrogen fixation in plants* (ed. P. S. Nutman) pp. 361–384. Cambridge University Press.
Date, R. A. (1965). Report No. 2012, Food and Agricultural Organization.
—— (1974). *Proc. Ind. natn. Sci. Acad. B. Biol. Sci.* **40,** 700–12.
—— (1975). In *Symbiotic nitrogen fixation in plants* (ed. P. S. Nutman) pp. 137–150. Cambridge University Press.
—— (1977). In *Exploiting the legume*–Rhizobium *symbiosis in tropical agriculture* (ed. J. M. Vincent, A. S. Whitney, and J. Bose) Workshop Proc. pp. 293–312. College of Tropical Agriculture, Misc. Pub. 145. University of Hawaii.
—— and Brockwell, J. (1978). In *Plant relations in pastures* (ed. J. R. Wilson) pp. 202–216. C.S.I.R.O., Australia.
—— and Halliday, J. (1979). *Nature, Lond.* **227,** 62–4.
—— and Roughley, R. J. (1977). In *A treatise on dinitrogen fixation,* Section 4. *Agronomy and Ecology* (ed. R. W. F. Hardy and A. H. Gibson) pp. 243–75. John Wiley, New York.
—— Burt, R. L., and Williams, W. T. (1979). *Agro-Ecosystems* **5,** 57–67.
Davidson, F. and Reuszer, H. W. (1978). *Appl. environ. Microbiol.* **35,** 94–6.
Dazzo, F. B., Napoli, C. A., and Hubbell, D. H. (1976). *Appl. environ. Microbiol.* **32,** 166–71.
De Polli, H., Franco, A. A., and Döbereiner, J. (1973). *Pesqui. Agropecu. Bras. Ser. Agron.* **8,** 133–8.
Deshpande, T. L., Greenland, D. J., and Quirk, J. P. (1964). *Trans. 8th. Int. Congr. Soil

Sci., Bucharest **3,** 1213–25.

Diatloff, A. (1969). *Aust. J. exp. Agric. Anim. Husb.* **9,** 357–60.

—— (1970). *Queensl. J. Agric. Anim. Sci.* **27,** 279–93.

—— (1977). *Soil Biol. Biochem.* **9,** 85–8.

—— and Brockwell, J. (1976). *Aust. J. exp. Agric. Anim. Husb.* **16,** 514–21.

Donald, C. M. and Williams, C. H. (1974). *Aust. J. Agric. Res.* **5,** 664–87.

Dudman, W. F. and Brockwell, J. (1968). *Aust. J. Agric. Res.,* **19,** 739–47.

Evans, J., Barnet, Y. M., and Vincent, J. M. (1979*a*). *Can. J. Microbiol.* **25,** 968–73.

—— —— —— (1979*b*). *Can. J. Microbiol.* **25,** 974–8.

Fehrmann, R. C. and Weaver, R. W. (1978). *Soil Sci. Soc. Am. J.* **42,** 279–81.

Foster, R. C. and Rovira, A. D. (1976). *New Phytol.* **76,** 343–52.

Franco, A. A. and Vincent, J. M. (1976). *Pl. Soil* **43,** 27–48.

Fred, E. B., Baldwin, I. L., and McCoy, E. (1932). *Root nodule bacteria and leguminous plants.* University of Wisconsin, Madison.

Garrett, S. D. (1965). In *Ecology of soil-borne plant pathogens* (ed. K. F. Baker and W. C. Soyder.) pp. 4–17. John Murray, London.

Gibson, A. H. (1967). *Aust. J. biol. Sci.,* **20,** 1087–104.

—— Date, R. A., Ireland, J. A., and Brockwell, J. (1976). *Soil Biol. Biochem.,* **8,** 395–401.

—— Currow, B. C., Bergersen, F. J., Brockwell, J., and Robinson, A. C. (1975). *Soil Biol. Biochem.* **7,** 95–102.

Gillman, G. P. (1974). *Aust. J. Soil Res.* **12,** 173–6.

—— and Bell, L. C. (1976). *Aust. J. Soil Res.* **14,** 351–60.

—— —— (1978). *Aust. J. Soil Res.* **16,** 67–77.

Giltner, W. and Longworthy, H. V. (1916). *J. agric. Res.* **5,** 927–42.

Graham, P. H. and Parker, C. A. (1964). *Pl. Soil* **20,** 383–96.

Gray, T. R. G., Baxby, P., Hill, I. R., and Goodfellow, M. (1968). In *The ecology of soil bacteria* (ed. T. R. G. Gray and D. Parkinson) pp. 171–97. Liverpool University Press, Liverpool.

Greenland, D. J. (1965). *Soil Fertilizers* **28,** 415–25.

—— Oades, J. M., and Sherwin, J. W. (1968). *J. Soil Sci.* **19,** 123–6.

—— (1971). *Soil Sci.* **111,** 34–41.

—— and Mott, C. J. B. (1978). In *The chemistry of soil constituents* (ed. D. J. Greenland and M. H. B. Hayes) pp. 321–354. John Wiley, New York.

Griffin, D. M. and Quail, G. (1968). *Aust. J. biol. Sci.* **21,** 579–82.

Gross, D. C., Vidaver, A. K., and Klucas, R. V. (1979). *J. gen. Microbiol.* **114,** 257–66.

Hagedorn, C. (1978). *Soil Sci. Soc. Am. J.* **42,** 447–51.

Ham, G. E., Frederick, L. R., and Anderson, I. C. (1971). *Agron. J.* **63,** 69–72.

Hambdi, Y. A. (1971). *Soil Biol. Biochem.* **3,** 121–6.

—— Yousef, A. N., Al-Tai, A., Al-Azawi, S. K., and Al-Baquari, M. (1978). *Soil Biol. Biochem.* **10,** 148–50.

Hardarson, G. and Jones, G. (1979). *Ann. appl. Biol.* **92,** 229–36.

Hingston, F. J., Posner, A. M., and Quirk, J. P. (1972). *J. Soil Sci.* **23,** 177–92.

Holding, A. J. and King, J. (1963). *Pl. Soil* **18,** 191–8.

—— and Lowe, J. F. (1971). *Pl. Soil* Special Volume, 153–66.

Hubbel, D. S. and Chapman, J. E. (1946). *Soil Sci.* **62,** 271–81.

Hunter, M. N., de Jabrun, P. L. M., and Byth, D. E. (1980) *Aust. J. exp. Agric. Anim. Husb.* **20,** 339–45.

Hutton, E. M. (1971). *Sabrao Newslett.* **3,** 75–81.

Ishizawa, S. (1953). *J. Soil Sci. Manure* **24,** 227–30.

Israel, D. W. and Jackson, W. A. (1978). In *Mineral nutrition of legumes in tropical and subtropical soils* (ed. C. S. Andrew and E. J. Kamprath) pp. 113–30. C.S.I.R.O., Australia.

Ireland, J. A. and Vincent, J. M. (1968). *Trans. 9th. Int. Cong. Soil Sci.* **2,** 85–93.

Jackson, M. L., Mackie, W. Z., and Pennington, R. P. (1946). *Soil Sci. Soc. Am. J.* **11,** 57–63.

Johnston, A. W. B. and Beringer, J. E. (1976). *J. appl. Bacteriol.* **40,** 375–80.

Jones, D. G. (1966). *Pl. Soil* **24,** 250–60.

Jones, R. C. and Uehara, G. (1973). *Soil Sci. Soc. Am. J.* **37,** 792–8.

Jones, R. M. and Date, R. A. (1975). *Aust. J. exp. Agric. Anim. Husb.* **15,** 519–26.

Keng, J. C. W. and Uehara, G. (1973). *Soil Crop Sci. Soc. Fl.* **33,** 120–6.

Keyser, H. H. and Munns, D. N. (1979*a*). *Soil Sci. Soc. Am. J.* **43,** 519–23.

—— —— (1979*b*). *Soil Sci. Soc. Am. J.* **43,** 500–3.

—— —— and Hohenberg, J. S. (1979). *Soil Sci. Soc. Am. J.* **43,** 719–22.

Kumarasinghe, R. M. and Nutman, P. S. (1979). *J. exp. Bot.* **30,** 503–15.

Labandera, C. A. and Vincent, J. M. (1975). *Pl. Soil* **42,** 327–47.

Lahav, N. (1962). *Pl. Soil* **17,** 191–208.

Lakshmi-Kumari, M., Singh, C. S., and Subba Rao, N. S. (1974). *Pl. Soil* **40,** 261–8.

Larsen, S. and Widdowson, A. E. (1968). *J. Sci. Fd. Agric.* **19,** 693–5.

Lee, B. (1980). *Rural Res.* **106,** 4–9.

Lie, T. A. (1969). *Pl. Soil* **31,** 391–406.

—— (1974). In *The biology of nitrogen fixation* (ed. A. Quispel). pp. 555–82. North Holland, Amsterdam.

—— (1981). In *Nitrogen fixation*, Vol. 1. *Ecology* (ed. W. J. Broughton) pp. 104–34. Clarendon Press, Oxford.

Lowther, W. L. and Loneragan, J. F. (1968). *Pl. Physiol.* **43,** 1362–6.

—— —— (1970). *Proc. 11th Int. Grassl. Cong.,* pp. 446–50.

Malcolm, W. M. (1966). *Bot. Rev.* **32,** 243–54.

Marques Pinto, C., Yao, P. Y., and Vincent, J. M. (1974). *Aust. J. agric. Res.* **25,** 317–29.

Marshall, K. C. (1964). *Aust. J. agric. Res.* **15,** 273–81.

—— (1967). *Aust. J. biol. Sci.* **20,** 429–38.

—— (1968*a*). *Biochim. biophys. Acta* **156,** 179–86.

—— (1968*b*). *Trans. 9th. Int. Cong. Soil Sci.* **3,** 275–80.

—— (1969*a*). *Biochim. biophys. Acta* **193,** 472–4.

—— (1969*b*). *J. gen. Microbiol.* **56,** 301–6.

—— (1971). In *Soil biochemistry* Volume 2 (ed. A. D. McLaren and J. J. Skujins) pp. 409–45. Marcel Dekker, New York.

—— (1975). *A. Rev. Phytopathol.* **13,** 357–73.

—— (1976). *Interfaces in microbial ecology.* Harvard University Press, Cambridge, MA.

—— and Cruickshank, R. H. (1973). *Archiv. Mikrobiol.* **91,** 29–40.

—— —— and Bushby, H. V. A. (1975). *J. gen. Microbiol.* **91,** 198–200.

—— Mulcahy, M. J., and Chowdhury, M. S. (1963). *J. Aust. Inst. agric. Sci.* **29,** 160–4.

—— Stout, R., and Mitchell, R. (1971). *J. gen. Microbiol.* **68,** 337–48.

Masterson, C. L. (1968). *Trans. 9th. Int. Congr. Soil Sci. Part 2,* 95–102.

Means, U. M., Johnson, H. W., and Date, R. A. (1964). *J. Bacteriol.* **87,** 547–53.

Mekaru, T. and Uehara, G. (1972). *Soil Sci. Soc. Am. J.* **36,** 296–300.

Minchin, F. R., Summerfield, R. J., Eaglesham, A. R. J., and Stewart, K. A. (1978). *J. agric. Sci.* **90,** 355–66.

Munns, D. N. (1965). *Aust. J. agric. Res.* **32,** 90–102.

—— (1968). *Pl. Soil* **28,** 129–46.

—— (1970). *Pl. Soil* **32,** 90–102.

—— (1977). In *Exploiting the legume*–Rhizobium *symbiosis in tropical agriculture* (ed. J. M. Vincent, A. S. Whitney, and J. Bose) pp. 211–36. University of Hawaii, College. Trop. Agr., Misc. Pub. 145.

—— (1978). In *Mineral nutrition of legumes in tropical and subtropical soils* (ed. C. S. Andrew and E. J. Kamprath) pp. 247–64. C.S.I.R.O., Melbourne.

—— and Fox, R. L. (1976). *Pl. Soil* **45,** 701–5.

—— —— (1977). *Pl. Soil* **46**, 533–48.

—— —— and Koch, B. L. (1977). *Pl. Soil* **46**, 591–601.

—— Keyser, H. H., Fogle, V. W., Hohenberg, J. S., Righetti, T. L., Lauter, D. L., Zaroug, M. G., Clarkin, K. L., and Whiteacre, K. W. (1979). *Agron. J.* **71**, 256–60.

Mytton, L. R. and de-Felice, J. (1977). *Annl. appl. Biol.* **87**, 83–94.

Newman, E. I. and Bowen, H. J. (1974). *Soil Biol. Biochem.* **6**, 205–9.

Nicol, H. and Thornton, H. G. (1941). *Proc. R. Soc. London Series* **B 130**, 32–59.

Norris, D. O. (1965). *Pl. Soil* **22**, 143–66.

—— (1967). *Trop. Grassl.* **1**, 107–21.

—— and Date, R. A. (1976). In *Tropical pasture research: principles and methods*. CAB Bull. No. 51. (ed. N. H. Shaw and W. W. Bryan) pp. 134–74. Comm. Agric. Bur., Farnham Royal, U.K.

Osa-Afiana, L. O. and Alexander, M. (1979). *Soil Sci. Soc. Am. J.* **43**, 925–30.

Ottow, J. C. G. (1975). *A. Rev. Microbiol.* **29**, 79–108.

Parfitt, R. L. and Greenland, D. J. (1970). *Soil Sci. Soc. Am. J.* **35**, 862–6.

Parker, C. A., Trinick, M. J., and Chatel, D. L. (1977). In *A treatise on dinitrogen fixation, Section IV. Agronomy and ecology* (ed. R. W. F. Hardy and A. H. Gibson) pp. 311–52. John Wiley, New York.

Parks, G. A. (1965). *Chem. Rev.* **65**, 177–98.

Paterno, E. S. (1979). Persistence of rhizobia in some Philippine soils. Ph.D. Thesis, University of the Philippines.

Peele, T. C. (1936). *Cornell Univ. Agric. Expt. Sta. Mem.* **197**, 1–18.

Pena-Cabriales, J. J. and Alexander, M. (1979). *Soil Sci. Soc. Am. J.* **43**, 962–6.

Philpotts, H. (1975). *Trop. Grassl.* **9**, 37–44.

—— (1977a). *Aust. J. exp. Agric. Anim. Husb.* **17**, 995–7.

—— (1977b). *Aust. J. exp. Agric. Anim. Husb.* **17**, 308–15.

Rand, B. and Melton, I. E. (1975). *Nature, Lond.* **257**, 214–16.

Rerkasem, B. (1977). Differential sensitivity to soil acidity of legume–*Rhizobium* symbioses. Ph.D. Thesis, University of Western Australia.

Reyes, V. G. and Schmidt, E. L. (1979). *Appl. environ. Microbiol.* **37**, 854–8.

Rice, W. A. (1975). *Can. J. Soil Sci.* **55**, 245–50.

—— Penney, D. C., and Nyborg, M. (1977). *Can. J. Soil Sci.* **57**, 197–203.

Riley, D. and Barber, S. A. (1969). *Soil Sci. Soc. Ann. J.*, **33**, 905–8.

Robinson, A. C. (1967). *J. Aust. Inst. agric. Sci.* **33**, 207–9.

—— (1969a). *Aust. J. agric. Res.* **20**, 827–41.

—— (1969b). *Aust. J. agric. Res.* **20**, 1053–60.

Robson, A. D. and Loneragan, J. F. (1970a). *Aust. J. agric. Res.* **21**, 435–45.

—— (1970b). *Aust. J. agric. Res.* **21**, 223–32.

Roper, M. M. and Marshall, K. L. (1974). *Microbial Ecol.* **1**, 1–13.

—— —— (1978a). *J. gen. Microbiol.* **106**, 187–9.

—— —— (1978b). *Microbial Ecol.* **4**, 279–89.

—— —— (1979). *Geomicrobiol. J.* **1**, 103–16.

Roughley, R. J. (1970). *Ann. Bot., Lond.* **34**, 631–46.

—— (1976). In *Symbiotic nitrogen fixation in plants* IBP No. 7. (ed. P. S. Nutman) pp. 125–36. Cambridge University Press.

—— and Dart, P. J. (1970). *Pl. Soil* **32**, 518–20.

—— Blowes, W. M. and Herridge, D. F. (1976). *Soil Biol. Biochem.* **8**, 403–7.

Rovira, A. D. (1961). *Aust. J. agric. Res.* **12**, 77–83.

—— (1978). In *Plant relations in pastures* (ed. J. R. Wilson) pp. 95–110. C.S.I.R.O., Melbourne.

Santoro, T. and Stotzky, G. (1967). *Arch. Biochem. Biophys.* **122**, 664–9.

—— —— (1968). *Can. J. Microbiol.* **14**, 299–307.

Sartain, J. B. and Kamprath, E. J. (1975). *Agron. J.* **67**, 507–10.

—— —— (1977). *Agron. J.* **69,** 843–5.

Shimshick, B. J. and Herbert, R. R. (1979). *Appl. Env. Microbiol.* **38,** 447–53.

Sprent, J. I. (1976). In *Symbiotic nitrogen fixation in plants* (ed. P. S. Nutman) pp. 405–20. Cambridge University Press.

Steinborn, J. and Roughley, R. J. (1974). *J. appl. Bacteriol.* **37,** 93–9.

—— —— (1975). *J. appl. Bacteriol.* **39,** 133–8.

Stotzky, G. (1972). *Critical Rev. Microbiol.* **2,** 59–137.

—— and Bystricky, V. (1969). *Bacteriol. Proc.* A93.

Stumm, W. and Morgan, J. J. (1970). *Aquatic chemistry* p. 478. Wiley Interscience, New York.

Tan, K. H. and Nopamornbodi, O. (1979), *Soil Sci.* **127,** 235–41.

Tazimura, K., Watanabe, I., and Shi, J. F. (1966). *Soil Sci. Plant Nutr.* **12,** 15–22.

Theng, B. K. G. (1979). *Formation and properties of clay–polymer complexes.* Elsevier, Amsterdam.

Trinick, M. J. (1970). *The Ecology of Rhizobium—interactions between Rhizobium strains and other soil microorganisms.* Ph. D. Thesis, University of Western Australia.

Tschapek, M., Tcheichvilli, L., and Wasouski, C. (1974). *Clay Minerals* **10,** 219–29.

Tsien, H. C. and Schmidt, E. L. (1977). *Can. J. Microbiol.* **23,** 1274–84.

Tweneboah, C. K., Greenland, D. J., and Oades, J. M. (1967). *Aust. J. Soil Res.* **5,** 247–61.

Uehara, G. (1977). In *Exploiting the legume—Rhizobium in tropical agriculture.* pp. 67–80. College of Tropical Agriculture, Misc. Pub. 145. University of Hawaii.

—— (1978). In *Mineral nutrition of legumes in tropical and subtropical soils* (ed. C. S. Andrew and E. J. Kamprath) pp. 21–36. C.S.I.R.O., Australia.

Upchurch, R. G. and Elkan, G. H. (1977). *Can. J. Microbiol.* **23,** 1118–22.

van Egeraat, A. W. S. M. (1975*a*). *Pl. Soil* **42,** 15–36.

—— (1975*b*). *Pl. Soil* **42,** 37–47.

—— (1975*c*). *Pl. Soil* **42,** 367–79.

—— (1975*d*). *Pl. Soil* **42,** 381–6.

—— (1975*e*). *Pl. Soil* **43,** 503–7.

—— (1976). *Pl. Soil* **44,** 501–3.

van Raij, B. and Peech, M. (1972). *Soil Sci. Soc. Am. J.* **36,** 587–93.

van Rensburg, H. J. and Strijdom, B. W. (1969). *Phytophylactica,* **1,** 201–4.

Vincent, J. M. (1965). In *Soil nitrogen* (ed. W. V. Bartholomew and F. C. Clark) pp. 384–435. Am. Soc. Agron., Madison, WI.

—— (1970). *A manual for the practical study of root-nodule bacteria.* IBP Handbook No. 15. Blackwell Scientific Publications, Oxford.

—— and Waters, L. M. (1953). *J. gen. Microbiol.* **9,** 357–70.

—— Thompson, J. A., and Donovan, K. O., (1962). *Aust. J. agric. Res.* **13,** 258–70.

Vittal Rai, P. and Patil, R. B. (1978). *Pl. Soil* **50,** 553–66.

Weaver, R. W. (1979). *Soil Biol. Biochem.* **11,** 545–6.

—— and Frederick, L. R. (1974). *Agron. J.* **66,** 229–32.

Webb, S. J. (1965). *Bound water in biological integrity.* Charles C. Thomas, Springfield, Ill.

Weber, D. F. and Miller, V. L. (1972). *Agron. J.* **64,** 796–8.

Wilkins, J. (1967). *Aust. J. agric. Res.* **18,** 299–304.

Williams, C. H. and Donald, C. M. (1957). *Aust. J. agric. Res.* **8,** 179–89.

Wilson, D. O. and Reisenauer, H. M. (1970). *J. Bacteriol.* **102,** 729–32.

Wilson, J. (1970). *Aust. J. agric. Res.* **21,** 571–82.

Wilson, J. K. (1931). *J. agric. Res.* **43,** 261–6.

Wilson, J. R. and Norris, D. O. (1970). *Proc. 11th. Int. Grassl. Congr.* 455–8.

Worrall, V. S. and Roughley, R. J. (1976). *J. exp. Bot.* **27,** 1233–41.

Zvyagintsev, D. G. (1962). *Sov. Soil Sci.* (Engl. trans.) 140–4.

3 Biology

M. J. TRINICK

3.1 Evolution and taxonomy

*Evolution of the legume–*Rhizobium *symbiosis*

The *Rhizobium*

Theorists have often suggested that nitrogen fixation occurred before the evolution of legume–*Rhizobium* symbiosis. It was considered likely that nitrogen fixation is a primitive character in micro-organisms because of its reductive nature, and that such micro-organisms were widely distributed in soil and water well before the appearance of the angiosperms (Parker and Scutt 1960; Parker 1968). Burns and Hardy (1975) summarized the concepts that life on earth originated under a reducing atmosphere after a long period of abiogenic, organic chemical evolution. They presumed that the primordial atmosphere contained methane, ammonia, and water and to a lesser extent hydrogen, nitrogen, hydrogen sulphide, carbon monoxide, and carbon dioxide, with oxygen being present only in trace amounts. After a millenia the environment changed, with the accumulation of oxidized components. They considered that ammonia was the major available form of nitrogen during the earliest period of successful life forms. It is unlikely that nitrogenase developed during this period. Then followed periods when various organisms obtained their energy by nitrification, when ammonia was oxidized to NO_2^- and NO_3^-. Other organisms developed to utilize nitrate and convert it back to ammonia, or to reduce it to molecular nitrogen or nitrous oxide (denitrification). Burns and Hardy (1975) point out that this opened up new sources of biological energy, a rare event in the still primarily reductive environment. Eventually, with the build-up of biologically inert nitrogen in the atmosphere, combined nitrogen became less rapidly available and perhaps, in many situations, combined nitrogen eventually became limiting for biological activity. Burris (1963) considered that nitrogen fixation originated in a photoautotroph, probably an antecedent of the present green sulphur bacteria and later (Burns and Hardy 1975) it was postulated and evidence was presented that other photosynthetic bacteria, the non-photosynthetic sulphur bacteria, and eventually the blue–green algae evolved. They suggested that the photosynthetic organisms which utilized solar energy were probably the dominant types in the immediate pre-oxidative period. It was suggested that *Clostridium* evolved later, since they require complex organic substrates for energy, and that clostridia were possibly the ancestors of the other (free living) diazotrophs, with *Azotobacter* being the most recent to evolve. Imshenetskii (1963), however, regards *Clostridium pasteurianum* as the origin of nitrogenase activity in the diazotrophs and suggested that the similarities between *Azotobacter* and certain blue–green algae indicated a more direct evolutionary link between them.

There is much disagreement on how new forms or species of life arose, and perhaps the proponents can be grouped as either those adhering to the populationist theory or to the theory of spontaneous change of individuals. In any environment an organism accumulates, over the course of time, inheritable variations which, perhaps, best fit or adapt it to its surroundings. It is not possible to understand evolution by just looking at single organisms, their populations, and their inherent variability that evolves, however. There is continual replacement of individuals within a population and evolution is a transformation of the group of individuals which have built in variation. The gene pool provides more suitable characteristics for survival and multiplication in the stable, or changing, or changed environment. Obviously, it is the behaviour of the individuals, as well as their relative numbers in the population, that have the variation and the need to use the attribute (i.e. ability to fix nitrogen).

These theories support the contention that the ability to fix nitrogen is a primitive character, originating in the primordial atmosphere described by Burns and Hardy (1975). The oxygen sensitivity of the nitrogenase also suggests that it evolved in organisms living in a reducing environment. Nitrogen-fixing, facultative, anaerobic bacteria fix nitrogen only under anaerobic conditions, and where fixation occurs with aerobic organisms, protective mechanisms have evolved. Similarities of nitrogenase (Dilworth 1974) and its reactions between the diazotrophs of diverse taxonomy suggests a common ancestry. Perhaps nitrogenase activity was at one time common amongst the diazotrophs and with changed environments, the advantage of having this attribute may have changed with subsequent loss of the required genetic material.

Others have suggested that nitrogenase activity was a relatively late arrival and that it only appeared in relatively few species that had retained a suitable internal environment or electron carriers for it to function (Dilworth 1974; Postgate 1974). They further support their contention by the very wide and unpredictable occurrence of nitrogenase in the bacteria and its very limited occurrence in the anaerobic groups. The similarities of the nitrogenase protein (Dilworth 1974) may also suggest its late arrival. Postgate (1974) considered that eukaryotes never acquired nitrogenase, since evolution had taken the eukaryote too far towards genetic rigidity and aerobiosis for it to be accommodated.

Prior to the findings of Kurz and LaRue (1975), McComb, Elliot, and Dilworth (1975) and Pagan, Child, Scowcroft, and Gibson (1975), *Rhizobium* was considered unable to fix nitrogen in the free-living state. It was soon reported that the levels of nitrogenase activity of the free-living *Rhizobium* were high under suitable conditions and approximated those of isolated bacteroids (Tjepkema and Evans 1975; Bergersen, Turner, Gibson, and Dudman 1976; Keister and Evans 1976). Demonstrations of intergeneric transformations between *Rhizobium* and *Azotobacter*, *Rhizobium* and *Escherichia coli*, as well as *Rhizobium* and *Klebsiella* (Sen, Pal, and Sen 1969; Venkataraman, Roychaudhury, Henriksson, and Henriksson 1975; Skotnicki and Rolfe 1978) support the probable movement of generic information between these and other bacterial species in nature. Studies of DNA homology have also indicated that *Rhizobium* is related to the genera *Azotobacter* and *Pseudomonas* (DeLey, Park, Tijlgat, and van Ermengem 1966), and to *Arthrobacter* (Hill 1966). Parker and Graham (personal communication) have also shown cross-agglutination between *Rhizobium* and these bacteria. These three genera contain nitrogen-fixing members (Paul and Newton 1961; Smyk and Ettlinger 1963). Transformants of *Rhizobium trifolii* with DNA from *A. chroococcum* were able to grow on nitrogen-free medium in the free-living state and were able to reduce acetylene to ethylene (Venkataraman *et al.* 1975). Page (1978) was able to transform mutant strains of *A. vinlandii* which were unable to fix nitrogen (Nif^- strain) to Nif^+ strains with DNA from *Rhizobium* spp. Free-living rhizobia, unlike the other bacteria, require specific conditions in laboratory culture to express their nitrogenase activity, but nevertheless their relationships with other nitrogen-fixing soil bacteria are evident.

Perhaps rhizobia were once able to fix equivalent quantities of nitrogen under similar conditions to other bacteria, such as *Azotobacter*, but lost many inheritable characteristics as a result of entering into a symbiotic association with higher plants. Inside the legume nodule, the rhizobia are supplied with an abundance of carbohydrates by the host. This nutritional environment of the nodule is in marked contrast to its other saprophytic environment. *Rhizobium* multiplies to large populations in the nodule giving ample opportunity for a selection process to operate, favouring a more-specialized environment for effective nitrogen fixation. The death and decay of the nodule produces an environment for the rhizobia that is rich in organic forms of nitrogen, eliminating the immediate requirement to continue to fix nitrogen. In the absence of the host, the soil rhizobial population tends to decrease and hence selective processes are probably restricted to the energy-rich host rhizosphere and within the nodule. In aerated soils, the presence of high enough pO$_2$ to suppress nitrogenase activity also favours the repression and perhaps encourages the elimination of the ability to fix nitrogen. However, the retention by *Rhizobium* of nitrogenase as an ecological advantage would only occur in soils low in combined nitrogen and oxygen. Nitrogenase activity has been detected in only a limited number, but not all, of cowpea-type rhizobia and *R. japonicum* tested (Bergersen and Gibson 1978), indicating that either the conditions imposed were unsatisfactory or that the non-fixing strains had lost part or all of the bacterial *nif* genes. Thus the nitrogen-fixing capacity of *Rhizobium* is, perhaps, a relic from its early non-symbiotic habitats. It seems most unlikely that it would have acquired the *nif* genes after entering the legume to form root nodules, which would have been ineffective and so serve no ecological advantage to the host or legume. It also seems logical for a plant to favour the development of an association with a bacterium able to supply combined nitrogen.

Evolution of symbioses between plants and micro-organisms

Parker (1957) suggested that symbiotic nitrogen fixation in plants developed in a gradual evolutionary process as a result of external environmental pressures acting on both partners. The soil micro-organisms, which required sources of energy-rich carbohydrates for nitrogen fixation and the plants, which required combined nitrogen, possibly grew initially in a loose association during a geological period when available combined nitrogen was inadequate. He postulated that the initial stage of a casual association between plants and nitrogen-fixing micro-organisms was in the rhizosphere region and on leaves and stems. This was followed by a symbiotic establishment within the host cortex and finally, in the case of nodulated plants, in the establishment of the organized nodular tissue. It is now well-established that nitrogen fixation occurs in the rhizosphere, by bacteria in the cortex of the root by *Spirillum* (Döbereiner and Day 1976; Döbereiner 1977), in stem glands at the bases of leaves of *Gunnera* by the blue–green alga *Nostoc punctiforme* (Schaede 1951) and nitrogen-fixing extracellular associations between blue–green algae and Pteridophyta (*Azolla*) (leaf space) (Saubert 1949) and in the root cortex of Cycadophyta (*Ceratozamia, Cycas* etc.) (Bergersen, Kennedy, and Wittman 1965). The most intimate association between higher plants and the blue–green algae occurs with *Gunnera* when the algae, initially extracellular, eventually becomes intracellular (Schaede 1951) and structures similar to the organized nodules formed by *Rhizobium* and Frankia (Becking 1970) have never evolved.

There were many major (known and unknown) obstacles (penetration, acceptance by the plant host, etc.) to be overcome before a micro-organism could enter into any sort of symbiotic association with a higher plant. It is well known that plants are resistant to almost all of the micro-organisms with which they come in contact and in response to their presence, they produce low molecular weight compounds, phytoalexins, which inhibit their growth. Like plant pathogens the rhizobia and actinomycetes of nodule-bearing angiosperms have presumably developed a mechanism which allows them to either avoid eliciting phytoalexin accumulation or the effects of the phytoalexins, if they have already accumulated (Albersheim, Ayers, Valent, Ebel, Hahn, Wolpert, and Carlson 1977). *Rhizobium* shares similarities or relationships with other soil bacteria (i.e. *Agrobacterium*) that invade plant tissue (see p. 85). *Agrobacterium* can enter the host through wounds, but not through root hairs. Many legume–*Rhizobium* symbiotic associations, however, depend on similar situations (Schaede 1940; Arora 1954) and recently Chandler (1978) has demonstrated the entrance of *Rhizobium* in *Arachis* through the root junction where laterals emerge. Both *Rhizobium* and *Agrobacterium* are able to stimulate gametophore in the moss *Pysisiella selwynii* (Spiess, Lippincott, and Lippincott 1977) and recently Skotnicki and Rolfe (1978) described a *R. trifolii* with characteristics of both *Rhizobium* and *Agrobacterium*. The control of infection is discussed in detail in later chapters. Once inside the plant and presented with a vastly different environment, the micro-organism and the host have to overcome an enormous array of physiological processes involved in infection, nodule initiation, and nodule development, as well as the basic problems associated with gene incorporation, maintenance, and expression for nitrogenase activity (e.g. control of pO_2). Parker (1957) considered it reasonable to suppose that the endophytic symbiosis was a result of mutual adaptation of the plant and the nitrogen-fixing microbe. This probably occurred over a very long period of time, with many intermediate types of nodule morphology and internal structures being involved. Perhaps the structure of nodules formed by *Rhizobium* on the non-legume *Parasponia* (Trinick 1979) is a relic and represents a more primitive nodule-type which may have occurred on many other plants, including members of the Leguminosae. Like the nodules on different non-legumes formed by other types of endophytes, the nodules on *Parasponia* resemble modified roots in that they have a central vascular bundle surrounded by an endophyte-infected zone. Perhaps a more advanced stage in evolution is represented by a nodule structure different from a root, as found in the legumes.

Therefore, the intimate association between *Rhizobium* and the host-nodule cell in *Parasponia*, where the *Rhizobium* is retained within continuous thread-like structures of varying thickness, maybe a more primitive stage than that in the legume, where the *Rhizobium* is released from the thread to be enclosed only by a thin membrane envelope.

There are many obvious differences between groups of legumes and non-legumes which may have evolutionary significance. For instance, the inner- (temperate legume) or outer- (tropical legume) cortical cells divide to provide a nodule meristem, possibly as a consequence of the release of cytokinins and indoleacetic acid by the rhizobia (Libbenga and Bogers 1974). In the non-legume, the cortical cells close to the penetrating hyphae of the actinomycete start to divide (as in the legume) but nodule formation commences only when cell divisions in the pericycle lead to a normal lateral-root primordium which becomes infected by the endophyte (Angulo, Dijk, and Quispel 1976). The nodule structure of *Parasponia* suggests a similar method of nodule formation by *Rhizobium* and perhaps this aspect of nodule development is due to the inherent genetic characteristics of the host plant prior to the establishment of the symbiotic system. The legume may have lost this involvement of the pericycle, if it ever had it, with consequential change in overall nodule structure.

The legume–*Rhizobium* symbiosis

The origin of the various symbiotic associations and the relative times that they evolved, are not known. However, it is generally thought that the legume and non-legume root-nodule symbiosis were the most recent to evolve. Bond, MacConnell, and McCallum (1956) have suggested that 'the possession of root nodules and the ability to use atmospheric nitrogen are very ancient characters, and mostly date from a time when only woody flowering plants existed and when conditions were apparently particularly favourable to the initiation of symbiotic associations'. The nodulated non-legumes with the symbiont *Frankia* (Becking 1970) are scattered amongst diverse and phylogenetically unrelated plant taxons covering seven orders, eight families, and 15 genera (Becking 1977). In the Leguminosae, the subfamily Caesalpiniaceae is largely un-nodulated, while the Papilionaceae and Mimosaceae are mostly nodulated (Allen and Allen 1961) by *Rhizobium*, which suggests that nodulation occurred early in the history of this family. Generally, relatively few plant families have been successful in establishing a nodule symbiosis and Bond *et al.* (1956) have suggested a common period of geological time, despite the very different endophyte involved. Perhaps the nodulation of *Parasponia* (Urticales) with *Rhizobium* (Trinick 1973) supports this contention of Bond *et al.* (1956). While no other substantiated excursion of *Rhizobium* outside the legumes is known, there are unconfirmed reports of it occurring in three genera of Zygophyllaceae (Sabet 1946; Mostafa and Mahmoud 1951). It may have happened that during this geological period there was a prevalence of free-living nitrogen-fixing bacteria, including actinomycetes, in soils that were generally lacking in combined nitrogen and low in available energy sources which encouraged the development of the root-nodule symbiosis. Most of the plants involved belong to woody families which are clustered around the Rosales (except for, perhaps, Rhamnales) and occur relatively low in the evolutionary tree according to Hutchinson's (1969) scheme.

Parker (1968) considered the literature relating to the origin and evolution of the legumes and discussed the belief of taxonomists (Andrews 1941; Tutin 1958) that the primitive legume 'was a large tree with rather stout branches and twigs, something like those of *Erythrina*, bearing large pulvinate bipinnate leaves'. However, with the present legumes he (Tutin 1958) concluded, 'It is, nevertheless, impossible to find even a single character by which a plant can infallibly be known to belong or not to belong to this family. . .'. According to Parker (1968) other taxonomists consider a polyphyletic origin from herbs and woody plants. Thus Hutchinson (1959) divides the dicotyledons into woody and herbaceous groups arising from hypothetical proangiosperms. Arber (1928), after considering the longer generation time of woody plants compared with herbs, suggested that the latter could have evolved more rapidly and also speculated that the Leguminosae were originally herbaceous. This also makes sense when visualizing the evolution of the symbiotic nodule with the far greater chance of variation occurring in the plant within a given time. This may have been necessary for the development of the multiple variations required within the plant to occur, in conjunction with the rapidly-multiplying soil nitrogen-fixing micro-organisms, and hence the variation within them. Arber (1928) also suggested that the adopted tree habit, with its reduced generation time, would 'put a brake' on its own evolution. Others have regarded low chromosome numbers to be the primitive state and Sen (1938) concluded that a large number of perennial leguminous species had arisen from annual types. The data also shows that woody legumes may have originated from herbaceous legumes with lower chromosome numbers. Parker (1968) concluded that 'We are uncertain whether the original legumes were woody or herbaceous, or whether any one of the three existing subfamilies of legumes can be regarded as more primitive than another'. Sporne (1956) also expressed his belief in the grave doubts of botanists about assessing the relative advancement of floral and other characters, and gave little hope that the angiosperms will ever be satisfactorily classified phylogenetically.

Undoubtedly, the legumes evolved during the Cretaceous (Andrews

1914; see references given by Parker 1968) period and Andrews (1914) claims that a 'mild and genial climate' was prevalent. Such an environment would hardly be a stimulus for evolution within the Leguminosae. Stebbins (1950) considers that 'a variable environment strongly promotes rapid evolution and may in fact be essential for speeding up evolutionary change'. Fossils of plants representing the three existing subfamilies have been reported for the early Tertiary (Axelrod 1958). Parker (1968) has reviewed the geological evidence and concluded that 'Major continental influences would have caused great climatic zonations over these land masses so that, as in the present day, a range of climates from humid through to subhumid to semi-arid and desert must have been present'. Parker (1968) suggested that since the Period was one of 'tetonic instability', the soils varied from low-nutrient acid soils as depicted by Norris (1956) to relatively fertile soils of neutral to alkaline reaction. Thus, he concluded that 'it seems unlikely that the legumes evolved in the wet tropics, but in subhumid areas, either tropical or temperate. Whether they developed on acid, neutral, alkaline or calcareous soils can only be guessed'.

Norris (1956) relied heavily on a previously neglected paper by Andrews (1914), who described a 'mild and genial climate' during the Cretaceous period and omitted consideration of opposing views on the various possible environments (climate and soil types). He formulated the concept that legumes and their symbiosis originated under wet, tropical conditions of low availability of essential soil nutrients and have subsequently adapted to arid and temperate climates, often with high soil fertility and pH. The hypothesis was also advanced 'that the cowpea-type of nodule bacteria that is associated with tropical Leguminosae represents the ancestral condition and is the type of genus *Rhizobium*' and the 'ancestral conditions' of the legume must have been one of cross-inoculation promiscuity. Thus all other more specialized legume–*Rhizobium*-type systems were evolved from this 'ancestral condition'. In support of this hypothesis, Norris (1956) described the present distribution of the three subfamilies of Leguminosae and the specialization of the *Rhizobium* symbionts and recorded the number of genera and species that occurred in the tropics, sub-tropics, and temperate zones. He found that the subfamilies Mimosoideae and Caesalpinioideae have essentially remained in the tropics and the Papilionatae have moved to the cooler temperate regions of the earth. From a consideration of the possible evolutionary trends from woody to herbaceous legumes as described by Tutin (1958), Norris and Date (1976) concluded that 'The annual species in the tribes Vicieae and Trifolieae (e.g. *Pisum, Vicia, Lens, Lathyrus, Trifolium, Medicago*) of traditional European and North American agriculture represent the most specialized species of the evolutionary tree. Their specialization extends even to their nutrition and associated symbiotic bacteria'. A further hypothesis was advanced by Norris (1965) that the slow-growing non-acid-producing ' "cowpea-type" of *Rhizobium* has persisted unchanged, because acid production during growth in the rhizosphere in acid soils would react unfavourably against survival'. It was also suggested that host groups with characteristically associated strong acid-producing *Rhizobium* are advanced types, which are strongly adapted to non-acid soils. This has been challenged by Parker (1968) and discussed by Parker, Trinick, and Chatel (1977) who said that production of acid or alkali depended on the energy source of the tested rhizobia. This was rediscovered by Date and Halliday (1979) for tropical rhizobia, and hence alkali production is not necessarily of ecological significance. Further, Trinick (1965*a*, 1980*a*) reports that 80 per cent of fast-growing rhizobia produce alkaline reactions. Munns (1977) stated that 'Taxonomy of the *Rhizobium* is no predictor of acid-tolerant associations which occur in some of the cowpea miscellany, some *R. lupini*, some *R. japonicum*, and some *R. trifolii*'.

Conclusions

Norris (1956, 1965) had a narrow approach to the evolution within the Leguminosae by only considering the views of Andrews (1914), by neglecting possible alternative conditions prevailing during the Cretaceous period, as well as the variable opinions on what is a primitive condition (see above). Parker (1968) has effectively summarized some alternatives. Many *Rhizobium* strains have been shown to fix nitrogen in the free-living state and show similarities with other genera of nitrogen-fixing bacteria commonly found in soil (see pp. 85–95). The plant pathogen *Agrobacterium* also has similarities with *Rhizobium* and both can invade plants and even stimulate gametophore formation in moss (see p. 80). This suggests that the predecessor or a primitive form of *Rhizobium* was probably abundant in the soils with other similar bacterial types, throughout the various regions of the earth when legumes appeared during the Cretaceous period. To-day rhizobia are still abundant in soils supporting legumes, but Brown, Jackson, and Burlingham (1968) and Schmidt (1978) consider that they are often poorly adapted to life in the soil in the absence of legume hosts or suitable plant rhizospheres. It seems reasonable, therefore, to speculate that legume–*Rhizobium* symbiosis evolved in different locations, simultaneously or at different times, with different nitrogen-fixing members of the soil population. The ability of very different physiological types of rhizobia to form effective nodules on *Vigna unguiculata* (Trinick 1965*a*, 1968, 1980*a*) also supports this hypothesis. The 'primitive' *Rhizobium* also ventured into nodule symbiosis with non-legume angiosperms, of which the non-legume *Parasponia* (Urticales, Ulmaceae) (Trinick 1973) is so far the only substantiated case (Becking 1977).

Rhizobial taxonomy

Introductory remarks

The sole criterion for the genus *Rhizobium* is the ability of the bacterium to form nodules on members of the Leguminosae (Allen and Allen 1950). This has resulted in bacteria with very different characteristics, and which have been isolated from an enormous range of legumes, being grouped into one instead of two or more genera. The attempt to group all such organisms into the one genus has caused confusion and much dissatisfaction with the taxonomy of *Rhizobium* (Norris 1956; Parker 1968). The placement of all nodule bacteria in the one genus occurred during the first part of this century, when only a small section of the family Leguminosae had been examined (Fred, Baldwin, and McCoy 1932). Recently, Trinick (1973) and Trinick and Galbraith (1980) found both fast- and slow-growing rhizobia nodulating a member of the Ulmaceae (*Parasponia* spp). Perhaps other non-legume, bacterial symbiotic nodulating species will eventually be found which will further emphasize the extensive host-range of the single genus *Rhizobium*. The loss of the ability of the organism to nodulate its host has been reported (Labandera and Vincent 1975) and by definition, this organism should no longer be called a *Rhizobium*. A recent edition of *Bergey's Manual* (Jordan and Allen 1974) places both *Rhizobium* and *Agrobacterium* in the family Rhizobiaceae because of their ability to cause cortical hypertrophies on plants, even though these structures are of very different types. Earlier editions (Breed, Murray, and Smith 1957) placed *Chromobacterium* in the same family as *Rhizobium* and *Agrobacterium*.

Plant infection

The infection of a legume by *Rhizobium* is an intimate association which depends on their specific mutual recognition. The discontinuous nature of nodulation amongst the legumes illustrates the failure in the evolution of the partners to reach a suitable stage for nodulation. Steps in the establishment of the symbiosis have frequently been ascribed to the colonization of the rhizosphere; entrance via the root hair, or during the emergence of lateral roots, or other means resulting in the formation of infection threads; the commencement and persistence of a nodule meristem; the release of the rhizobia from the infection thread followed by their multiplication within membrane envelopes of the nodule cell before their conversion to bacteroids; the establishment and continuance of a shared metabolism between the plant and the bacterium. In the non-legume *Parasponia*, the rhizobia are retained within the infection thread in all infected nodule cells, without any loss of symbiotic effectiveness (Trinick 1976, 1979, 1980b). Both the bacterial and plant partners are infinitely variable in their nature and so their effectiveness

in nitrogen fixation between them will depend on their mutual com-
patability. Genetic engineering is revealing the complexity of symbiotic
systems. Usually a single strain of *Rhizobium* is represented in the nodule
but there are numerous reports (Purchase and Nutman 1957; Škrdleta
and Karimová 1969; Means, Johnson, and Erdman 1961; Lindemann,
Schmidt, and Ham 1974; Franco and Vincent 1976) of dual infection by
similar rhizobia, and recently both fast- and slow-growing rhizobia
have been shown to share the same nodule (Franco and Vincent 1976;
Trinick 1981). Non-nodulated plants probably failed during the initial
stages of infection whilst the commonly observed ineffective-association
failed at a later stage(s).

Early studies on legume–*Rhizobium* symbiosis were concerned with
the annual species of agricultural importance in the tribes Vicieae and
Trifolieae (Papilionideae) (e.g. *Pisum, Vicia, Lens, Lathyrus, Trifolium,
Medicago*) of traditional European and North American agriculture. It
was soon noticed that the isolates from one genus would not necessarily
nodulate a plant in another, and early workers soon found that the
legumes studied formed groups based on the ability of their associated
rhizobia to be mutually interchangeable. This led to the concept of the
cross-incoculation groups. The rhizobia nodulating each group was
named according to the dominant host of that group (Table 3.1). Fred *et
al.* (1932) listed 22 plant–bacterial groups and recognized six species of
Rhizobium but gave no name to the rhizobia-producing nodules on
plants of the so-called 'cowpea group'. The 'cowpea-group' is considered synonymous with the promiscuous slow-growing rhizobia commonly encountered in the nodules of tropical legumes; similar rhizobia
were isolated from the temperate indigenous legumes of Western

TABLE 3.1

Some cross-inoculation groups within the Leguminosae

Group	Representative hosts	*Rhizobium* species
Pea	*Pisum, Lathyrus, Lens, Vicia, Cicer*	*R. leguminosarum*
Bean	*Phaseolus vulgaris*	*R. phaseoli*
Clover	*Trifolium*	*R. trifolii*
Medic	*Medicago, Melilotus, Trigonella*	*R. meliloti*
Lupin	*Lupinus, Ornithopus*	*R. lupini*
Soyabean	*Glycine max*	*R. japonicum*
Cowpea	A great variety of genera from the three subfamilies of Leguminosae	*Rhizobium* spp
Others	Small groups of one or more genera or species which are 'specific' in their *Rhizobium* requirement, i.e. *Lotus, Sesbania, Leucaena, Lotononis* etc.	—

Australia (Lange 1961). Unfortunately, there has been a tendency to use this group as a dumping ground for all unspecialized and specific groups of legumes not falling into the other cross-inoculation groups.

This classification was developed on the premise that each species of *Rhizobium* would only nodulate plants within the particular 'cross-inoculation' group and that within such groups rhizobia from one plant will nodulate all other plants, and vice versa (Fred *et al.* 1932). It soon became very evident that these groups were not discrete as there were many reports of boundary jumping between groups (Allen and Allen 1939; Wilson 1944; Lange 1961, 1966; Trinick 1965*b*, 1968, 1980*a*; Trinick and Galbraith 1980). There also became an increased awareness of specific subgroups which included distinct plant groups within plant genera (Vincent 1974, 1977). These findings have affected the usefulness and credibility of this classification.

Examples of *Rhizobium* strains nodulating plants in other host groups are illustrated in Table 3.2. The interchange between the lupin–soyabean–cowpea group was reviewed by Lange (1961), and that between the pea and clover groups by Kleczkowska, Nutman, and Bond (1944). The nodulation of *Phaseolus vulgaris* by many slow-growing rhizobia was reported by Lange (1961); the reciprocal infection between the two groups of *Lotus* species was reviewed by Vincent (1974); the nodulation of *Medicago sativa* by *Leucaena* fast-growing isolates and a similar fast-growing strain from *Lablab* was shown by Trinick (1965*b*, 1980*a*); the non-legume *Parasponia*, like cowpea, was nodulated with both the 'normal' slow-growing cowpea-type rhizobia as well as the fast-growing rhizobia of the *Leucaena–Mimosa–Sesbania* group (Trinick 1980*a*). The limited number of *Leucaena* (and *Lablab*) strains able to ineffectively nodulate *Medicago sativa*, is really insufficient to invalidate the usual discrete nature of the medic group. Frequently the out-of-group nodulations are ineffective, except those between the slow-growing lupin, soyabean, and cowpea groups. The slow-growing organisms from the non-legume *Parasponia* were invariably ineffective and occasionally partially effective on hosts belonging to the cowpea group, but the symbiosis between *Parasponia* and slow-growing rhizobia from tropical legumes was often highly effective. The fast-growing rhizobia of the *Leucaena–Mimosa–Sesbania* type were variable in their effectiveness.

Vincent (1974) has reviewed the various host groups and subgroups listed in Table 3.1 and illustrated the high degree of specialization between the host and its symbiont that often occurs within a plant group. Attention was drawn to the several studies made with *Trifolium* spp of mid-Africa, which showed them to be poorly compatible with the rhizobia from European species, and the extremely heterogeneous nature of the symbiosis amongst themselves. Further differences in *Trifolium* were noted by Brockwell and Katznelson (1976) studying the

TABLE 3.2
Examples of 'boundary jumping' between different plant–
Rhizobium *groups*

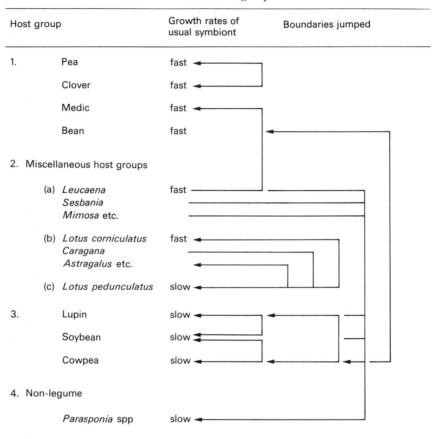

Host group		Growth rates of usual symbiont	Boundaries jumped
1.	Pea	fast	
	Clover	fast	
	Medic	fast	
	Bean	fast	
2. Miscellaneous host groups			
(a)	*Leucaena* *Sesbania* *Mimosa* etc.	fast	
(b)	*Lotus corniculatus* *Caragana* *Astragalus* etc.	fast	
(c)	*Lotus pedunculatus*	slow	
3.	Lupin	slow	
	Soybean	slow	
	Cowpea	slow	
4. Non-legume			
	Parasponia spp	slow	

symbiotic characteristics of ten species of *Trifolium* and their symbionts from Israeli soils. Significant differences in rhizobial strain specificity between varieties of the same *Trifolium* species was also shown by Gibson and Brockwell (1968). Vincent (1974) has concluded that the *Medicago* group is discrete, but there is incomplete cross-nodulation within the group; *M. sativa* and *Melilotus alba* are promiscuous while *M. laciniata* is very selective and other hosts occupy an intermediate position. The complex situation found in the *Lupin–Lotus* group involving both fast- and slow-growing types of rhizobia (Vincent 1974) further emphasizes the difficulties of grouping hosts and their symbionts.

The 'cowpea group' should, perhaps, be preserved for those legumes with non-specific slow-growing symbionts and all specialized host

× *Rhizobium* legumes placed in groups indicating the specific nature of their symbiosis. The list of legumes with specific *Rhizobium* requirements that do not belong to the groups listed in Table 3.1 is growing. All *Stylosanthes* species were once considered to be nodulated with the usual non-specific or promiscuous cowpea-type *Rhizobium*, but recent reports (Date and Norris 1979; Date, Burt, and Williams 1979; Souto, Coser, and Döbereiner 1972) have shown varying degrees of specificity between species and even varieties and often the specificity was related to phytogeographic and soil pH backgrounds of both the host and the strains of *Rhizobium*. Some other tropical legumes with slow-growing rhizobia showing marked symbiotic specificities include *Lotononis bainesii* (Norris 1958*b*), species of *Centrosema* (Bowen 1959), *Aspalathus linearis* (Staphorst and Strijdom 1975), *Desmodium* species demonstrate varying specificities (Diatloff 1968; Diatloff and Luck 1972), and lines of *Glycine wightii* (Diatloff and Ferguson 1970; Nicholas 1971). Date (1977) and Burton (1977) list tropical legumes showing varying degrees of specificity within species and genera. A number of tropical legumes are effectively nodulated with highly specific fast-growing rhizobial types; these include *Leucaena* (Ishizawa 1955; Trinick 1968), *Mimosa* (Ishizawa 1955; Trinick 1968, 1980*a*; Campêlo and Döbereiner 1969), *Acacia farnesiana* (Trinick 1968, 1980*a*), and *Sesbania* (Johnson and Allen 1952*a*). A fast-growing *Rhizobium* strain has been isolated from *Lablab* and this isolate has similar bacteriological features to isolates from *Leucaena* (Trinick 1980*a*). This unusual strain, free from slow-growing rhizobia, was promiscuous and highly effective on a number of tropical legumes previously shown to have only slow-growing symbionts, but it could produce only ineffective nodules on *Leucaena* and *Mimosa*. Thus the infective characteristics of this organism are akin to the typical promiscuous cowpea-type.

The complexities of the infective characteristics of the rhizobia are enormous and Wilson (1944) was probably first to suggest that the cross-inoculation groups were not clear-cut; he wrote the paper 'Over five hundred reasons for abandoning the cross-inoculation groups of legumes'. However, the degree of promiscuity between rhizobia and the legumes that Wilson (1944) thought to occur has not been the experience of others. It is clear that the present grouping of plants and their symbionts (see Table 3.1) according to their infection patterns is not adequate or satisfactory and has therefore led to dissatisfaction. Is it practical or even desirable however, to attempt to classify the presently known legumes and their symbionts into a longer list of groups?

The taxonomy of *Rhizobium* should, perhaps, provide an indication of the inter-relationships between the different species of *Rhizobium*, as well as imparting as much practical information on its host range. Vincent (1974, 1977) suggested that there is 'a case for looking at any grouping on the basis of "preferred" hosts'. He also summarized the plant tests that permit reasonable distinction between the present species. There is a need for a concerted effort to collate the available data on a similar basis with other specific host–*Rhizobium* associations which would probably produce further useful and possible *Rhizobium* species or groups. There need not be any limit to the number of *Rhizobium* species designated, provided that the new species convey the necessary information. Of course, there will always be those rhizobia that are either extremely promiscuous, jumping across many plant groups, or highly specific and these should be recognized separately for this characteristic. An attempt should always be made to incorporate the irregularities. Perhaps further studies on cross-inoculation patterns are required before further groupings can be suggested. It is evident that nodulating patterns do exist and that taxonomic systems should use them. This taxonomic system provides meaningful information on the types of *Rhizobium* infecting different groups of plants, as well as giving an indication of the degree of the relationship between the groups.

The 'Symbiotic Rating' system suggested by Norris (1956) has not been developed further, possibly because it only divides the rhizobia according to his 'Ancestral' and 'Calcicole' types. It does not indicate nodulating patterns of rhizobia and hence imparts very little practical information from the *Rhizobium* side. Any further classification scheme based on plant infection should not conform to any particular evolutionary theory, especially when evolution may have progressed along various paths.

Adansonian approach

The principle of classifying organisms giving equal weight to all ascertainable characters is attributed to Adanson (1727–1806), who also claimed that minute and superfluous characteristics should be ignored (Cowan 1968)! The problem of which characters to include or exclude in bacterial classification is difficult and must, to a degree, reflect the priorities of the research worker. Lange (1961) first suggested classifying the nodule bacteria using a system based on non-biased over-all similarities, and recommended the Adansonian classification as proposed by Sneath (1957).

Graham (1964*b*) applied the approach using 83 strains of the genus *Rhizobium* and 38 strains belonging to the genera *Agrobacterium*, *Chromobacterium*, *Beijerinckia*, and *Bacillus* and by applying 100 features including vitamin, carbohydrate, and nitrogen nutrition, antibiotic sensitivities, morphological and colonial, and infective attributes. *Chromobacterium*, *Bacillus*, and *Beijerinckia* remained distinct and unrelated to the other genera. Graham (1964*b*) concluded that the results indicated a need for major taxonomic changes for *Rhizobium* and *Agrobacterium* and proposed the following: (i) The homogeneous nature of *R. trifolii*, *R. leguminosarum*, and *R. phaseoli* indicated their consolidation into the single species *R. leguminosarum*; (ii) *R. meliloti* should remain unchanged; (iii) the fast-growing rhizobia are closely related, especially *R. meliloti* to *Agrobacterium* which should be included in the genus *Rhizobium* as *R. radiobacter*; (iv) the slow-growing *R. lupini*, *R. japonicum*, and cowpea miscellany form a group distinct from the fast-growers. It was proposed that a new genus, *Phytomyxa*, should be created for them.

A later study of Graham's data by 'tMannetje (1967) found that a more satisfactory or perhaps different grouping of the genera was obtained by using the 'furthest neighbour' and 'flexible' sorting techniques, rather than either 'nearest neighbour' or the modified method of Sneath (1957) used by Graham (1964b). The results supported the division between the fast- and slow-growing rhizobia and that *Agrobacterium* was more closely related to the fast-growers. However, it was postulated that the genus *Rhizobium* should not be split at the generic level and that *Agrobacterium* should remain a separate taxon. 'tMannetje (1967) pointed out that some fast-growing and slow-growing strains had very high similarity values and with different sorting techniques these strains could be grouped together and perhaps be linked between the two groups of rhizobia.

The Adansonian analysis of the Rhizobiaceae by Moffett and Colwell (1968), although the numbers of rhizobial strains were small, supported Graham's (1964b) conclusion. They grouped the fast growing rhizobia into *R. leguminosarum* (incorporating *R. trifolii* and *R. phaseoli*) and *R. meliloti* and included two other species, *R. radiobacter* and *R. rubi*. All the slow-growing rhizobia were ranked, like Graham, into the genus *Phytomyxa*. Other species of *Agrobacterium* were not sufficiently similar to *Rhizobium* for inclusion, and it appeared from their results that *Rhizobium* and *Phytomyxa* were more closely related to Pseudomonadaceae.

Despite the relatively small numbers of characters (37) considered by Jarvis,McLean, Robertson, and Fanning (1977), the Gower metric similarity-index and flexible sorting strategy (Lance and Williams 1967, 1968) on 110 strains of rhizobia generated clusters of varying similarity. The New Zealand indigenous rhizobia from native legumes had only a low level of similarity with exotic strains of the *R. trifolii–leguminosarum–phaseoli* complex, *R. meliloti* and non-acid-producing *R. lupini*. However, the acid-producing *R. lupini* were clustered with the indigenous strains. Further, *R. meliloti* was found, as in previous reports, to have a low level of similarity with the *trifolii–leguminosarum–phaseoli* group. Prior to this study, New Zealand rhizobia were considered to belong to two minor subgroups according to plant-infection tests and having some relationship to the *R. trifolii–leguminosarum–phaseoli* complex (Jarvis *et al.* 1977). These results illustrate the potential of using overall similarities to show the development or evolution of different rhizobial groups in areas of geological isolation.

Using the principal component (PC) analysis of 65 tests to examine binary descriptive data for 38 cultures of *Arthrobacter*, 16 of *Agrobacterium* and 27 of *Rhizobium*, and without including data for morphological or symbiotic characteristics, or for pathogenicity, Skyring, Quadling, and Rouatt (1971) showed that each genus was separable and that the rhizobia and agrobacteria were most alike. However, their work indicated similarities between the three genera. In a preliminary Adansonian analysis they supported the earlier finding of Graham (1964*b*) of mutual similarities of greater than 70 per cent between the fast-growing strains of rhizobia and agrobacteria. However, their limited number of strains of rhizobia prevented them from drawing taxonomic relationships between fast- and slow-growing rhizobia.

Unfortunately the range of rhizobia examined by overall similarity methods does not represent the true diversity of nodulated legumes, and perhaps further examinations, covering a greater range, may reveal an uncertain number of distinct groups of rhizobia.

DNA base composition, hybridization, and base sequence homology

DeLey and Rassel (1965) found a correlation between the base composition of pure DNA from 35 strains of *Rhizobium* and their type of flaggelation. The peritrichously flagellated strains which are usually fast-growers have a lower percentage (G + C) composition in the range 58.6–63.1 per cent than the subpolarly flagellated slow-growing strains, whose range is mostly between 62.8 and 65.5 per cent. They considered that their fast-growing, peritrichous, low (G + C) groups coincide with *R. meliloti* and *R. leguminosarum* as first proposed by Graham (1964*b*). The technique of DNA base composition could not distinguish between these species. The subpolarly flagellated, high (G + C), slow-growing group was recognized as a single genetic group, which could be either another species of *Rhizobium* (*R. japonicum*) or another genus (*Phytomyxa japonicum*). The relationship between both groups was evident by their closely related percent (G + C) values.

Hill (1966) listed the DNA base composition for a large number of bacterial species which emphasized the relative homogeneity of the two groups of rhizobia described by DeLey and Rassel (1965) and pointed to a similarity with *Agrobacterium* and many species of *Pseudomonas*. This study also showed that there is a wide range of variability of the DNA base composition between, and sometimes within, the genera of bacteria. DeLey *et al.* (1966) also concluded that the genera *Rhizobium*, *Azotobacter*, and *Azomonas* appeared to be rather closely related to *Pseudomonas*, since they shared 40–56 per cent of their DNA. The heterogeneity of *R. japonicum* in DNA base composition was shown by Elkan (1971) with 26 strains, which contained three statistically significant but overlapping clusters. At the extremes of these clusters there was only about 70 per cent homology. The lack of overall sensitivity of the method was pointed out by Elkan and Usanis (1971) by its failure to correlate with phenotypic characteristics such as nutrient requirement. They also recorded a much higher percentage (G + C) content (68–69 per cent) for the exceptional pink *Rhizobium* first isolated by Norris (1958*b*). DNA base composition lacks sensitivity and can only distinguish between fast- and slow-growing rhizobial types, but it can further relate the agrobacteria to rhizobia.

From DNA hybridization experiments, Heberlein, DeLey, and Tijtgat (1967) found that strains of *Agrobacterium* species, *R. leguminosarum* and *R. meliloti* exhibited a mean percentage of DNA homology of greater than 50 per cent with their two reference strains (*A. tumefaciens* and *R. leguminosarum*). They found that *A. tumefaciens*, *A. radiobacter*, and *A. rubi* were indistinguishable, with strain variations involving up to 30 per cent of their base sequences. The remainder of their test organisms fell into six distinct genetic groups: (i) *A. rhizogenes* located closer to the rhizobia than to the typical members of its own genus, (ii) *R. leguminosarum*, (iii) *R. meliloti*, (iv) *R. japonicum* which was less related to *R. leguminosarum* than *R. meliloti*, (v) *Chromobacterium*, like *R. japonicum*, was close to two members of the Pseudomondales, and (vi) *A. pseudotsugae* which only had a DNA homology of about 10 per cent with the reference strains. The nucleic acid hybridization experiments of Gibbons and Gregory (1972) failed to distinguish between *R. trifolii* and *R. leguminosarum* and their two strains of *R. phaseoli* were more closely related to *R. meliloti* than to *R. leguminosarum*.

A comparison between phenetic similarity and DNA base sequence homology of *Rhizobium* was made by Jarvis *et al.* (1977) using the rhizobia from native legumes of New Zealand and representatives from a number of established *Rhizobium* groups. Base sequence homologies failed to substantiate the distinctions that were made between the primary clusters of the native rhizobia found by the phenetic classification, but they confirmed that native rhizobia were more closely related to the acid-producing *R. lupini* than to *R. trifolii*.

With 27 strains of *R. trifolii*, four of *R. leguminosarum*, and four of *R. phaseoli*, Jarvis, Dick, and Greenwood (1980) found that DNA homologies correlated with serological relationships and that there was no detectable change in homology when rhizobia lost their ability to nodulate their host. They considered the possibility that the ability to nodulate its host is contained on a relatively short DNA base sequence, and hence only a very small portion of the genome. The rhizobia which effectively nodulated *T. repens, T. subterraneum, T. ambiguum,* and *Vicia hirsuta* formed one population with respect to reference strains able to nodulate the first two clover species. They proposed the combining of *R. trifolii* with *R. leguminosarum* and retaining *R. leguminosarum*. Within this species biovars should be designated according to their plant specificity. *R. phaseoli* showed sufficient genetic divergence not to be included in this group. Jarvis and Crow (personal communication) examined a larger group of fast-growing rhizobia from a comprehensive range of legume genera and found that their DNA homology supported the recognized distinction between 'fast'- and 'slow'-growing rhizobia and that DNA homology can divide their 'fast-growing' rhizobia into four main genetic races which are sufficiently different to rank species level within the genus *Rhizobium*.

Serological relationships

The serology of *Rhizobium* is reviewed in detail in Chapter 8. The serological characteristics of *Rhizobium* have limited application to their taxonomy. Serological differences between rhizobia isolated from different leguminous species, and even within species, were recognized by early contributors such as Stevens (1923) and Jimbo (1930). Most of the early work was restricted to agglutination techniques. Fred *et al.* (1932) summarized the work to show that cultures isolated from different nodulating groups were serologically distinctive, cross reactions outside the plant groups were uncommon, and that reactions were often not obtained with cultures obtained from within a particular group. Vincent (1977) considered that this early work suffered through the limited number of strains tested and the poor sensitivity of the agglutination methods.

Vincent (1941) was the first to increase the sensitivity by differentiating between the flagella (H) (relatively non-specific) and somatic (O) (relatively specific) agglutinations, and he also made use of agglutinin absorption tests. Using these methods he postulated three flagella and seven somatic antigens to account for the cross reactions of six strains from *Medicago* spp. Studies on isolates from *Trifolium*, *Medicago*, and *Pisum* soon followed and showed different antigenic constitutions within groups of isolates and their serological groupings (Vincent 1942; Hughes and Vincent 1942; Kleczkowski and Thornton 1944; Kleczkowski *et al.* 1944; Purchase, Vincent, and Ward 1951). Later gel diffusion techniques (Dudman 1964) which depend on soluble or solubilized antigens (depending on preparation techniques), were developed. These techniques have permitted further differentiation of rhizobia.

Graham (1963*b*), tested 113 strains of a range of rhizobial species for both flagella and somatic agglutination with 58 rhizobial antisera, as well as testing 16 against *Agrobacterium*. The rhizobia were divided into the following three broad serological groups: (i) *R. trifolii, R. leguminosarum,* and *R. phaseoli,* (ii) *R. meliloti, Agrobacterium radiobacter,* and *A. tumefaciens,* (iii) *R. japonicum, R. lupini,* and the cowpea rhizobia. Cross-reactions were common within these groups but absent between them. Recently, Date (1974) reported that relations between rhizobial strains for tropical legumes, based on those for soyabean, reveal a high degree of diversity and specificity in somatic antigen composition. Individual strains of *Rhizobium* within each group can be very specific and show little or no serological affinity with many other strains of the same group, particularly amongst the fast-growing strains. These groupings are in agreement with those reported using the Adansonian approach (p. 90) and DNA base composition and homology (p. 92). Purchase et al. (1951) reported minimal antigenic constitution in *R. trifolii* and *R. meliloti* and the heterogeneity of other collections (Drozanska 1966; Scheffler and Louw 1967), as well as the multiplicity of somatic antigens of *R. japonicum* (Date and Decker 1965; Koontz and Faber 1961; Škrdleta 1965*b*; Elkan 1971). Similarly, immunodiffusion has been used extensively for strain identification and antigenic analysis rather than taxonomic studies of rhizobia (Dudman 1977).

The application of immunodiffusion techniques to study the internal antigens of both fast- and slow-growing rhizobia by Vincent and Humphrey (1970) and Vincent, Humphrey, and Škrdleta (1973) has produced valuable taxonomic information, since internal antigens show broader group specificities. They found that lines produced with fast-diffusing thermo-labile antigens in gel diffusions of broken cells of antiserum for *R. trifolii* strain SU 329 revealed a relatedness between *R. trifolii, R. leguminosarum,* and *R. phaseoli* and a less close relationship between these and *R. meliloti,* as well as a relationship between the latter and agrobacteria. The slow-growing rhizobia failed to react with the antisera. Conversely, when Vincent *et al.* (1973) tested both fast- and slow-growing rhizobia against antisera to three strains of *R. japonicum,* at least one, usually two common antigens were found in *R. japonicum, R. lupini,* cowpea rhizobia, and slow-growing strains from *Lotus.* Later Vincent (personal communication) found similar bands for the slow-growing rhizobia from the non-legume *Parasponia.* Sixty-four fast-growing rhizobia were either wholly negative or gave a much weaker non-identical line, and the agrobacteria again grouped with the fast-growing rhizobia. The fast-growing strains from *Lotus* and *Leucaena* grouped with *R. meliloti* and the agrobacteria, and so occupied an intermediate position between the *R. leguminosarum–R. trifolii* and the slow-growers. Vincent (1977) found that *Lotononis bainesii* isolates failed to react with antisera to either the fast- or slow-growers. A similar investigation by Graham (1971) with antiserum for *Agrobacterium* showed a relatedness with the fast-growing rhizobia.

Carbon utilization

Early investigations into the carbon utilization were reviewed by Allen and Allen (1958) who generalized that the fast-growing rhizobia are less fastidious in their carbon requirements than the slow-growers. Mannitol and sucrose are the preferred carbon sources for the former; arabinose and xylose are preferred by the latter (Fred *et al.* 1932). The slow-growing rhizobia utilize mannitol, mannose, glycerol, lactose, sucrose, and erythritol sparingly, or not at all (Neal and Walker 1935). Graham (1964*a*) confirmed the more specific requirement of the slow-growing rhizobia, with most strains utilizing only glucose, sodium citrate, xylose, mannitol, arabinose, and fructose and he also concluded that the fast-growing root-nodule bacteria and agrobacteria were similar in reaction, giving vigorous growth on most of the carbon compounds tested. Vincent (1977) tabulated the similar results of Graham (1964*a*), Elkan and Kwik (1968) and Abdel-Ghaffar (1966). A consistent feature of the slow-growing organisms was their failure to use rhamnose, sucrose, trehalose, raffinose, and dulcitol; and their response to organic acids, maltose, and lactose varied considerably between strains. Growth on a range of carbon sources further confirms the major division between the fast- and slow-growing rhizobia. Individual strains in a *Rhizobium* species may be distinguished by their inability to utilize the substrate or by a change in colony morphology (Kuykendall and Elkan 1976; Trinick 1980*a*).

The Entner–Doudoroff, the Embden–Meyerhoff glycolytic, and the oxidative pentose–phosphate pathways have been reported in the fast-growing rhizobia (Katznelson 1955; Katznelson and Zagallo 1957; Jordan 1962; Johnson, Evans, and Ching 1966; Martinez-de Drets and Arias 1972) and only the Entner-Doudoroff pathway occurred in the slow-growing rhizobia (*R. japonicum*) (Elkan 1971). Similarities between a collection of strains of *R. japonicum*, representing the slow-growing rhizobia and pseudomonads, which includes dependence on Entner–Doudoroff and other catabolic pathways (Martinez-de Drets and Arias 1972; Elkan 1971), has pointed to a relatively-close genetic relationship between the two genera (Elkan 1971).

After examining the literature, Skotnicki and Rolfe (1978) concluded that a general feature shared by both fast- and slow-growing rhizobia is an apparently defective tricarboxylic acid cycle (TCA). They also found that some tricarboxylic acid intermediates and related compounds stimulated growth of *R. trifolii* strains in the presence of sucrose and arabinose, while others inhibited growth partially or completely. They were able to differentiate between the various species of fast- and slow-growing rhizobia. Although some variations between strains and species of the fast-growing rhizobia (*R. trifolii*, *R. leguminosarum*, and *R. meliloti*) occurred, they did show characteristic patterns of growth stimulation and inhibition by TCA cycle intermediates and related compounds. The *R. meliloti* strains tested could be differentiated by the marked stimulation rather than inhibition with citrate, lactate, and acetate. The slow-growing organisms were completely inhibited by the same tricarboxylic acid cycle intermediates but were also inhibited by 20 mM succinate and fumarate in contrast to the fast-growing ones, which were stimulated. *R. japonicum* and *Vigna* rhizobia were more closely related than to *R. lupini*, since they were stimulated by all tricarboxylic acid cycle intermediates, by lactate, and by pyruvate. However *R. lupini* strains were not stimulated by these compounds and were inhibited by pyruvate. These studies are the first to show distinct differences between the various species of *Rhizobium*, irrespective of growth rate. Unfortunately, only relatively few strains were examined, but the authors have produced interesting correlations between the existing species of *Rhizobium*. Further studies along these lines could enhance the relative importance of carbon utilization in the taxonomy of *Rhizobium*.

Growth rate

The fast-growing rhizobia (colonies $\geq$ 1 mm in 3–5 days on yeast–mannitol agar) are *R. trifolii, R. leguminosarum, R. phaseoli,* and *R. meliloti* as defined according to the host of isolation (pp. 85–90 and Tables 3.1 and 3.2). In addition, there are un-named fast-growing strains which nodulate various other genera of both temperate and tropical legumes. These include *Leucaena leucocephala, Acacia farnesiana, Mimosa invisa, M. pudica* (Trinick 1968, 1980*a*), *M. caesalpiniaefolia* (Câmpelo and Döbereiner 1969), *Sesbania grandiflora* (Johnson and Allen 1952*b*), *Lotus corniculatus, Lupinus densiflorus, Anthyllis vulneraria, Astragalus glycyphyllus, Caragana aborescens,* and *Ononis repens* (Gregory and Allen 1953; Brockwell and Neal-Smith 1966; Brockwell, Hely, and Neal-Smith 1966). Some of these have been termed as the Lupin–Lotus group (Vincent 1974) and comparisons between various fast-growing isolates from tropical legumes have been made (Trinick 1980*a*). Perhaps these rhizobia are intermediate forms between the named fast- and the slow-growers. The slow-growing rhizobia (colonies < 1 mm in 10 days) are traditionally designated *R. lupini* and *R. japonicum* and appear to be reasonably well-defined (Vincent 1977). However the rest of the slow-growing organisms from other legumes include a vast array of rhizobia with variable host specificities (see pp. 85–90), often with greater specificities than found within the *Lupinus* and *Glycine* groups (e.g. *Centrosema pubescens*). No attempt has been made to extend the number of *Rhizobium* species in this group.

Generally, the species within a genus are nodulated by either fast- or slow-growing rhizobia but there are notable exceptions; *Lupinus* spp, excluding *L. densiflorus*, (see Vincent 1974) and *Acacia* spp are generally considered to nodulate with slow-growing strains but Trinick (1980*a*) had found that *A. farnesiana* was only nodulated with fast-growing strains of the *Leucaena* type. A legume host is usually nodulated with either a slow- or a fast-growing strain, but there are a number of hosts which nodulate with either type. They include *Lotus* spp, *Ornithopus sativus*, *Anthyllis vulneraria*, *Astragalus glycophyllus*, and many hosts of the cowpea miscellany normally nodulated with slow-growing organisms. Both groups of *Rhizobium* have an extensive host-range and have managed to form nodules on the non-legume, *Parasponia* (Trinick and Galbraith 1980). However, the *Rhizobium* association is usually only effective in nitrogen fixation with either the fast- or the slow-growing type but the following exceptions occur. The *Leucaena–Mimosa–Acacia farnesiana* fast-growing rhizobia are usually effective on hosts such as *Vigna unguiculata* and *Vigna unguiculata* spp *sesquipedalis* which are normally effectively nodulated with slow-growing isolates. (Trinick 1968, 1980*a*). Recently, an effective fast-growing *Leucaena*-type *Rhizobium* was isolated from *Lablab purpureus* which also effectively nodulated a number of other tropical legumes known to nodulate in the field only with slow-growing organisms (*Calopogonium caeruleum*, *Flemingia congesta*, *Macroptilium atropurpureum*, *M. lathyroides*, *Tephrosia candida*, *Vigna unguiculata* and *V. unguiculata* ssp *sesquipedalis*, *Glycine max*) (Trinick 1980*a*). Dual infection by both fast- and slow-growing rhizobia have been observed in *M. atropurpureum* (Franco and Vincent 1976) and *V. unguiculata* (Trinick 1981). Great care was taken to ensure that the fast-growing strain (NGR 234) was free from *Agrobacterium*-like contaminants and from accompanying slow-growing rhizobia. Certain hosts such as *Vigna* spp, *Pueraria* spp, *Desmodium* spp and *Centrosema* spp have repeatedly been reported to have slow-growing symbionts and reports of fast-growing organisms isolated from such legumes must always be suspected of contamination with *Agrobacterium* (Lange 1960) and must be tested accordingly (Kleczkowska, Nutman, Skinner, and Vincent 1968). Recent reports from tropical regions have stated that only fast-growing strains were isolated from many legumes previously reported to have slow-growing symbionts (Lim and Ng 1977). Repeated isolations by the author from these same legumes grown in various parts of Papua New Guinea have always yielded slow-growing isolates. Nevertheless, the number of hosts with both fast- and slow-growing rhizobial symbionts is continually being increased. As with the other criteria for dividing the nodule bacteria, the usual boundaries are broken.

Vincent (1974, 1977) has tabulated the cultural, serological, and biochemical characteristics of the fast- and slow-growing rhizobia to compare them with those of *Agrobacterium*. These attributes have linked *R. trifolii*, *R. leguminosarum*, and *R. phaseoli* together and have tended to separate *R. meliloti*. However, *R. phaseoli* has not been researched as much as the other fast-growing organisms and perhaps its low DNA homology with *R. leguminosarum* could support its retention as a separate species. Also, the frequent nodulation of its host, *Phaseolus vulgaris*, with slow-growing rhizobia (Lange 1961) adds further weight to this. The fast-growing rhizobia from the *Lotus–Lupinus* group (Abdel-Ghaffar and Jensen 1966) and *Leucaena* (Tan and Broughton 1981) have subpolar flagella rather than the peritrichous flagella of the other fast-growing *Rhizobium* spp and have a general inability to utilize dulcitol (Abdel-Ghaffar and Jensen 1966; Trinick 1980*a*). In many other respects, such as internal antigens, they resemble the others. Norris (1965) regarded these faster growing organisms, possibly also including those from *Leucaena, Mimosa, Sesbania* etc., as forms intermediate between *R. trifolii* and the slow-growing cowpea rhizobia. Perhaps the effective nodulation of tropical legumes such as *Vigna* spp (Trinick 1968, 1980*a*) by such fast-growing organisms support this view, but further data may permit their inclusion with existing species. Vincent (1974) does not consider specific ranking between the slow-growing *R. japonicum* and *R. lupini* to be justified, even though they can be distinguished from each other on the basis of DNA homology, a greater frequency of cross-agglutination within than between them, as well as their host preference. He suggests, however, a symbiotype distinction such as *R. japonicum* symb. *lupini*. The rest of the slow-growing organisms were considered to be extremely heterogeneous, many with very specialized host requirements (see p. 89) and more work is required before any attempt to group them could be contemplated.

Rhizobium × phage susceptibility

The usefulness of *Rhizobium* phage for taxonomic purposes is limited. Rhizobial phage commonly occurs in nodules and in field soils supporting legumes, and are generally more prevalent in nodules showing early senescence than in young healthy tissue (Allen and Allen 1950; Kowalski, Ham, Frederick, and Anderson 1974). Bacteriophage were early demonstrated for all the main groups of rhizobia and these were reviewed by Allen and Allen (1950) who reported the variability in behaviour of the various phages from rhizobia. Conn, Bottcher, and Randall (1945), employing 22 phages and 33 cultures representing six cross-inoculation groups, defined four groups of rhizobia, (i) clover-pea-bean, (ii) alfalfa, (iii) soyabean, and (iv) lima bean. Early reports (Conn *et al.* 1945; Cook, Watson, and Allen 1949) indicated that bacteriophage from other genera of Rhizobiaceae did not attack *Rhizobium*. However, Roslycky, Allen, and McCoy (1962) reported lysis of rhizobia by bacteriophage of *Agrobacterium*, thus giving further evidence to the link between *Rhizobium* and *Agrobacterium* discussed in earlier sections. The host range of phage from rhizobia is very variable; many studies on phage typing on different species of *Rhizobium* have shown that the phage were mostly polyvalent (Conn *et al.* 1945; Staniewski, Kowalski, Gogaez, and Sokolowska 1962). Staniewski (1970) was able to establish 31 phage types for 157 strains belonging to various species of *Rhizobium*, a high proportion of the phage types proving to be multivalent. In contrast to these results, Kowalski *et al.* (1974) found bacteriophages against *R. japonicum* serogroups highly specific (i.e. 45 of 51 phage isolates). A detailed study with phage by Bruch and Allen (1957) supported the distinction between fast- and slow-growing *Lotus* group of rhizobia. Phage from *L. corniculatus* lysed (22–77 per cent) rhizobia from the same host group but failed to do so with rhizobia from *L. uliginosus*. Apart from one strain of *R. meliloti*, susceptible rhizobia to the phage were only isolated from legumes with specialized *Rhizobium* symbionts such as *Anthyllis* and *Astragalus* (their relationships with the *Lotus* rhizobia are discussed on pp. 97–9). Other fast- and slow-growing rhizobia, including *R. lupini* and the cowpea type, were not attacked, Ziemiecka (1963) was able to show a much wider cross-reactivity between the phage for fast- and slow-growing rhizobia.

Close serological relationships between phages of *R. lupini* and *R. trifolii* were shown by Staniewski, Kowalski, and Lomanska (1963). Occasional correlations between phage and the *Rhizobium* serotypes attacked have also been reported. Kowalski *et al.* (1974) found that among 51 isolates, 45 lysed only rhizobial strains from the same serological group as the strain from which the phage was isolated. However up to 50 per cent of *Rhizobium* strains belonging to a serological group were not sensitive to the phage isolated. This showed that the phage could further differentiate between the rhizobia. Barnet (1972) found that some *R. trifolii* phage of different morphological groups were restricted to one or a few serotypes, while others attacked a wide range of rhizobial hosts of distinct serological characteristics. A connection between phage sensitivity and the serology of *Rhizobium* was noted by Barnet and Vincent (1970), who found a loss of absorbing capacity was associated with lysogenic modification of *R. trifolii*.

The knowledge of *Rhizobium* susceptibility to bacteriophage is not really sufficient to be useful in *Rhizobium* classification, although correlations with other criteria for classifying rhizobia have sometimes been made.

Other features of *Rhizobium* used for differentiation of types

The Adansonian approach (p. 90) used as a large number of characteristics as possible to group rhizobia. However, a number of single attributes of the *Rhizobium* have been used to separate them from other soil organisms, as well as to substantiate similarities or differences between the different species of *Rhizobium*.

Acid production. Some workers, like Norris (1956, 1965), have emphasized, the importance of the production of acid or alkali by the various rhizobia when considering *Rhizobium* taxonomy, survival in the soil, and evolution. However, the production of acid or alkali depends on the composition of the medium, the presence of various carbohydrates and organic nitrogen compounds, and the preferential requirement of the particular *Rhizobium* strain (Parker *et al.* 1977). In the same study, both fast- and slow-growing rhizobia were unable to alter the pH of the rhizosphere of the hosts as well as the water extracts of the soil. Eighty-five strains of slow-growing rhizobia from native legumes of Western Australia produced acid on arabinose and xylose, an acid reaction or no change of pH on rhamnose, and alkaline reactions on maltose and sucrose (Lange 1961). Similarly, Jarvis *et al.* (1977) found that many of New Zealand's indigenous rhizobia grew slowly on yeast–mannitol agar and produced acid only from a limited range of carbohydrates (e.g. glucose). The production of acid or alkali by rhizobia is thus of limited value and Date and Halliday (1979) have further challenged its significance by again stating that results can be altered by minor changes in the test medium. Surely the resulting pH change of the media is due to substances produced as by-products, or total or partial digestion of the substrates, or through secretion by the rhizobia. Organic acids such as isobutyrate, butyrate, acetate, proprionate, and pyruvate which Holding and Lowe (1971) found in the growth medium of *R. trifolii* would produce acid reaction of the medium, while end products from the breakdown of various nitrogenous compounds would be alkaline. Tan and Broughton (1981), like Parker *et al.* (1977), concluded that the pH of the medium is a result of selective utilization of available substrates. The importance of acid or alkali production (Norris 1956, 1965) in *Rhizobium* taxonomy, as well as being a valuable ecological characteristic, is very doubtful.

General bacteriological features. Incorporation of congo red into media separates most strains of rhizobia from related genera such as *Agrobacterium* and other soil organisms (Hahn 1966). Rhizobia absorb the dye weakly compared with other bacteria. Other tests such as growth in litmus milk and on calcium glycerophosphate allow very limited grouping of rhizobia. Rhizobia produce slow changes in litmus milk, often towards alkalinity, but a few strains of *R. meliloti* (Parker and Graham 1964) and fast-growing *Leucaena* type rhizobia (Trinick 1965*a*, 1980*a*) can give an acid reaction. Many strains can form a 'serum' zone and acid producing organisms can sometimes digest sufficient to give the appearance of 'a clot of blood' in the tube (Trinick 1965*a*, 1980*a*). Hofer (1941) and Graham and Parker (1964) found that only strains of *R. meliloti* and *Agrobacterium radiobacter* produced a precipitate in calcium glycerophosphate, whilst slow-growing rhizobia grew poorly or not at all. Other features showing limited differentiation (Graham and Parker 1964) were production of penicillinase, growth at pH 9.5 and in 340 mM sodium chloride, and the production of hydrogen sulphide in certain media. The peritrichate flagellation of fast-growing rhizobia compared with subpolar flagellation of slow-growing ones has already been mentioned. The occurrence of predominantly subpolar flagella with fast-growth in the *Lotus–Lupinus densiflorus* rhizobia (Vincent 1974) and *Leucaena* rhizobia (Tan and Broughton 1981) perhaps represents intermediate evolutionary stages. Vincent (1974, 1977) has tabulated cultural and biochemical characteristics of the major groups of rhizobia.

Chemical composition. Studies have usually been limited to only a few species or even a few isolates within a rhizobial species, but attempts have been made to differentiate rhizobia on polyacrylamide-gel electrophoresis of protein extracts (Pechy and Szende 1974; Shemakhanova and Oleinikov 1971), carbohydrate composition (Nalbandyan 1976), phospholipid composition and fatty acid composition (Bunn and Elkan 1971; Gerson and Patel 1975; Mackenzie, Lapp, and Child 1979). However, Mackenzie *et al.* used 42 isolates belonging to the major plant-affinity groups of *Rhizobium* and used 15 major fatty-acid components for numerical taxonomic analysis. They found that the rhizobia constituted a uniform group, but that clusters comprising *Glycine–Vigna* isolates and *Pisum–Phaseolus* isolates were evident, which was consistent with conclusions formed by others (see p. 95) using a variety of features. Two-dimensional polyacrylamide-gel electrophoresis was used by Roberts, Leps, Silver, and Brill (1980) to identify and classify 57 strains of *Rhizobium*. They found all the slow-growing rhizobia to be closely related and distinct from the fast-growing strains. *R. leguminosarum* and *R. trifolii* formed a distinct group and *R. meliloti* were discrete. The other fast-growing organisms, *R. phaseoli*, *Sesbania* (perhaps *Leucaena* type), and *Lotus* rhizobia which had similarities, seemed much more diverse. Peterson, Greenwood, Belling, and Bathurst (1971), used electrophoretic mobility of proteins to place isolates of *Rhizobium* into various groups, and concluded that these methods gave little information on the relatedness of the *Rhizobium* groups. Craig, Greenwood, and Williamson (1973) considered the significance of polyphosphate inclusions and associated nuclear material which divided rhizobial bacteroids from *Lotus* nodules into two groups corresponding to acid and alkali production in culture. Vincent (1977) has summarized patterns of isoenzymes amongst the groups of rhizobia and concluded that they have made only a limited contribution towards taxonomy and clarification of rhizobial relationships.

Concluding remarks

Perhaps the ideal taxonomic system for *Rhizobium* should reveal as much practical information as possible to the user and indicate groups for invasiveness and effectiveness in nitrogen fixation, as well as showing relationships and even evolutionary trends between the different rhizobia. The bacterium is *Rhizobium* because it nodulates a legume and its early classification was built solely on phenotypic data, often with very few plant-host range tests. This only accounted for a minute fraction of the total genome which led DeLey *et al.* (1966) to consider numerical analysis and DNA homology to be better approaches to relatedness between bacterial genera. In addition, several groups (Jordan and Allen 1974; Nuti, Ledeboer, Lepidi, and Schilperoort 1977) have suggested that infectiveness and effectiveness are plasmid-borne and if this should be demonstrated, then the present cross-inoculation groupings would show meaningless distinctions and relationships between the present *Rhizobium* groupings. Recently Johnston, Beynon, Buchanan-Wollaston, Setchell, Hirsch, and Beringer (1978) have obtained a high-frequency transfer of nodulating ability from *R. leguminosarum* to *R. trifolii*, *R. phaseoli*, and a member of the cowpea miscellany. Jarvis *et al.* (1980) also consider that the loss of nodulating ability was contained on a relatively short DNA base sequence which could be lost without a significant effect on overall genetic homology with an independent *Rhizobium* reference strain. Such happenings do not support the various evolutionary trends for *Rhizobium* hypothesized, including the 'symbio-taxonomy' proposed by Norris (1965) and Norris and 't Mannetje (1964).

The various attempts at including a greater part of the rhizobial genome (Adansonian or overall similarities, DNA base composition, or homology) have not advanced *Rhizobium* taxonomy and possible evolutionary trends as far as they were first envisaged. These attempts have further confirmed the broad grouping first defined with the possible consolidations already indicated on pp. 90–5. Grouping rhizobia along these lines fails to indicate the host preferences of individual strains of rhizobia.

Perhaps a classification system combining as much of the bacterial genome as possible, with indication of host specificity or preference, could be devised. Before such a goal could be achieved, a tremendous effort is required to study the enormous array of organisms now grouped as the cowpea miscellany, the presently known and often host-specific fast-growing rhizobia of the *Leucaena–Mimosa–Sesbania– Lotus–Cicer* etc. type(s) and other distinctive rhizobia like the isolates from *Lotononis bainesii* (Norris 1958*b*; Godfrey 1972). A range of phenotypic and genotypic tests designed to show similarities and differences between rhizobia should be formulated and standardized. In the meantime, more attention could be given to the suggestions of Vincent (1977) by developing the 'symbiotype' concept. Thus if *R. leguminosarum* should include the present *R. trifolii* and perhaps *R. phaseoli*, then the nomenclature for clover rhizobia would be *R. leguminosarum* symbiotype *trifolii*, and similarly *R. japonicum* symbiotype *vignae* for the cowpea strain (Vincent 1977). This system would, with present information, indicate rhizobial relationships as well as specificities in infection characteristics.

3.2 General biology

Morphology of Rhizobium

The free-living cell and growth on solid media
Rhizobium growing on yeast–mannitol agar (YMA) are small to medium-sized Gram-negative rods, varying in width from 0.5 to 0.9 μm, and having rounded ends. They occur singly or in pairs and can often appear in groups with a side by side arrangement (e.g. *Leucaena* type cultures, Trinick 1965*a*). Size of the rhizobial cells can vary according to the physiological age of the culture.

Young cells are frequently motile and have peritrichous, polar, or subpolar flagella. DeLey and Rassel (1965) made use of flagella arrangement as a distinguishing feature between the fast- and slow-growing rhizobia; the peritrichous organisms are usually fast-growing while the subpolarly flagellated organisms are slow (Leifson and Erdman 1958). Flagella are easily lost (DeLey and Rassel 1965) during specimen preparation for light and electron microscopy and the presence of subpolar multiple flagella would indicate a peritrichous arrangement, the other flagella being lost during preparation. The fast-growing rhizobia from *Lotus corniculatus*, *Lupinus densiflorus*, and *Leucaena leucocephala* differ from other fast-growing strains in having subpolar flagella (Abdel-Ghaffar and Jensen 1966; Tan and Broughton 1981).

Heat resistant endospores are absent from *Rhizobium*. The reports of their presence by Bisset and Hale (1951) and Bisset (1952) remain unsupported. Graham, Parker, Oakley, Lange, and Sanderson (1963) were unable to find heat-resistant endospores in 164 strains of *Rhizobium*. They used media designed to promote endospore production, but all strains failed to survive at 60 °C for 5 min.

Young cells are evenly stained with basic stains. However Graham and Parker (1964) found that strains of *R. leguminosarum* and *R. trifolii* commonly contained metachromatic granules. As the cell ages, there is an accumulation of granules of poly-β-hydroxybutyrate (PHB) which do not stain (Vincent 1974, 1977) and often give the cell a banded appearance. This is due to the stainable cytoplasmic material being compressed between large granules of PHB. This substance, although not a true lipid, is stained with lipophilic dyes and is extracted with chloroform (Vincent 1974, 1977). The presence of PHB in older rhizobial cells complicates the embedding and sectioning of material for electron microscopy but the cell envelope, the various membranes, the nuclear material including ribosomes and other organized material have been described. Details of the ultrastructural components of the free-living cell are given in Chapter 6.

The fast-growing strains of rhizobia produce an abundance of water-soluble gum (exopolysaccharide) which seems to have no morphological role. Dudman (1968) found capsular-like structures which became more prominent with age. Only one strain of *R. trifolii* produced capsules under all conditions and capsulation in other strains varied between wide limits with the nature of the medium, and with shaking or static conditions, and appeared independent of polysaccharide production. Colonial variants of *R. trifolii* contained different proportions of capsulated cells (Dudmam 1968). Humphrey and Vincent (1969) thought that the capsular material seemed to bear no relationship to gum production in *R. trifolii*. Vincent (1977) has concluded that the capsules may be interpreted as extracellular accumulations of normal antigenic material of the cell wall, since fluorescent-labelled antibody developed (Dudman 1968) against cell-wall lipopolysaccharide reacted with capsules. The cell-surface chemistry will be described in detail in Chapter 7.

The general morphology of the rhizobial cell can be changed according to the imposed environmental, including nutritional conditions. Allen and Allen (1958) summarized the morphological changes observed when rhizobia were grown in media containing alkaloids, glucosides, high acidity, thiamine, blood, and substances such as caffeine, strychnine, and quinine compounds. More recently, Skinner, Roughley, and Chandler (1977) reported that levels of yeast extract (Difco, Oxoid, Vegemite) usually used in media (> 0.35 per cent) to grow rhizobia, produced enlarged and distorted cells. These cells often contained several granules of PHB and whorls of intracytoplasmic membranes, indicating greater international disorganization than that seen in root nodule bacteroids. The morphology of the rhizobia within soft agar was found by Pankhurst and Craig (1978) to change with location of the cells within the agar. The larger than normal vegetative *Rhizobium* occurred at constant depths and hence at similar oxygen levels and were pleomorphic and similar to nodule bacteroids. They attributed nitrogen fixation of the colonies to these rhizobial forms. Schwinghamer (1964, 1968) suggested that soil microbial metabolites could induce irregular forms of nodule bacteria in the soil. Recently it has been claimed (Koleshko 1975), using a method of capillary microscopy, that rhizobia in the soil are present in the form of rods, 'bacteroids', and cocci. The term 'bacteroid' has been loosely used to describe all pleomorphic rhizobial forms, but perhaps should be restricted to the nodule-inhabiting forms. Jordan (1962) makes the distinction 'artificially produced'.

The rhizobia form colourless, white, milky, or cream colonies that are circular, convex (especially when young), glistening, and with an entire margin. The red isolate from *Lotononis* (Norris 1958*b*) is the only known exception. The fast-growing isolates with a mean generation time of 2–4 h form relatively large colonies (2–4 mm in diameter) in about five days. *R. trifolii*, *R. leguminosarum*, *R. phaseoli*, and some fast-growing strains from tropical legumes (i.e. *Leucaena*, *Mimosa*, etc.) produce large amounts of extracellular gum which is often clear or misty, with or without opaque areas. The colonies can appear to be very large with extended growth. Other fast-growing strains, *R. meliloti* and some strains from *Leucaena*, *Mimosa* etc. as well as the fast-growing isolate from *Lablab* (Trinick 1980*a*), tend to produce less gum and their growth is more opaque and white. The slow-growing strains, with a mean generation time of 6–8 h, form smaller colonies, and reach a maximum size $\leqslant 1$ mm after 10 days' growth. These produce less gum which is usually dense and sticky. An 'autoplaque' condition resembling self-lysis and phage plaques was found in *R. trifolii* (Barnet and Vincent 1969), which appeared to be triggered under certain undefined circumstances on the surface of solid media.

A strain of *Rhizobium* is established from an isolated colony obtained by streaking a surface sterilized squashed nodule. The colony type within a *Rhizobium* strain can change by mutation, so that streaking a culture on agar plates will produce more than one colony type. Vincent (1954, 1962*a*) isolated an effective strain of *R. trifolii* that produced three variants having high, medium, and low gum production; the small colony variants, though fully invasive, invariably lost their effectiveness. Herridge and Roughley (1975*a*) examined 17 cultures of the cowpea-type *Rhizobium* CB 756 and found they varied in symbiotic effectiveness and all contained a number of colony types. They indicated that colony characteristics of most isolates were unstable and that the most effective sub-strains were all isolated pinpoint, dry colonies. The large, gummy colonies were ineffective. A similar variation was noted by Kuykendall and Elkan (1976) and Upchurch and Elkan (1977) with *R. japonicum*. Roughley (1976) lists similar variations occurring in a number of commercially used inoculant strains of *Rhizobium*.

This 'spontaneous' mutation that can occur frequently in some strains of *Rhizobium* raises the question of the definition of a strain and the purity of cultures. It makes the task of ensuring the purity of a culture very difficult and emphasizes the desirability of continuing *Rhizobium* strains by subculturing from general agar growth and not from an isolated colony. The identity and purity of a culture can be established and checked using serological techniques, antibiotic resistance, phage resistance, and biochemical and nutritional markers. Cultures should also be examined under the microscope for cell type variation.

Bacteroid formation and variability

The term bacteroid should refer only to the *Rhizobium* forms found in the central tissue of the nitrogen-fixing nodules of legumes, and perhaps the non-legume *Parasponia* (Trinick 1979). They represent a specialized form which has differentiated with respect to their physiology and in many cases in their morphology. It was considered that nitrogen fixation was confined to the bacteroid state, but recent detection in free-living rhizobia suggests that fixation may start before bacteroids are formed, as suggested by Bergersen (1974).

The fast-growing rhizobia in nodules of plants belonging to the *Trifolium, Medicago*, and *Pisum* cross-inoculation groups may develop considerable pleomorphy, often referred to as the club, X, and Y shapes. Pleomorphic bacteroids are also formed in *Astragalus glycyphyllus* (Dart 1969) but other fast-growing rhizobia such as those from *Leucaena* do not form such bacteroids (Trinick unpublished). The slow-growing organisms do not generally display such dramatic changes in cell morphology. In *Lupinus* spp, bacteroids are smaller and retain the rod shape (Jordan and Grinyer 1965; Dart and Mercer 1966; Kidby and Goodchild 1966), while in *Lotus* spp the rod shaped bacteroids are slightly larger (Dart 1969). With *Glycine, Phaseolus*, and *Vigna* cross-inoculation groups, there is little pleomorphy and the bacteroid size increases are usually in length (Dart 1975). Van Rensburg, Hahn, and Strijdom (1973) described the morphological development of *Rhizobium* bacteroids in *Arachis hypogaea* and found that the rod shaped *Rhizobium* cells changed into polymorphous bacteroids and eventually into spheroplasts; the bacterial cell-walls gradually disappear with age and the spheroplasts are converted into protoplasts. There are instances where the host induces morphological change depending on the infecting strain of *Rhizobium*. Thus Van Den Berg (1977) reported the morphological changes of two strains of rhizobia (*R. leguminosarum*) on *Pisum*, and *Vicia*. One strain produced branched bacteroids on both hosts while the other formed a few branched bacteroids on broad beans and first developed regular bacteroids which later changed to spherical forms in *Pisum*. Details of the bacteroids and their function within the legume nodule are discussed in Volume 3.

Briefly, the bacteroids have altered nuclear material (Dart and Mercer 1963; Dilworth and Williams 1967; Bergersen 1955), ribosomes are reduced in numbers (Bergersen 1968; Sutton and Robertson 1974), cell walls are altered (Mackenzie, Vail and Jordan 1973) and levels of PHB, glycogen, and polyphosphate generally increase in size and number with age (Goodchild and Bergersen 1966; Craig and Williamson 1972). Bacteroids were once considered incapable of cell division (Almon 1933; Bergersen 1968, 1974; Dart 1977), which was supported by the reduced amount of DNA and altered cell walls previously reported. However, recent nucleic acid studies contradict earlier reports with regard to DNA loss during bacteroid development and higher levels of DNA for bacteroids have been reported by Sutton (1974), Reijnder, Vesser, Aalbers, van Kammen, and Houwers (1975), Paau, Lee, and Cowles (1977) and Bisseling, van den Bos, van Kammen, van den Ploeg, van Duijm, and Houwers (1977). This supports more recent studies showing that bacteroids are able to divide and form bacterial colonies if given a suitable environment.

Sutton, Jepsen, and Shaw (1977) reported only 0.3–10 per cent (depending on nodule age) variability of bacteroids from *Lupinus angustifolius* nodules. They found that recovery of viable bacteroids depended upon the osmolarity of the plating and extracting medium with optimum results at 0.3–0.4 M mannitol. Ninety per cent viability of *R. japonicum* and *R. phaseoli* bacteroids was shown by Tsien, Cain and Schmidt (1977) irrespective of nodule age or nodule environment. Similar findings were also reported by Gresshoff, Skotnicki, Eadie, and Rolfe (1977) and Gresshoff and Rolfe (1978) who established 90 and 100 per cent viability of bacteroids derived from nodule protoplasts of *Trifolium* and *G. max* respectively. Full recovery of *R. trifolii* bacteroids depended upon the isolation of the free bacteroids in an osmotically protected medium and with *R. japonicum*, the plating medium required at least 0.2 M mannitol. The precise mode of action of the extra mannitol was not known. Bergersen (1961), using an identical medium without the extra mannitol, was unable to obtain colony formation from bacteroids.

Nutritional and environmental requirements for growth

If *Rhizobium* (chemo-organotroph) evolved as suggested on p. 79, then the bacterium would be expected to be able to utilize an enormous range of mineral and organic compounds of varying degrees of complexity, in different environments. Their survival and persistence (see pp. 107–8) inside nodule tissue also depends on their ability to cope with extremes of changed physical environments, as well as their source of energy-yielding organic compounds (for nitrogen fixation) synthesized by the host plant. Their attraction to (chemotaxis) and penetration of the legume/non-legume root requires them to be sensitive to specific organic complexes, and for them to secrete enzymes capable of breakdown of 'resistant' plant tissue.

Mineral nutritional requirements

The few critical studies on the mineral nutrition of rhizobia have been restricted to a limited range of rhizobial types and strains. Even more limited is the data on the requirements for phosphorus, sulphur, and potassium, which is surprising when it is known that soils are frequently deficient of these minerals.

Since the reports of Kurz and LaRue (1975), McComb *et al.* (1975) and Pagan *et al.* (1975) a large amount of information on the utilization of molecular nitrogen by free-living *Rhizobium* strains has accumulated. The level of nitrogenase activity has often been reported to approximate that of isolated bacteroids (Tjepkema and Evans 1975; Bergersen *et al.* 1976; Keister and Evans 1976). Early reviews (Gibson, Scowcroft, and Pagan 1977; Bergersen and Gibson 1978) have summarized the conditions required for the expression of nitrogenase activity. The effect of various carbon sources varied, but best activity occurred with a range of pentoses and hexoses with high activity, supported with supplements of tricarboxylic acid cycle intermediates (Kurz and LaRue 1975; McComb *et al.* 1975; Pagan *et al.* 1975; Gibson *et al.* 1977). A low concentration of combined nitrogen was also shown to be essential for induction and continued nitrogen fixation with glutamine promoting the highest activity (Gibson *et al.* 1977), but a range of sources can be used including NO_3^- and NH_4^+ salts, casamino acids, urea, and asparagine. As would be expected, nitrogenase expression depends on low oxygen tensions and Bergersen (1977) found 1 μm oxygen or less in chemostat cultures permitted activity. In agar cultures grown at 0.2 atm of pO_2, nitrogenase activity was thought to occur in a restricted part of the colony (Gibson *et al.* 1977; Bergersen and Gibson 1978) where low concentrations of oxygen would be expected. If the culture is disturbed, the physiological state of the individual rhizobial cell is changed (oxygen diffusion rates to individual cells etc.) halting or disturbing nitrogenase activity. Pankhurst and Craig (1978) found a band of morphological distinct cell types in *Rhizobium* colonies on agar which suggested that they contained the nitrogenase. These cells were larger and pleiomorphic and were similar in morphology to the nitrogen-fixing bacteroids formed by the same strain in root nodules. Other morphological and ultrastructural comparisons between the asymbiotic nitrogen-fixing forms have been made (Wilcockson and Werner 1978; van Brussel, Costerton, and Child 1979).

Most studies have shown that nitrogenase activity in free-living rhizobia is restricted to the slow-growing organisms but Bednarski and Reporter (1978) have shown activity with rhizobia from *Lupinus*, *Pisum*, *Trifolium*, and *Vicia* as a result of their growth in 'plant cell-condition medium'. Their report indicated an exchange of materials across dialysis membranes between the endosymbionts and the host legume (soyabean) cell culture. Child and Kurz (1978) reported studies on the inducing effect of plant cell tissue cultures of both legume and non-legume origin on the nitrogen-fixing activity of both *Spirillum* and *Rhizobium* sp. Both organisms had similar nutritional requirements for induction of nitrogenase activity. In pure culture they required a pentose sugar and a tricarboxylic acid for induction of high activity, but with tissue cultures, the plant cell callus tissue appeared to supply only the tricarboxylic acid, not the sugar component. However, the plant tissue seemed to satisfy a nutritional deficiency in the bacterial cultures. The successful induction of nitrogenase activity in association with culture cells of non-legumes was first described by Child (1975) and Scowcroft and Gibson (1975).

Rhizobia are also able to utilize inorganic forms of nitrogen. Most strains can utilize both nitrate and ammonium as the sole or supplementary form of nitrogen. Norris (1959) used nitrate as the source of nitrogen in his synthetic medium. Bergersen (1961) obtained relatively poor growth with *R. trifolii* on nitrate as the nitrogen source, in contrast to Vincent's (1962*a*) experience. However, both authors reported poor growth with ammonium [NH_4NO_3 or $(NH_4)_2SO_4$] which can exercise an inhibitory effect on the good growth usually obtained with nitrate. An inhibitory change in pH of cultured media using either nitrogen source could occur in lightly buffered media (Vincent 1977). Jordan and San Clemente (1955) have reported the use of nitrite by rhizobia. Rajagopalan (1938) and Wilson (1947) claim that the legume bacteria can liberate gaseous nitrogen from nitrate. Daniel and Appleby (1972) noted the disappearance of nitrate and nitrite from anaerobic cultures of *R. japonicum*. Some strains of slow-growing rhizobia are able to reduce nitrate under anoxic conditions in one of three ways: (a) to nitrous oxide and nitrogen (denitrification), (b) reduction to and subsequent accumulation of nitrogen dioxide (nitrate respiration) and (c) no reduction (Zablotowicz, Eskew, and Focht 1978). The fast-growing strains of *R. leguminosarum*, *R. phaseoli*, and *R. trifolii* tested were unable to reduce nitrate by dissimitatory means.

An early study of the effects of phosphorus on rhizobia indicated to Truesdell (1917) that there was a positive growth response. Unfortunately, he did not assess the phosphate requirement of the host and hence the real response of the rhizobia could not be ascertained. Kamata (1962) related the ability to nodulate phosphate-deficient soyabeans with the rhizobial strains relative response to phosphorus in culture media. Werner and Berghauser (1976) showed that rhizobia could utilize phosphorus from very low concentrations in solution. Sixty-five strains of slow-growing cowpea-type rhizobia were limited to a total attainable population density of 5×10^7 cells ml^{-1} by low phosphorus (5–10 μM) when compared with high phosphorus (1000 μM) (Keyser and Munns 1979*a*). The only data on the quantitative requirement for potassium is given by Vincent (1974, 1977), who found that *R. trifolii* and *R. meliloti* responded up to 0.006 mM.

Both calcium and magnesium are essential for the growth of rhizobia (Vincent 1962*a*, *b*) despite the conflicting claims by Norris (1958*a*, 1959). Vincent (1962*b*) concluded that there was a more general requirement for divalent cation (magnesium or calcium) which is met by a total concentration of 0.5 mM. When levels for each element fell below 0.1 mM for magnesium and 0.025 mM for calcium, specific deficiency effects were noted (Vincent and Colburn 1961; Vincent 1962*b*). Rhizobial viability was effected by deficiency, particularly for magnesium. Cells deficient in magnesium became elongated and sometimes branched, while calcium-deficient cells were irregularly swollen and roughly spherical and were osmotically stable (Vincent and Colburn 1961; Vincent 1962*b*). The calcium effects suggested, together with other evidence, that calcium is involved in normal wall structure (Humphrey and Vincent 1962, 1965; Vincent and Humphrey 1963, 1968). Strontium, but not barium or magnesium, could partially overcome the growth restriction and abnormal morphology. Steinborn and Roughley (1975*a*) found *R. trifolii* and a fast-growing *Leucaena* isolate were unaffected by Mg:Ca ratios from 30:1 to 1:10, but *R. meliloti* (SU 47) grew faster at the higher ratio.

Rhizobia have been shown to respond to iron (0.005–0.2 mM; Thorne and Walker 1936) evidently required for haemoprotein synthesis, zinc (0.1–1.0 μM) and manganese (0.1 10 μM) (Wilson and Reisenauer 1970), molybdenum (Pillai and Sen 1970) and to cobalt (0.1 μM) (Lowe, Evans, and Ahmed 1960). Fast-growing *R. meliloti* is especially dependent on cobalt (Cowles, Evans, and Russell 1969). Although growth response was greater at 0.03 μM added cobalt than at 3 μM, greater quantities of vitamin B_{12}—coenzyme synthesis occurred at the higher level. Recently, Wilson and Reisenauer (1970) found that when supplied, cobalt could prevent inhibition of rhizobia by other heavy metals such as nickel and copper. Inukai, Sato, and Shimizu (1977) reported that methionine could substitute for cobalt ion, and promote the growth of *R. meliloti* in response to its concentration. High levels of manganese in artificial media are tolerated by rhizobia (Masterson 1968; Holding and Lowe 1971) and Keyser and Munns (1979*b*) found that 200 μM manganese slowed the growth rate of slow-growing rhizobia. Continued cultivation in the presence of high levels of manganese resulted in a marked loss of symbiotic effectiveness without any adaptation to manganese (Holding and Lowe 1971). Also 50 μM aluminium increased the lag time of many rhizobia and stopped the growth of 40 per cent of the strains (Keyser and Munns 1979*a*). The heavy metals (mercury, uranium, nickel, copper) are toxic to rhizobia at low levels and are discussed by Fred *et al.* (1932).

Rhizobia respond to low levels of sodium ions, but salts of both sodium and potassium can be toxic at higher concentrations. Chlorides can be more toxic than sulphates of sodium, potassium, and magnesium (Yadav and Vyas 1973). Magnesium chloride was not stimulatory like magnesium sulphate (Pandher and Kahlon 1978). Maximum sodium chloride concentrations tolerated by rhizobia seem to vary between strains and between species. Yadav and Vyas (1971) reported some strains of *R. japonicum* tolerant of 510 mM sodium chloride and some other slow-growers tolerated up to 445 mM. Pillai and Sen (1970) found that *Lablab* isolates had optimum levels at about 170 mM, while Steinborn and Roughley (1974) showed that *R. trifolii* viability was reduced by only 34 mM sodium chloride in peat. Differences between *Rhizobium* species were reported by Steinborn and Roughley (1975*b*) with *R. meliloti* tolerating 600 mM, slow-growing strains 86 mM, and the clover and pea group 170 mM sodium chloride. Adaptation of *Rhizobium* to high levels of salts have been reported (Steinborn and Roughley 1975*b*; Méndez-Castro and Alexander 1976). Inhibitory effects of low concentrations of sodium carbonate and sodium bicarbonate have also been reported (Yadav and Vas 1973; Singh, Lakshmi-Kumari, Biswas, and Subba Rao 1973; Pandher and Kahlon 1978). Although nodulation of lucerne was inhibited at 0.1–0.2 per cent carbonate/bicarbonate, rhizobia could tolerate up to 0.75 per cent (Subba Rao, Lakshmi-Kumari, Singh, and Biswas 1974).

General organic nutritional sources

In the soil and the rhizosphere, rhizobia have the difficult task of contending with a vigorously competitive general microflora for energy substrates which are often in very limited supply. Rhizobia succeeds in the rhizosphere but is often less successful in the soil. Despite the obvious influence of nutrition upon survival in natural habitats, very little is really known about what is available and what can be used. This type of knowledge would assist in manipulating the environment to promote rhizobial growth. However, in the laboratory, information has been accumulating on the requirements of rhizobia for a range of carbon and nitrogen sources of varying molecular complexities.

Differences between the fast- and slow-growing rhizobia in their carbon utilization and the various metabolic pathways (see also Chapter 4) have already been discussed. Recently, compounds of increased molecular complexity have been tested with various rhizobia. Mino (1970) has tested fatty-acid utilization by *R. trifolii* and found that many compounds, including unsubstituted aliphatic dicarboxylic, substituted mono- and dicarboxylic acids, and aromatic acids, with the exception of acetic acid, inhibited growth. Their action increased with increasing length of the carbon chain. The inhibitory effects of glycollic and propionic acids increased with decrease in pH of the media. At the concentrations used, Tween 40 and Tween 60 also inhibited growth, whereas Tween 80 and Tween 85 promoted growth. Others have used the detergent Tween 40 at 0.1 per cent (v/v) to separate and wash rhizobial cells without inhibiting them (Kuykendall and Elkan 1976; Gaur and Sen 1979). Organic acids were found to be used less than carbohydrates (Nalbandyan, Ovsepyan, Ovsepyan, and Stepanyan 1976). Some tricarboxylic-acid cycle intermediates and related compounds were found by Skotnicki and Rolfe (1977) to stimulate *Rhizobium* growth in the presence of sucrose and arabinose, while others inhibited growth partially or completely. Enzyme differentiation between the fast- and slow-growing rhizobia and their difference in sucrose utilization and invertase activity have been reported (Martinez-de Drets and Arias 1972; Martinez-de Drets, Arias, and Rovira de Cutinella 1974). In the presence of sodium glutamate, (Hussien, Tewfik, and Hamdi 1974), all rhizobia tested degraded catechol (99–100 per cent), *p*-hydroxybenzoic acid (79–99 per cent), protocatechuic (81–97 per cent), and salicyclic acid (20–83 per cent). The level of sodium glutamate used was important because at levels higher than 1 mM, it stimulated degradation of *p*-hydroxybenzoic and salicyclic acid, and inhibited protocatechuic acid.

Rhizobium is also able to utilize various organic forms of nitrogen. A large collection of strains and species of rhizobia have been shown to utilize urea and many showed moderate degrees of growth on biuret (Jensen and Schroder 1965). It has been shown frequently that amino acids support *Rhizobium* growth, but sodium glutamate was shown to be superior (Bergersen 1961); other amino acids used by Bergersen (1961) singly and/or in combination were asparagine, aspartic acid, arginine, histidine, tyrosine, lysine, and serine. Some strains were found to prefer other amino acids than glutamic acid. The requirement of an amino acid may be replaced provided the *Rhizobium* strain is supplied with a suitable carbon such as α-ketoglutarate and a source of combined inorganic nitrogen (Elkan and Kwik 1968; Jordan and San Clemente 1955; Jordan 1952). Glycine (Longley, Berge, van Lanen, and Baldwin 1937; Wolf and Baldwin 1940; van Egeraat 1976) and homoserine (Kaszubiak 1965) have shown toxicity towards rhizobia at higher levels. *R. leguminosarum* is less affected by levels of homoserine that are toxic to other species of *Rhizobium* (van Egeraat 1975). Toxic effects of amino acids are dealt with further on pp. 121–2. Vincent (1977) considered that vitamin-free casein hydrolysate was possibly superior to other amino acids due to the presence of a peptide growth factor. Also peptone was shown to be a poor source of nitrogen but a possible substitute for yeast extract for the growth of the *Lotononis* strains of *Rhizobium* (cited by Vincent 1977).

Allen and Allen (1950) summarized the early reports of the vitamin requirements of the various species of *Rhizobium*, particularly for biotin and thiamine. Graham (1963c) examined the vitamin requirements of 63 strains of *Rhizobium*. Twenty six out of 31 strains of the *R. leguminosarum*, *R. trifolii* and *R. phaseoli* group responded to the addition of one or more of biotin, thiamine and calcium pantothenate. Strains of *R. meliloti* and the slow-growing organisms sometimes responded only to biotin. Elkan and Kwik (1968) also found that the slow-growing *R. japonicum* utilized only biotin, out of ten vitamins tested, and only three out of 39 strains responded. Generally nicotinic acid, pyridoxin, folic acid, *p*-amino benzoic acid, inositol, B_{12}, and riboflavin have no effect on rhizobial growth (Vincent 1977). Vitamin B_{12} was shown by Burton and Lochhead (1952) to be synthesized in excess of requirements of rhizobia. Recently, Schwinghamer (1970) produced an ineffective auxotroph, (derived from an effective strain [T1] of *R. trifolii*), which had a specific requirement for riboflavin in culture and this auxotroph only produced an effective response on *Trifolium pratense* when riboflavin was added to the plant growth medium.

In culture media, rhizobia are usually supplied with nitrogen and vitamins in the form of yeast extract. However, unfavourable effects (reduced growth and induced ineffectiveness) have been reported (Jordan and Coulter 1965; Date 1972; Sherwood 1972; Skinner *et al.* 1977). This led Norris and Date (1976) to recommend the preparation of fresh yeast-water from bakers yeast in laboratories working with *Rhizobium*.

In the soil situation, very little is known about the nutritional sources for *Rhizobium*. Parker *et al.* (1977) state that minute quantities of energy-yielding substances in the soil can give rise to significant populations and estimated that 1 μg of organic material can produce approximately 1–4 × 10^6 bacterial cells. Natural soil populations of rhizobia rarely exceed 1 × 10^5 g^{-1} soil and would account for less than 0.05 μg of the available substrate. Soil water extracts have been shown to contain vitamins, amino acids, phenolic compounds (such as cinnamic, ferulic, *p*-amino benzoic, *p*-hydroxy benzoic, *p*-coumaric, syringec, and valic acids), and organic acids (see above) (Whitehead 1964; Kononova 1966; Patrick and Toussour 1970; Alexander 1971). Humic acid as sodium humate, other humic extracts, and fulvic acid obtained from soils, and well decomposed farmyard manure were all equally effective at stimulating rhizobia when used at optimum concentrations (Bhardwaj and Gaur 1972). Parker *et al.* (1977) selected 15 phenolic compounds which were possible breakdown products of lignin and humates and found that *R. meliloti*, *R. japonicum*, and *R. lupini* gave a growth response to them. However, Purushothaman and Balaraman (1973) reported the isolation from four soils (and toxic effects on rhizobia) of the phenolics *p*-hydroxy benzoic, *p*-amino benzoic, *p*-amino benzoic, *p*-coumaric, syringic, and cinnamic acid. Low levels of pectolytic enzymes have been reported in rhizobia (Hubbell, Morales and Umali-Garcia 1978) but their usefulness in the soil environment for obtaining nutrients is not known.

The understanding of the organic nutrition of rhizobia in natural soils is very limited, but it is clear that the soil can contain inhibitory, as well as growth-promoting substances.

Oxygen, temperature, pH, and moisture requirements

Wilson (1940) obtained excellent growth of rhizobia at less than 0.01 atm oxygen, but was unable to obtain growth in its complete absence. Graham and Parker (1964) did not obtain anaerobic growth with 79 strains of rhizobia representing the major *Rhizobium* groups. Comparable growth curves for *R. meliloti* were obtained by Ertola, Mazza, Balatti, Cuevas, and Daguerre (1969) over a 40-fold range of oxygen and it was found that growth rate was affected only at low availability (i.e. 29 ml O_2 l^{-1} h^{-1}). Rhizobia growing in colonies on agar are at differing oxygen tensions. Broth cultures of rhizobia, when shaken, grow more rapidly, possibly through better aeration as well as the supply of nutrients to individual cells. Roughley (1968) found aeration of nodule bacteria important in peat culture with rapid reduction in numbers occurring with complete restriction of aeration.

Allen and Allen (1950, 1958) state that the optimum temperature for most rhizobia occurs between 25 and 30 °C, with 35 °C preferred by *R. meliloti* isolates. Various strains of *R. leguminosarum* and *R. phaseoli* exhibit temperature optimum of 25 °C, with a rapid decline in growth with temperatures on either side of this (Roponen, Valle, and Ettala 1970). Broughton, Chan, Padmanabhan, and Tan (1975) concluded that tropical rhizobia optima fell into two groups; one at about 20 °C, the other at about 25 °C. Growth of *R. lupini* and *R. trifolii* in various sterilized soils was optimum between 25 and 30 °C with a mean about 28 °C (Chowdhury, Marshall, and Parker 1968).

Studies involving 87 strains of rhizobia were shown by Bowen and Kennedy (1959) to have maximum temperatures for growth on agar-slopes of 31–38.4 °C, 32–32.7 °C, 36.5–42.5 °C and 30–42.0 °C for *Trifolium*, *Pisum*, *Medicago*, and tropical legume strains of *Rhizobium* respectively. The maximum temperature of individual strains within each *Rhizobium* group (except for *Pisum*) showed a marked variability of high temperature tolerance, with some strains, even those from the tropical legumes, sensitive to temperatures as low as 30 °C. In another study with 79 strains of rhizobia, Graham and Parker (1964) obtained growth at 39 °C with only eight strains, all belonging to the *Medicago* group; other rhizobial types failed to survive. The growth of tropical rhizobia was severely impaired by temperatures in excess of 30 °C and some isolates showed a negative growth pattern at 40 °C (Broughton *et al*. 1975). Fred *et al*. (1932) recorded survival of *Rhizobium* at between 0 and 50 °C, but growth and survival at 50 °C in laboratory culture is unlikely. However, Graham and Parker (1964) found 22 rhizobial strains out of 79 to survive heating at 50 °C for 10 min. Another study, Graham *et al*. (1963) subjected a large collection of *Rhizobium* to 5 min at 60 °C without any survivors, and attempts to increase heat resistance by training were not successful. However, natural development of a resistance or tolerance to high temperatures has been demonstrated. El Essawi and Abdel-Ghaffar (1967) found that Egyptian isolates from *Trifolium alexandrinum* grew at 35 °C and Wilkins (1967) found strains of *Acacia, Lotus,* and *Psoralea* rhizobia from western New South Wales which survived higher temperatures than strains from the cooler New England tablelands of New South Wales.

Rhizobia are tolerant of temperatures below 4 °C but very little growth occurs at this temperature. Vincent (1958, 1977) found storage of agar cultures at 5 °C resulted in decimal reduction times for 18 weeks; similarly the reduction rate was 8 weeks at 25 °C and for liquid cultures, 8–21 weeks at 2 °C and 23–40 weeks at −15 °C, according to strain and substrain.

The pH requirements for rhizobial growth on agar and in liquid media have been examined for strains representing all the major groups. Unfortunately, most studies do not define the control of the medium obtained during growth of the organism, and hence care should be exercised when interpreting results, especially as there is also a likelihood that pH interacts with other factors (Munns 1977). A comprehensive survey of the rhizobial groups was made by Graham and Parker (1964) who found that all strains examined grew at pH values between 5.5 and 7.5. Only 12 of the slow-growing rhizobia grew at pH 4 and one survived at pH 9. All *R. meliloti* strains survived and grew at pH 9.5 but could not grow at pH 4.5. All 97 strains failed to survive at pH 10.

Researchers have consistently found that strains of *Rhizobium* within a group vary greatly in their sensitivity to low or high pH values (Graham and Parker 1964; Ham, Frederick, and Anderson 1971; Jensen 1942, 1943; Loneragan 1972; Okafor and Alexander 1975). For instance, the minimum pH value for *R. japonicum* can vary as much as three units (Damirgi, Frederick and Anderson 1967; Ham *et al*. 1971). The greater overall sensitivity of *R. meliloti* than *R. trifolii, R. leguminosarum,* and *R. phaseoli* and tolerance of slow-growing rhizobia to low pH has often been reported (Jensen 1942; Fred *et al*. 1932; Munns 1977; Okafor and Alexander 1975; Parker and Graham 1964). Growth, although poor, was detected with three strains of *R. trifolii* at pH 3.5 (van Schreven

1972) but this is exceptional. Yadav and Vyas (1973) obtained some growth with five out of ten cowpea-type rhizobia at pH 3.5; most strains grew well at pH 4 and higher. Rhizobia were found by Cabezas de Herrera (1956) to be modified into unusual forms when grown at pH 4.

Abundant growth usually occurs over a wide range of pH as shown by Parker and Graham (1964). Yadav and Vyas (1973) obtained maximum growth for slow-growing organisms between pH 4.5 and 8.5 for many, and with others satisfactory growth extended to pH 10. The ability of rhizobia to grow at pH 10 is supported by Bhardwaj (1975) who found *R. leguminosarum* strains to survive in a saline–alkaline soil with a pH 10, but not at pH 10.5, the rhizobia possessing a greater tolerance for alkalinity than their host legumes. Broughton *et al.* (1975) obtained clear optima and found most slow-growing strains to have maximal growth rates at about pH 6, while others displayed increased growth at pH 8. Diatloff (1970*b*) obtained an optimum at about pH 6 for an isolate of *R. japonicum*. Generally, optimum pH values are of the order of 6.8–7.2 for clover rhizobia, 7.0–7.5 for medic rhizobia, and 5.8–6.2 for the slow-growing rhizobia.

Brayan (1923) observed that the critical pH in soil was approximately the same as that in pure culture for *Medicago* nodule bacteria. Other data has accumulated relating natural occurrence in soils to pH tolerance in agar or liquid media. Thus *R. meliloti* is rarely abundant in soils with pH less than 5, and *R. trifolii* in soils below 4.5 (Jensen 1969; Peterson and Gooding 1941; Robson, Edwards, and Loneragan 1970). An interesting result was recently reported by Keyser, Munns, and Hohenberg (1979), who successfully identified 65 per cent of cowpea-type rhizobial strains that were symbiotically sensitive to low pH by testing their ability to grow in liquid media containing aluminium at pH 4.5 (50 μM). Perhaps the pH requirements for growth of most useful strains may be close to the soil pH value required for nodulation of their respective host. The effect of pH on *Rhizobium* ecology and on nodulation is reviewed in detail elsewhere (see Chapter 2 and Volume 3).

Surprisingly little definitive work has been done on the tolerance of rhizobia to moisture stress. Similar studies to those made with the fungus *Rhizoctonia solani* (Dubé, Dodman, and Flentjc 1971) describing the minimum water activity for growth should be done with *Rhizobium* so that a better understanding of the effect of desiccation on rhizobia in soil could be realized. Vincent (1958, 1962*a*) and Vincent, Thompson, and Donovan (1962) have studied the survival of *R. trifolii* suspended in simple media, applied to glass beads, and seeds of *Trifolium subterraneum* held in an atmosphere of controlled relative humidity from 0 to 100 per cent. Improvement in survival occurred due to the incorporation of sucrose (McLeod quoted by Vincent 1958, 1962*a*). Results were improved further with maltose (Vincent *et al.* 1962*a*), which was superior to other substances including cellobiose and gum arabic. They also found that drier conditions (0–20 per cent) favoured survival. More detailed examinations are not available for other species of *Rhizobium*.

Inhibitory agents of Rhizobium

Antibiotics

Generally, rhizobia are sensitive to wide-spectrum antibiotics and many authors (Davis 1962; Graham 1963*a*; Skrdleta 1965*a*) have tested large numbers of fast- and slow-growing rhizobia against them. The strain to strain variation within a rhizobial group is large enough to make generalizations unreal, but Graham (1963*a*) was able to conclude that strains of the slow-growing rhizobia appear less susceptible to antibiotics than the fast-growing, *Pisum–Trifolium* type organisms. He found that the slow-growing strains were generally more resistant to streptomycin, aureomycin, and penicillin G. After examining 48 strains of *R. japonicum*, Cole and Elkan (1979) concluded that antibiotic resistance was the norm for this species. They considered this unusual because soil bacteria are considered susceptible to several commonly used antibiotics (Stout 1962). Isolates from tropical legumes (mainly slow-growing) were resistant to lincomycin, most were moderately resistant to clindamycin, erythromycin, and penicillin and most were sensitive to kanamycin, streptomycin, and tetracycline (Broughton *et al.* 1975); some rhizobia were resistant to all antibiotics tested and others were sensitive to most. Davis (1962) found the rhizobia tested from all cross-inoculation groups were resistant to the sulpha drugs, isoniazid, nystatin, and penicillin. Kecskés and Manninger (1962) found a collection of temperate rhizobia resistant to penicillin and sensitive to chloramphenicol, streptomycin, tetracycline, and kanamycin. Schwinghamer (1967) tested 17 strains of fast-growing rhizobia, including *R. trifolii*, against 15 antibiotics and found them to be generally sensitive to antibiotics including penicillin, the highest levels of resistance occurring against spectinomycin and chloramphenicol. Abdel-Wahab, Rifaat, Ahmed, and Hamdi (1976) found that several *R. trifolii* strains were resistant to penicillin and sensitive to streptomycin and chloramphenicol. Cross-resistance has often been reported (Schwinghamer 1967; Brockwell, Schwinghamer and Gault 1977; Hagedorn 1979) for rhizobia and it appears largely associated with those antibiotics possessing similar modes of action. Full cross-resistance was noted for strains selected for resistance to viomycin and neomycin. Furthermore, kanamycin resistance conferred resistance to neomycin, but not to viomycin (Schwinghamer 1964). Cole and Elkan (1979), working with *R. japonicum*, found that most common grouping of resistances in strains was simultaneous resistance to tetracycline, penicillin G, neomycin, chloramphenicol, and streptomycin (25 per cent of strains tested). The various reports of sensitivity and resistance to antibiotics for specific species of *Rhizobium* shows much diversity amongst the strains which makes generalization unwise. Perhaps the previous history of the *Rhizobium* strain is important in that previous exposures to toxic elements within the natural soil/rhizosphere environment may help determine its pattern of sensitivity.

Antibiotic resistance markers and natural resistance have frequently been used for strain identification. Unfortunately in developing the markers, the *Rhizobium* often changes in other respects such as alteration in growth rate or cell permeability, increased or decreased effectiveness of the symbiosis, and the possibility that the resistance can be transferred between strains in the soil as well as in the nodule (Johnston and Beringer 1976; Pariiskaya and Gorelova 1976). Schwinghamer (1964) reported a loss of effectiveness of *R. trifolii*, *R. leguminosarum*, and *R. meliloti* closely associated with mutation to viomycin and neomycin resistance. In contrast, loss of effectiveness occurred only infrequently in clones resistant to kanamycin or polymyxin and did not occur with streptomycin-resistant mutants. Later, Schwinghamer (1976) concluded that the frequency and degree of such changes in effectiveness depends on the type of antibiotic resistance involved in the mutation. Thus resistance to streptomycin, spiramycin, chloramphenicol, or the tetracycline group of antibiotics was associated with little or no change, while resistance to D-cycloserine, novobiocin, vancomycin, bacitracin or penicillin resulted in partial or full loss of effectiveness in about one-half of the mutants. Schwinghamer (1967) postulated an alteration of cell-wall characteristics being responsible for the decrease of effectiveness, particularly D-cycloserine. In examining spectinomycin resistance as a marker, Schwinghamer and Dudman (1973) found that partial or full loss of effectiveness occurred in only about 20 per cent of the resistant mutants of effective strains, representing four species of *Rhizobium*. Pankhurst (1977) reported similar results for fast- and slow-growing strains of *Rhizobium* nodulating *Lotus* species. Resistance to streptomycin, spectinomycin, chloramphenicol, and tetracycline (inhibition of protein synthesis) was associated with little or no loss of effectiveness, but resistance to nalidixic acid and rifampicin (inhibitors of nucleic acid synthesis), and to D-cycloserine, novobiocin, and penicillin (inhibitors of cell wall–cell membrane synthesis) was associated with loss of effectiveness in 20–100 per cent of the mutants. Pankhurst (1977) suggests that the symbiotic effectiveness of the resistant mutants is influenced by characteristics of both the rhizobial strain and the host plant. Similar results with streptomycin-resistant mutants were reported by Law and Strijdom (1974), Gollobin and Levin (1974), and Abdel-

Wàhab *et al.* (1976) but Abdel-Wahab *et al.* (1976) found no changes in effectiveness with penicillin-resistant mutants. Different results were recorded by Levin and Montgomery (1974) who were unable to find dramatic differences in infectivity or effectiveness between antibiotic-sensitive strains and their resistant mutants. Hagedorn (1979) did not agree with the above reports and found that losses of effectiveness in *R. trifolii* occurred at random throughout a large collection of resistant mutants and was not associated with any particular group or class of antibiotics.

Viomycin-resistant ineffective mutants of *R. meliloti* have been examined in some detail (Hendry and Jordan 1969; Jordan, Yamamura, and McKague 1969; Alexander, Jordan, and McKague 1969; MacKenzie and Jordan 1970; Yu and Jordan 1971). The inhibitory action of the antibiotic was prevented by divalent and monovalent inorganic ions (Mg^{2+}, Na^+, K^+, Li^+) and it was interpreted in terms of competition for sites of action. They concluded that the resistance was likely to be due to selective reduction in the permeability of cells towards viomycin. They thought that this might occur as a result of greater accumulation of phospholipid at the expense of neutral lipid in the cell envelope. The antibiotic appeared to complex with the phospholipid *in vitro*.

Some changes in other characteristics of antibiotic-resistant strains have been noted. Schwinghamer and Dudman (1973) noted minor antigenic changes in two strains of rhizobia, and nutritional changes were noted by Abdel-Wahab *et al.* (1976). Similarities in colony morphology, nodulation competitiveness etc., between mutants and parent strains have often been commented upon.

Seed-coat toxins

The presence of water-soluble toxins in seed coats were first demonstrated independently by Thompson (1960) and Bowen (1961) in *Trifolium* sp, *M. sativa*, and *Centrosema*. The inhibition with the water-soluble extract from *Trifolium subterraneum* was more active than that from *Centrosema*, and there was little activity from alfalfa; the degree of inhibition varied with the micro-organisms tested (various rhizobia, other Gram-negative and Gram-positive organisms). Vincent *et al.* (1962) substantiated the inhibitory effect by comparing the death of rhizobia on seeds with glassbeads, and on glassbeads impregnated with a water extracts containing the toxin. The list of legumes having seeds with inhibitors of *Rhizobium* is increasing, and includes *Trifolium* spp, *Pisum sativum*, *Medicago sativum*, *Glycine max*, *Phaseolus* spp, *Vigna* spp, *Cajan cajan*, *Cicer arietinum*, *Macrotyloma uniflorum*, *Lablab purpureus*, *Cyanopsis tetraconoloba* and others. The toxicity produced from the seed often varied from crop to crop, as well as from variety to variety (Dadarwal and Sen 1973; Kandasamy, Kesavan, Ramasamy, and Prasad 1974; Faizah, John, and Broughton 1976; Jain and Rewari 1976). The sensitivity of the different rhizobia to the inhibitors also varied markedly. The inhibitors affected other soil and rhizosphere organisms to varying extents. Bowen (1961) found that *Agrobacterium* was only slightly inhibited, but that the *Xanthomonas–Flavobacterium* group was more affected than *R. trifolii*. Similar inhibitory effects were noted on *Azotobacter* and a *Bacillus* by Dadarwal and Sen (1973). Bacteria producing large quantities of extracellular gums in the test medium were less susceptible to the toxic effects (Dadarwal and Sen 1973). On the other hand, Hale and Mathers (1977) found the diffusate from Huia *T. repens* seed was limited to *Rhizobium* species, some being inhibited more than others. Fottrell, O'Connor, and Masterson (1964) obtained two toxins from *T. repens* seeds, which they identified as myricetin and a tannin or a mixture of tannins. Inclusions of various salts into the test media reduced the toxicity and the addition of vanadium increased the survival of rhizobia on *T. repens* seed. Treatments of seeds with phenolic absorbents such as polyvinyl pyrrolidone or activated charcoal removed the toxin (Hale and Mathers 1977) and it was considered that the toxic factors has a phenolic structure. However, Kandasamy *et al.* (1974) found no correlation between total phenolic content of seed exudates and their inhibitory activity towards *Rhizobium* spp.

Amino acids

Although amino acids are used readily by rhizobia, certain amino acids such as glycine can be inhibitory. Wilson (1940) first reported the inhibition of *Rhizobium* with 100 ppm glycine and the effect of its inhibition has been often studied. Depressive effects on rhizobial growth with increasing levels of the amino acids L-arginine (1 per cent) and L-glutamic acid (0.5 per cent) have been demonstrated, but not with L-aspartic acid up to 1 per cent in the medium (Mino, Mino, Hayashi, and Chiji 1968) or by biotin at 10 μg l^{-1} (Elkan and Kwik 1968). Strijdom and Allen (1966) observed repression of growth with histidine (0.075 per cent) (either the L- or D- form) Hamdi (1969) found that *Medicago Rhizobium* were inhibited by D- and DL- methionine at 0.125 per cent and stimulated by L-methionine at the same concentration. Skinner *et al.* (1977) reported varying degrees of inhibition of *Rhizobium* sp (CB 756), *R. japonicum* (CB 1809), *R. lupini* (WU 425), *R. meliloti* (SU 47), and *R. trifolii* (TA 1) to four different mixtures of amino acids. *R. meliloti* was only slightly affected by each mixture and CB 756 was unaffected by one of the mixtures. TA 1, the strain most susceptible to amino acids was not inhibited by aspartate, cystine, leucine, proline, serine, or glutamate.

Strijdom and Allen (1966) trained *R. meliloti* to tolerate 750–1500 ppm of amino acids such as glycine, L-cysteine, and D-alanine and

Hamdi (1968) found strains tolerant of high D-methionine. The tolerant organisms often developed a loss of effectiveness and even their ability to invade their host. Further work by Strijdom and Allen (1969) showed that there was an amino axid×species×strain interaction in that amino acid varied in the effect, strains of *Rhizobium* responded differently and there were marked differences between *R. trifolii* and *R. meliloti*. Similarly, Schwinghamer (1968) obtained mutants resistant to analogues of L-amino acids and found that the mutants of *R. trifolii* and *R. leguminosarum* were typically non-nodulating on their respective hosts. Vincent (1977) has suggested that there is an apparent similarity to the situation with selected resistance to antibiotics such as neomycin, and viomycin, but the available data (Vincent 1977) do not indicate what common mechanism might be involved.

Cell distortion in bacteria inhibited by glycine has been known (Gordon and Gordon 1943) for a long time. Skinner *et al.* (1977) found that all their amino acids produced some distorted cells of *Rhizobium*, and the proportion of distorted cells was least with those acids that caused little or no inhibition of growth. They found that large spheroplasts were especially abundant with the amino acid valine. 'Bacteroid-like' forms of *R. leguminosarum* growing on 10 μg ml^{-1} glycine were observed by van Egeraat (1976).

Pesticides and herbicides

Pesticides and herbicides, including fumigation, are frequently used in agricultural practice and hence rhizobia often come into close contact with potentially potent chemicals. The need for their use is well established and is reviewed by Chatel and Graham (Volume 3). It is difficult to make firm generalizations concerning the toxicity of the agents towards *Rhizobium* since tests are often undertaken under different conditions using various levels of the chemical. In agar tests the inhibition recorded is a reflection of the solubility of the chemical and the size of the molecule. In field tests the results are complicated by many unknown factors. No attempt is made here to list the various agents and their possible relative effect on *Rhizobium*.

Some workers have compared the toxicity of the pesticide/herbicide towards *Rhizobium* under laboratory conditions (agar plates and/or plant tube culture) with field experiments, and often found the results do not correlate (Diatloff 1970*a*; Lakshmi-Kumari, Biswas, Vijayalakshmi, Narayana, and Subba Rao 1974; Curley and Burton 1975; Tewfik, Embaba, and Hamdi 1975; Murthy and Raghu 1976; Carg, Tauro, and Grover 1973; Brockwell and Robinson 1976). Diatloff (1970*a*) found that both fungicides and insecticides, which were toxic in agar tests, permitted early nodulation in most plants, even with Ceresan which *in vitro* was highly toxic. Seed pelleting and inoculation with *R. meliloti* and *R. trifolii* seed previously treated with organo-phosphorus insecticides the previous day, was followed by a high rhizobial mortality, particularly with *R. trifolii* (Brockwell and Robinson 1976). Tewfik *et al.* (1975) found nodulation of seedlings in agar tubes completely inhibited by herbicides and nematocides and similar observations were made with pot plants treated with the chemical at the time of planting. Aldrin, which is relatively non-toxic to rhizobia, decreased nitrogen fixation and yield of *Cicer arietinum* at levels of 1, 5, and 10 ppm in soil (Kapoor, Singh, Khandelwal, and Mishra 1977). The differences noted by various workers in reaction to insecticide/herbicide treatment are probably due to various soil factors. Soil moisture, as suggested by Milthorpe (1945) and Ruhloff and Burton (1951), and soil texture may be important. Dilution of toxic fungicides by soil moisture (Isaac and Heale 1961), and the addition of organic matter in the form of chickpea straw, reversed the toxicity of Ceresan (Mishra and Gaur 1975). Methods of seed pelleting have often protected *Rhizobium* and recently Iswaran (1975) added activated charcoal to the list of suitable materials.

The various pesticides, as would be expected, show different degrees of toxicity. Thiram, Dexon, Dithane, and PCNG are less toxic than most of the other fungicides, particularly those containing toxic metals (mercury, copper, and zinc) and halogen-substituted compounds (Kecskés and Vincent 1969; Diatloff 1970*a*; Khatri and Choksey 1973; Curley and Burton 1975; Backman 1977). Less work has been done in the laboratory with insecticides but inhibition of rhizobia for *Cajanus cajan* has been noted (Oblisami, Balaraman, Venkataramanan, and Rangaswami 1973). Inhibition has also been observed in the field (Vincent 1958; Goss and Shipton 1965). Swamiappan and Chandy (1975) found stimulation of nodulation by phorate, endrin, and chlor-fenvinphos. Malathion was reported not to influence the nodulation of peanuts (Kulkarni, Sardeshpande, and Bagyaraj 1975) but fumigants used for insect pests during seed storage adversely affected nodulation. Aldrin and lindane at low levels (1–5 ppm) had no effect on the growth of *R. japonicum*, but altered the incorporation of radio-carbon (^{14}C-glucose) into different constituents of the growing cells (Balasubramanian 1976). Some insecticides (Dexon, simazine, atrozine, and protetryne) non-toxic to rhizobia have affects on the plant metabolism (Karanth and Vasantharajan 1974; Paromenskaya 1975; Smith, Funke, and Schulz 1978). Hormone herbicides usually have no effect on *Rhizobium* when used at the recommended levels (Fletcher and Alcorn 1958). Vincent (1977) has concluded that it is pointless to arrive at any generalization about the relative sensitivities of these herbicides. Dalapon at 500 ppm and paraquat at 1000 ppm inhibit rhizobia in culture (Namdeo and Dube 1973). Camugli and Gomez Etchart (1974) found 2,4-dichlorophenoxacetic acid was non toxic to rhizobia while Brominal and Tordon were very toxic. At lower levels Tordon was sometimes stimulatory. Hamdi (1977) compared the relative potency as protectants and the inhibitory action towards rhizobia of pesticides commonly used in Egypt. *Rhizobium* strains within and between species vary in the degree of their sensitivity to the various pesticides and herbicides, but no generalizations can be made to differentiate between the species of *Rhizobium*. Care should be taken to differentiate between the effects due to the chemical and those due to the carrier or solvent.

Odeyemi and Alexander (1977*a*) obtained isolates of *Rhizobium* resistant to high levels of the fungicides Spergon, Thiram, and Phygon. They compared these resistant strains of *Rhizobium* with the non-resistant types on their hosts and when the host had been treated with the fungicides, the resistant rhizobia, unlike the non-resistant ones, were able to nodulate and fix nitrogen as well as the plants without the pesticides. It was suggested that the findings have practical application. Odeyemi and Alexander (1977*b*) showed that the pesticides were destroyed by the resistant bacteria, but not by the susceptible parent. Staphorst and Strijdom (1976) were unable to isolate Thiram-resistant mutants. At high levels, Dalapon and paraquat can act as mutagenic agents (Namdeo and Dube 1973).

Microbial interactions

Competition between rhizobial strains for nodule formation

Rhizobium strains growing in the rhizosphere of their host compete with each other for nodule sites. This phenomenon has been observed with many host × *Rhizobium* combinations, and some attempts have been made to elucidate the mechanisms involved in these interactions (Vincent and Waters 1953; Marques Pinto, Yao, and Vincent 1974; Labandera and Vincent 1975; Franco and Vincent 1976). The successful inoculation of legumes in field soils depends upon the inoculant strain of *Rhizobium* competing for nodule sites with the naturally-occurring soil population of *Rhizobium*, which can often be poorly effective with the introduced host (Holland 1970; Ireland and Vincent 1968; Vincent and Waters 1954).

Field trials testing the relative success of one strain of *Rhizobium* compared with another in their competitive ability at nodule formation on the host have frequently been reported. Vest, Weber, and Sloger (1973) reviewed the work on strain competition with *R. japonicum*. Using two chlorosis-inducing strains (as markers), Means *et al.* (1961) showed that one (USDA 76) was more competitive than nine 'normal' strains; the other was a poor competitor against eight of these nine strains. Subsequently Johnson and Means (1964) ranked three other strains

USDA 110 > USDA 76 > USDA 38 in competitive ability which compared with their effectiveness rankings of USDA 110 > USDA 38 > USDA 76 (Caldwell 1969). In another experiment, Caldwell and Weber (1970), found that different serogroups were able to dominate nodule samples according to planting date which was later attributed to root temperature (Weber and Miller 1972). The competitive ability in nodule formation between single *Rhizobium* strain inoculum and the naturally occurring populations of rhizobia has shown two *R. trifolii* strains (WA 67, TA 1) to occupy a higher proportion of the nodules than a third strain (UNZ 29) in competition with each other in a mixed inoculant, WA 67 > TA 1 > UNZ 29 (Brockwell and Dudman 1968). Gibson (1968) demonstrated that a poorly nodulating *R. trifolii* could lower the proportion of nodules formed by a second strain in competition with the naturally-occurring soil population. It was found that in a soil containing 100 000 ineffective rhizobia per gram, the proportion of effectively nodulated plants was approximately doubled for a 10-fold increase in the size of the inoculum, and at least one million effective rhizobia per seed were required to obtain 90 per cent effectively nodulated plants (Ireland and Vincent 1968). Herridge and Roughley (1975*b*) found that CB 756 (cowpea-type *Rhizobium*) was less able to compete for nodules on *Lablab purpureus* than other slow-growing strains from native legumes in low and high competition sites in kraznozen soils of northern New South Wales.

The complexity of the factors involved in the competition between strains of *Rhizobium* have not been defined, but may involve inherent characteristics of both the host and the *Rhizobium*. These include compatibility as measured by effectiveness, the relative numbers of the competing strains in the inoculum and on the root surface, the relative growth rate of the rhizobia, the physiological state of the rhizobia at the time of inoculation, temperature, moisture, oxygen, and nutritional levels. Most studies have related the results of competition back to the relative numbers applied in the original inoculum, without consideration of their subsequent establishment in the rhizosphere in the presence of the natural soil population.

Effective symbionts are often, but not always, the most competitive strains in nodule formation (Marques Pinto *et al.* 1974; Russell and Jones 1975; Vincent and Waters 1953; Robinson 1969*a*; Singer, Holding, and King 1964; Israilsky and Rushkova 1967). No correlation between effective and ineffective *R. japonicum* was noted by Ruiz-Argueso, Cabrera, and Santa Maria (1977). Effective rhizobia can vary in their ability to compete for nodule sites (Rensburg and Strijdom 1969; Franco and Vincent 1976). Many effective rhizobia have been reported to be poor competitors (Rensburg and Strijdom 1969; Ireland and Vincent 1968; Johnson and Beringer 1976; Franco and Vincent 1976) and form very few nodules in the presence of certain ineffective strains. Recently, Winarno and Lie (1979) reported that the nodulation of pea cv. Afghanistan and cv. Iran by a *Rhizobium* strain was suppressed by the presence of a non-nodulating strain. The cultivar, host-line, and species can have an effect on the competitive ability of strains of *Rhizobium* (Roughley, Blowes, and Herridge 1976; Vincent and Waters 1953; Lahiri 1974; Masterson and Turner 1968) and thus it is well known that the rhizobia selected from the native soil-population differs according to the host species. Robinson (1969*b*) obtained isolates from *T. subterraneum* and *T. pratense* grown together in the field, and then tested them in tube culture against each host and found that the isolates were more effective with their homologous host. However, the initial selection of the isolates as the most prominent nodule on a plant may have biased the results, especially when it is appreciated that strains effective on *T. subterraneum* are often ineffective on *T. pratense*, and vice versa. A similar finding was made by Masterson and Sherwood (1974) with *T. repens* and *T. subterraneum*.

Various attempts have been made to relate nodule representation of the rhizobia with the population levels in the rhizosphere of the host (Vincent and Waters 1953; Marques Pinto *et al.* 1974; Robinson 1969*a*; Israilsky and Rushkova 1967; Franco and Vincent 1976; Labandera and Vincent 1975). Investigators have often applied high inoculum levels of mixed inocula and not considered the population dynamics on the root surface after inoculation. However, Vincent and Waters (1953) found that rhizobial strains gave unequal growth in the plant rhizosphere. The dominant strain varied according to the *Trifolium* species used and the distribution of strains in the nodules was unrelated to that occurring outside the plant. Marques Pinto *et al.* (1974) studied paired effective, ineffective, and effective-with-ineffective strains of rhizobia, and found that the proportion of nodules formed by a strain was related to its proportionate representation on the root surface. From this study, they defined the term 'Competitive Index' as the ratio of nodules originating from each strain under conditions of equal representation on the root. Labandera and Vincent (1975) found the 'competitive index' of variable value when studying competition between introduced and native Uruguayan strains of *R. trifolii*; the success of some strains was not very dependent on relative representation on the root surface. In another study (Franco and Vincent 1976) five out of seven cases of nodulation success could be related quantitatively to root-surface representation and so a 'competitive index' calculated. Franco and Vincent (1976) were unable to attribute any single feature of the symbiotic capacity of the interacting strains to their competitive ability. The dominant root colonizer of paired slow-growing organisms was a strain characteristic not affected by host, and often unrelated to the inoculum ratio applied. Growth of the fast- (NGR 8, from *Leucaena*) and slow-growing isolate in the rhizosphere of *Macroptilium* in agar was independent of each other.

The above deals with competition for nodule sites between similar types of *Rhizobium* and does not compare the competitive performance between rhizobia of different species on the one host. This is probably because there is no direct application for such information. In attempting to define the relationships between the degree of root surface colonization and the order of appearance of the nodules by the individual members of paired inocula, Franco and Vincent (1967) used a number of paired inocula on *Macroptilium atropurpureum*, including the fast-growing isolate from *Leucaena leucocephala* (ineffective on this host) with an effective slow-growing strain. In this case, the fast-growing strain was overwhelmed by the effective slow-growing one; significant numbers of nodules were formed by NGR 8 only when the ratio of slow- to fast-growing organisms in the inoculum was greater than 1.2×10^5. They found that the slow-growing strain was able to grow in the root region without restriction in the presence of the fast-growing one. Trinick (in preparation) compared the competitive ability of a number of paired fast- with slow-growing rhizobia on tropical legumes and found, like Franco and Vincent (1976), that the effective *Leucaena* and *Mimosa* isolates, ineffective on other tropical legumes, were not able to compete with the effective slow-growing strains on hosts like *Vigna unguiculata*. A fast-growing strain isolated from *Lablab purpureus* was found to dominate in nodule formation in competition with slow-growing rhizobia when temperatures were 25 °C/23 °C (day/night soil temperatures) but not when the temperature was raised to 30 °C/25 °C. It was not surprising, therefore, that fast-growing rhizobia from *L. purpureus* were not isolated from other nodules collected from this host growing in the same soil where soil temperatures would equal 30 °C/ 25 °C or higher. It is surprising, therefore, that fast-growing rhizobia from *L. purpureus* were not isolate from other nodules collected from this host growing in the same soil.

Škrdleta and Karimová (1969) found that one of two strains of *R. japonicum* dominated and that their relative success in forming nodules was influenced by the ratio of the two in the applied inoculum. It was also noted that the delayed application of one of the competing strains put it at a competitive disadvantage since, presumably, the other strain achieved dominance (Škredleta 1970). A change in soil pH from acid to neutral and alkaline conditions favoured an ineffective strain, the effective strain dominating under the more acid conditions (Russell and Jones 1975). The proportion of effective isolates was found by Masterson and Sherwood (1974) to be affected by soil moisture. Infection thread numbers produced by *R. trifolii* were found to be influenced by the physiological condition of the cells of the prepared inocula; there was an effect due to cell age, growth media (including root extract), and whether or not the cells were flocculated (Napoli and Hubbell 1976). Perhaps when more is known of the infection of legumes by rhizobia (see Dazzo and Hubbell, Chapter 9), the mechanisms determining the competitive ability of rhizobia will be better understood.

Until recently, it was generally accepted that only one serological type is found in a given nodule (Waksman 1927; Allen and Allen 1958). Purchase and Nutman (1957) considered dual occupancy of nodules as a rare occurrence. One soyabean nodule out of 150 examined by Means *et al.* (1961) was found to have two strains of *R. japonicum*. A slightly higher proportion of nodules with two strains of rhizobia were found by Škrdleta and Karimová (1969). Thirty-two per cent of soyabean nodules were found by Lindemann *et al.* (1974) to have double infection when the inoculation density was 10^8 rhizobia per plant; dual infections decreased with lower levels of inoculation. Most of their doubly infected nodules had 75 per cent or more of one strain and only 25 per cent or less of the other. Trinick (in preparation) found both fast- and slow-growing rhizobia occupying the same nodule and like Lindermann *et al.* (1974) the frequency of dual occupancy of nodules decreased with lower inoculation levels. In one paired fast- slow-growing combination on *Vigna*, the ineffective *Leucaena* isolate formed a very low proportion ($<$ 0.01 per cent of the nodule *Rhizobium* population). Two to twenty-seven per cent of the nodules formed on *P. sativum* exposed to a mixture of *R. leguminosarum* with a strain of *R. trifolii* or *R. phaseoli* (which do not normally nodulate peas on their own) contained two species of *Rhizobium* (Johnston and Beringer 1976).

Stimulation, neutral, and inhibition interactions between micro-organisms

In the soil and rhizosphere environment, micro-organisms survive by being best adapted to the specific environment or ecological niche provided. The environment supplies them with challengers that permit certain members of the community to be more successful at competing for the available nutrients, space, and the different types of surfaces supplied. The soil microflora best able to utilize the available nutrients (including available oxygen, carbon dioxide, etc.) have an ecological advantage that could be offset by other members of the population producing toxic metabolites, or by the alteration of the immediate environment in some other way (i.e. pH change), or by direct predation (i.e. by protozoa, myxobacteria, *Bdellovibrio*). Some organisms of the population may produce metabolites that are specifically required by others and hence stimulate some members of the community; other organisms may not interfere in any way with the activities of others.

It is not surprising that micro-organisms isolated from soil show many types of interactions on laboratory media. The various interactions between micro-organisms that may occur in the soil and rhizosphere are discussed by Parker *et al.* (1977). Recently Trinick and Parker (1981*a*) found that standard laboratory-strength media can produce different interactions from those observed at one-quarter or one-tenth strength. Sometimes interactions observed at the lower concentrations were different, reduced, or increased in intensity at the higher nutritional status of the medium. Thus, it cannot be assumed that the interactions observed on laboratory media do, in fact, occur in the soil and rhizosphere environment. There are reports (Chowdhury 1977) where very positive interactions have been observed in the soil as a result of the addition of energy sources such as glucose, but these are considered unnatural. The inhibition of rhizobial growth by toxic soil-water extracts (Chatel and Parker 1972) of unamended soil suggest the possible role of microbial antagonisms in *Rhizobium* establishment. Culture filtrates of micro-organisms that are able to influence (stimulate or inbibit) rhizobia have a similar effect when they are substituted for the growing organism (van Schreven 1964; Trinick and Parker 1981*a*).

Amongst Rhizobium strains. An inhibitory amino acid, rhizobitoxine, is produced by *R. japonicum* (Keith, Tortora, Ineichen, and Leimgruber 1975). Medium that has supported the growth of some rhizobia has been shown to be unfavourable to the growth of the same organisms (Demolon 1952), and this was not considered to be due to the depletion of essential nutrients but possibly to the accumulation of a toxic factor. Trinick (unpublished) found media that had supported *R. meliloti*, were unsuitable for other micro-organisms. Trinick and Parker (1981*a*) showed that *R. meliloti* could inhibit *R. trifolii* and *R. lupini* on agar as well as strongly inhibiting themselves when streaked at right-angles to each other. *R. lupini* was unable to interfere with the growth of other rhizobia and *R. trifolii* showed only weak inhibition of *R. meliloti*. A relatively heat-stable polypeptide-type of antibiotic from an ineffective *R. trifolii* strain (T 24) was isolated by Schwinghamer and Belkengren (1968). This antibiotic was active against a number of strains of different species of *Rhizobium*. When T 24 was present in a mixed inoculum with antibiotic-sensitive effective strains of *R. trifolii*, nodulation of *Trifolium* spp by the effective strains was strongly suppressed. A further survey of 270 isolates from *Trifolium* nodules at five localities reported by Schwinghamer (1971) showed 35 per cent were mildly antibiotic towards two of six indicator strains; 8 per cent of the cultures were shown to be lysogenic or produced bacteriocin-like substances. The potent antibiotic produced by T 24 may have ecological significance in the field, but the mildly antibiotic isolates probably do not produce large enough quantities in the soil or host rhizosphere to have an effect. Similarly, nothing is known about the *R. meliloti* antibiosis (Trinick and Parker 1981*a*) and its influence on other rhizobia and soil bacteria in the field.

Between Rhizobium *and other bacteria.* Soil and rhizosphere bacteria isolated onto laboratory media can stimulate, have no effect, or inhibit rhizobia in culture. Fred *et al.* (1932) reviewed early work dealing with the stimulatory effects of bacteria on the growth of rhizobia in laboratory media. Hattingh and Louw (1966*a*) found that only some of the Gram-negative non-spore-forming rhizosphere bacteria from *Trifolium* were stimulatory towards *R. trifolii.* Trinick and Parker (1981*b*) found the percentage of bacteria isolated from natural environments were generally less frequently stimulatory towards rhizobia than actinomycetes and fungi; 47 per cent (454 total) showed some degree of stimulation of *R. lupini*, approximately 30 per cent towards *R. meliloti* and approximately 20 per cent towards *R. trifolii.* Two rhizosphere isolates from *Aristida oligantha* showed stimulation of *R. japonicum* (Leuck and Rice 1976) and Bhalla and Sen (1973) observed that about 50 per cent of isolates from the rhizosphere of *Cicer arietinum* requiring yeast extract, soil-extract amino acids, and growth-promoting factors stimulated the growth of *Rhizobium*, and that most of these organisms were Gram-negative. The inoculation of rhizobia in association with other isolates of legume hosts have on occasions produced beneficial results. For example Krasil'nikov and Korenyako (1944) presented evidence for the ability of 'activating' bacteria to hasten and increase nodulation by *R. trifolii*, and Harris (1953) found that the infectivity of a weakly-invasive strain of *R. trifolii* was stimulated in the presence of other soil bacteria.

Other bacterial isolates are neutral on agar in their effects towards *Rhizobium.* Bhalla and Sen (1973) found that 25–30 per cent of the isolates had no effect and Leuck and Rice (1976) observed that 77 per cent failed to influence the growth of rhizobia on agar. A large variation in this type of interaction was found between different species of *Rhizobium* (*R. meliloti*, 50 per cent; *R. lupini*, 43 per cent; *R. trifolii*, 15 per cent) using the same group of soil/rhizosphere isolates (Trinick and Parker 1981*b*). This probably has little or no ecological significance but in the soil and rhizosphere environment the organisms could influence rhizobial growth through competition for scarce resources.

Most attention has been directed towards the inhibitory effects of soil and rhizosphere isolates towards *Rhizobium.* Most reports have shown a high incidence of Gram-negative-type cells inhibiting rhizobia. Krasil'nikov and Korenyako (1944) found the most inhibitory bacteria of those studied to be a strain of *Pseudomonas* and one of *Achromobacter*. In a later study, Hattingh and Louw (1966*b*) found that the majority of the inhibitory isolates from the rhizoplane of *Trifolium* spp were Gram-negative non-spore forming rods. The percentage of bacterial inhibitions from problem (with nodulation), non-problem soils, and the rhizosphere of *Trifolium* varied towards different species of *Rhizobium* (*R. trifolii*, 40 per cent; *R. meliloti*, 12 per cent; *R. lupini*, 5 per cent). This was reflected in their ability to survive in soil (Trinick and Parker 1981*b*). Inhibitory bacteria ranged from small Gram-negative rods resembling *Pseudomonas* spp, abundant in the rhizosphere to the various Gram-negative and Gram-variable forms frequently encountered in the soil, including the globiform and sporulating bacteria. Only one sporulating *Bacillus* was isolated from the *Trifolium* rhizosphere and it was antagonistic to *R. trifolii* but not to *R. meliloti* or *R. lupini*. Leuck and Rice (1976) found strains of *Xanthomonas axonopodis*, *Arthrobacter citreus*, *Arthrobacter simplex*, *Enterobacter aerogenes*, and *Micrococcus luteus* inhibitory to *Rhizobium*. Virtanen and Linkola (1948), and Afrikyan and Tumanyan (1958) have indicated the importance of spore-forming bacteria as antagonists of rhizobia.

The antagonists of rhizobia have a range of morphology and Gram reaction, but the frequency of inhibitors is greater amongst the Gram-negative rods. However, these inhibitors do not necessarily inhibit nodulation by the sensitive rhizobia (Smith and Miller 1974), but Trinick (1970) found that agar inhibitors and stimulators often had a similar effect on rhizobial numbers in both the soil and rhizosphere without themselves being affected. Pereira, Abrahao, and Oliveira (1974) claim that *Bacillus polymyxa* was responsible for the inhibition of *R. japonicum* in soyabean crops grown after potatoes and Rao and Dhala (1978) found the nodulation of *Lablab purpureus* in pot culture was markedly suppressed by *Pseudomonas aeruginosa*.

Between Rhizobium and actinomycetes. Antibiotic production by actinomycetes and their effects on *Rhizobium* are well known. Very little attention or emphasis has been placed on their possible stimulation or neutral effects on *Rhizobium* however. Antoun, Bordeleau, Gagnon, and Lachance (1978) indicated that 70 per cent of actinomycetes isolated exhibited no inhibitory effects and Robison (1945) found certain streptomyces spp beneficial. Some stimulatory effects towards *R. trifolii* and *R. meliloti* were noted with 40–69 per cent of 551 actinomycetes from West Australian soils but *R. lupini* only responded to 8 per cent (Trinick and Parker 1981*b*).

Van Schreven (1964) demonstrated the strong toxic effects of the metabolic products of actinomycetes. Some reports have shown that a minority of soil actinomycetes have an antibiotic effect (Holland and

Parker 1966; Thornton, DeAlencar, and Smith 1949). Trinick and Parker (1981*b*) reported numbers of actinomycetes inhibiting rhizobia varied with different species of *Rhizobium* (30–50 per cent against *R. trifolii*, 15 per cent for *R. meliloti*, and 12 per cent for *R. lupini*). In five soils examined by Patel (1974), 23–70 per cent inhibited their rhizobia, representing many cross-inoculation groups. Many of the antibiotics produced were specific for certain strains of the rhizobia tested. Actinomycetes were also able to affect the *Rhizobium* population in sterilized soil and the rhizosphere of plants growing in otherwise sterile conditions; both stimulation and inhibitory effects, as with bacteria, were observed (Trinick 1970). Damirgi and Johnson (1966) reported a decrease in nodulation of 35 per cent by *R. japonicum* in the presence of a strain of an actinomycete that strongly inhibited the *Rhizobium* strain in agar culture.

Rhizobium can influence the population of certain actinomycetes (and bacteria) growing in sand culture and on agar medium, both stimulatory and inhibitory effects being observed (Trinick 1970). This aspect of rhizobial behaviour is rarely considered (see also pp. 130–1 for *Rhizobium* × *Rhizobium* interactions) and perhaps rhizobia best able to establish in soil/rhizosphere environments have attributes which permit them to influence others in these environments.

Between Rhizobium *and fungi.* Brown *et al.* (1968) reported that certain fungi (e.g. *Giocladium*, *Paecilomyces*, and *Humicula*) increased *Rhizobium* infection and hastened nodulation, but they did not determine the actual cause of this stimulation. From a sample of 495 fungi isolated from West Australian soils, 33–47, 66 and 8 per cent stimulated (on agar) the growth of *R. trifolii*, *R. meliloti*, and *R. lupini* respectively (Trinick and Parker 1981*b*). Only one strain of *R. trifolii* (TA 1) was strongly inhibited (5 per cent of the fungi), the other rhizobia tested showed either a weak or no inhibition by the other fungi. Hattingh and Louw (1966*b*) noted that relatively few fungi were inhibitory against any of five strains of *R. trifolii*. These results are different to those of Holland and Parker (1966) who reported strong inhibition due to the indigenous fungi in areas of Western Australia displaying 'first-year mortality' symptoms. Fungal species reported to be most active against rhizobia are *Penicillium* and *Aspergillus* (Robison 1945; Thornton *et al.* 1949; Vyas and Prasad 1960; Holland 1966; Holland and Parker 1966; Gupta 1973). Other species showing inhibitory effects on *Rhizobium* belong to *Alternaria*, *Cephalosporium*, *Fusarium*, *Paecilomyces*, *Thielvia* and others (Subba Rao and Vasantha 1965; Chonkar and Subba Rao 1966; Sethi and Subba Rao 1972; Gupta 1973).

Antifungal activity of three different *Rhizobium* isolates on agar was reported by Drapeau, Fortin and Gagnon (1973). More recently, Tu (1979) showed that when cultures of the root-rot fungi *Phytophthora megasperma*, *Pythium ultimum*, *Aschochyta imperfecta*, and *Fusarium oxysporum* were inoculated with a culture of *R. japonicum*, fungal sporulation was reduced by 75, 65, 35 and 47 per cent respectively. The mycelium surface of the first two fungi were extensively colonized by rhizobia.

Between Rhizobium *and* Bdellovibrio/*Protozoa*. *Bdellovibrio bacteriovorus* strains which parasitize *Rhizobium* were first reported by Parker and Grove (1970) who found that they were capable of destroying large numbers of rhizobia in the laboratory. Keya and Alexander (1975) detected *Bdellovibrio* in 32 of the 90 soils examined. *Bdellovibrio* required host densities of about 10^8 ml^{-1} for multiplication in liquid culture and in the soil situation large numbers of rhizobia were required for bdello-vibrios to increase appreciably. They concluded therefore that there was a need for a critical host-cell frequency for *Bdellovibrio* replication.

Like *Bdellovibrio*, protozoan predators in the soil and in culture are host-cell dependent. Danso, Keya, and Alexander (1975) suggested that the 'size of the prey population diminishes until a density is attained at which the energy used by the predator in hunting for the survivors equals that obtained from the feeding'. *Paramecium* sp caused the rapid decline in soil of *Vigna* rhizobia from 10^9 g^{-1} to 10^7 g^{-1} before stabilization (Habte and Alexander 1977). Further studies on the mechanisms of the persistance of low numbers of bacteria in soil and solution culture preyed upon by protozoa were reported by Habte and Alexander (1978), who concluded that their maintenance is governed by their capacity to reproduce and replace the cells consumed by predation. Protozoa studied by the above authors and Sardeshpande, Balasubramanya, Kulkarni, and Bogyaraj (1977) include *Colpoda*, *Uroleptus*, *Tetramitus*, *Hartmanella*, *Naegleria*, *Vahlkampfia*, *Tetrahymena* and *Paramecium*. However, nothing is known of their effect, if any, on the natural soil and rhizosphere populations of *Rhizobium*.

Between Rhizobium *and bacteriophage/bacteriocins*. Bacteriophage and bac-teriocins and their lytic action have long been known and described for the main groups of rhizobia. Their biology, including host specificity (see p. 99), has recently been reviewed in detail by Vincent (1974, 1977) and will not be discussed further.

Bacteriophages specific for rhizobia are commonly found in soils and nodules, especially when legumes are grown regularly (Allen and Allen 1958; Staniewski *et al.* 1962). For instance Kleczkowska (1957) found Trifoliphage in all ten extracts from *Trifolium* roots and nodules, and in all the ten soils in which the clover had been grown. Kowalski *et al.*

(1974) found bacteriophages lytic to several host strains of *R. japonicum* in nearly all samples of soil and nodules examined. The possibility that bacteriophage could have adverse effects on the survival of rhizobia was reviewed by Allen and Allen (1958) who described the early work of Demolon and Dunez (1936) on 'alfalfa fatigue'. They postulated that bacteriophage may directly interfere with the legume–*Rhizobium* symbiosis. Kleczkowska (1957) indicated that the effect of rhizobiophages on nodulation by *Rhizobium* may be indirectly increasing the proportion of ineffective strains. However, she suggested that the span of phage soil is relatively short and their effects can be localized so that phage susceptible bacteria and the phage can exist close to each other without any apparent interaction. Vincent (1965) concluded that 'despite a relatively voluminous literature, the practical significance of rhizo-biophage as a factor militating against the survival and functioning of rhizobia in the soil has yet to be unequivocally demonstrated'. However, Evans, Barnet and Vincent (1979a,b) have shown that the presence of bacteriophage can reduce the population of susceptible strains in the root zone of *Trifolium* growing in seedling agar. Evens *et al.* (1979b) found that the reduction in rhizoplane population led to the appearance of variant substrains which were susceptible to the bacteriophage and mostly ineffective in symbiotic nitrogen fixation. Schwinghamer and Brockwell (1978) also reported the competitive advantage of bacteriocin and phage-producing strains of *R. trifolii* in mixed broth and peat culture.

Lysogeny in *Rhizobium* was first reported by Marshall (1956) for *R. trifolii* and since then it has been widely reported and studied in several fast-growing species of *Rhizobium* (Ordogh and Szende 1961; Takahashi and Quadling 1961; Schwinghamer and Reinhardt 1963; Staniewski and Kowalski 1965; Kowalski 1966). Perhaps the lytic action of temperate phage from lysogenic strains upon other strains of rhizobia might be important in the survival of rhizobia in various ecological situations.

Bacteriocins are protein-containing substances, ranging widely in molecular weight, which are produced by bacteria and kill specific cells of some other bacteria, but do not reproduce in those cells. These occur in the various cross-inoculation groups (Roslycky 1967). 'Bacteriocin-like' non-phage substances were found with *R. leguminosrum* in an earlier study by Schwinghamer and Reinhardt (1963) dealing with lysogenic strains of *Rhizobium*. Schwinghamer (1971) found potent bacteriogenic strains were encountered in a study of antagonism between strains of rhizobia isolated from clover in south eastern Australia. Bacteriocins of *R. lupini* were characterized by Lotz and Mayer (1972) as defective bacteriophage particles since they resembled the tails of certain *E. coli* phages, but had no nucleic acid. Further descriptions and properties of bacteriocins produced by rhizobia have been reported (Schwinghamer, Pankhurst, and Whitfeld 1973; Schwinghamer 1975; Gissmann and Lotz 1975). Schwinghamer *et al.* (1973) have suggested that from the standpoint of competition between strains of rhizobia in the field environment, the known bactericidal function of the bacteriocin could be significant in providing the bacteriocinogenic strain with some competitive advantage over other rhizobia.

References

Abdel-Ghaffar, A. S. and Jensen, H. L. (1966). *Arch. Mikrobiol.* **54**, 393–405.

Abdel-Wahab, S. M., Rifaat, D. M., Ahmed, K. A., and Hamdi, Y. A. (1976). *Abl. Bakt. Abt.* II **131**, 170–8.

Akrikyan, E. K. and Tumanyan, V. G. (1958). *Izv. Akad. Nauk armyan. SSR ser. biol. s-kh Nauki* **11**, 37–46.

Albersheim, P., Ayers, A. R., Valent, B. S., Ebel, J., Hahn, M., Wolpert, J., and Carlson, R. (1977). *J. Supramolec. Struct.* **6**, 599–616.

Alexander, D., Jordan, D. C., and McKague, M. (1969). *Can. J. Biochem.* **47**, 1092–4.

Alexander, M. (1971). *Microbial ecology*. John Wiley, New York.

Allen, E. K. and Allen, O. N. (1950). *Bacteriol. Rev.* **14**, 273–330.

—— —— (1958). In *Encyclopedia of plant physiology*, (ed. W. Ruhland) pp. 48–118. Springer-Verlag, Berlin.

—— —— (1961). In *Recent advances in botany*, Vol. 1, pp. 585–8. University of Toronto Press.

—— —— (1976). In *Symbiotic nitrogen fixation in plants* (ed. P. S. Nutman) pp. 113–21. Cambridge University Press.

Allen, O. N. and Allen E. K. (1939). *Soil Sci.* **47**, 63–76.

Almon, L. (1933). *Zentralblatt für Bakteriol. Parasitenkd. Infektionskr. Hyg. Abt. Orig.* **87**, 289–97.

Andrews, E. C. (1914). *J. R. Soc. N.S.W.* **47**, 333.

Angulo, A. F., Dijk, C. van, and Quispel, A. (1976). In *Symbiotic nitrogen fixation in plants*, (ed. P. S. Nutman) pp. 475–83. Cambridge University Press.

Antoun, H., Bordeleau, L. M., Gagnon, C., and Lachance, R. A. (1978). *Can. J. Microbiol.* **24**, 558–62.

Arber, A. (1928). *New Phytol.* **27**, 69–84.

Arora, N. (1954) *Phytomorphology* **4**, 211–6.

Axelrod, D. I. (1958). *Bot. Rev.* **26**, 431–509.

Backman, P. A. (1977). *Am. Phytopathol. Soc. Proc.* **4**, 221.

Balasubramanian, A. (1976). *Zbl. Bakt. Abt.* II **131**, 517–21.

Barnet, Y. M. (1972). *J. gen. Virol.* **15**, 1–15.

—— and Vincent, J. M. (1969). *Aust. J. Sci.* **32**, 208.

—— —— (1970). *J. gen. Microbiol.* **61**, 319–25.

Becking, J. H. (1970). *Inst. J. Syst. Bacteriol.* **20**, 201

—— (1977). In *A Treatise on dinitrogen fixation, Section III. Biology* (ed. R. W. F. Hardy and W. S. Silver) pp. 185–275. John Wiley, New York.

Bednarski, M. and Reporter, M. (1978). *Appl. environ. Microbiol.* **36**, 115–20.

Bergersen, F. J. (1955). *J. gen. Microbiol.* **13**, 411–19.

—— (1961). *Aust. J. biol. Sci.* **14**, 349–60.

—— (1968). *Trans. 9th Int. Congr. Soil Sci. Soc.* Vol. 2, pp. 49–63.

—— (1974). In *The biology of nitrogen fixation* (ed. A. Quispel) pp. 473–98. North Holland, Amsterdam.

—— (1977). In *Recent developments in nitrogen fixation* (ed. W. Newton, J. R. Postgate, and C. Rodriguez-Barrueco) pp. 387–417. Academic Press, New York.

—— and Gibson, A. H. (1978). In *Limitations and potentials for biological nitrogen fixation in the tropics*, Basic Life Sciences (ed. J. Döbereiner, R. H. Burris, A. Hollaender, A. A. Franco, C. A. Neyra, and D. B. Scott) Vol. 10, pp. 263–74. Plenum Press, New York.

—— Kennedy, G. S., and Wittman, W. (1965). *Aust. J. Biol. Sci.* **18**, 1135–42.

—— Turner, G. L., Gibson, A. H., and Dudman, W. F. (1976). *Biochim. biophys. Acta* **444**, 164–74.

Bhalla, H. and Sen, A. N. (1973). *Sci. Culture* **39**, 191–3.

Bhardwaj, K. K. (1975). *Pl. Soil* **43**, 377–85.

Bhardwaj, K. K. R. and Gaur, A. C. (1972). *Ind. J. Microbiol.* **12**, 19–21.

Bisseling, T., van den Bos, R. C., van Kammen, A., van den Ploeg, M., van Duijm, P., and Houwers, A. (1977). *J. gen. Microbiol.* **101**, 79–84.

Bisset, K. A. (1952). *J. gen. Microbiol.* **7**, 233–42.

—— and Hale, C. M. F. (1951). *J. gen. Microbiol.* **5**, 592–5.

Bond, G., MacConnell, J. T., and McCallum, A. H. (1956). *Ann. Bot. N.S.*, **20**, 501–12.

Bowen, G. D. (1959). *Qd. J. agric. Sci.* **16**, 267–81.

—— (1961). *Pl. Soil* **15**, 155–65.

—— and Kennedy, M. M. (1959). *Qd. J. agric. Sci.* **16**, 177–97.

Brayan, O. C. (1923). *Soil Sci.* **15**, 37.

Breed, R. S., Murray, E. G. D., and Smith, N. R. (1957). *Bergey's Manual of Determinative Bacteriology*, 7th edn. Williams and Wilkins, Baltimore.

Brockwell, J. and Dudman, W. F. (1968). *Aust. J. agric. Res.* **19**, 749–57.

—— and Katznelson, J. (1976). *Aust. J. agric. Res.* **27**, 799–810.

—— and Neal-Smith, C. A. (1966). *Aust. C.S.I.R.O. Div. Pl. Ind. Fld. Stn. Rec.* **5**, 9–15.

—— and Robinson, A. C. (1976). *Aust. C.S.I.R.O. Div. Pl. Ind. Fld. Stn. Rec.* **15**, 15–26.

—— Hely, F. W., and Neal-Smith, C. A. (1966). *Aust. J. exp. Agric. Anim. Husb.* **6**, 365–70.

—— Schwinghamer, E. A., and Gault, R. R. (1977). *Soil Biol. Biochem.* **9**, 19–24.

Broughton, W. J., Chan, P. Y., Padmanabhan, S., and Tan, I. K. P. (1975). *Mal. Agric. Res.* **4**, 141–53.

Brown, M. E., Jackson, R. M., and Burlingham, S. K. (1968). In *The ecology of soil bacteria* (ed. T. R. G. Gray and D. Parkinson) pp. 531–51. University of Toronto Press.

Bruch, C. W. and Allen, O. N. (1957). *Can. J. Microbiol.* **3**, 181–9.

Brussel, A. A. N., van Costerton, J. W., and Child, J. J. (1979). *Can. J. Microbiol.* **25**, 352–61.

Bunn, C. R. and Elkan, G. H. (1971). *Can. J. Microbiol.* **17**, 291–5.

Burns, R. C. and Hardy, R. W. F. (1975). *Nitrogen fixation in bacteria and higher plants.* Springer-Verlag, Berlin.

Burris, R. H. (1963). In *Evolutionary Biochemistry* (ed. A. I. Oparin) pp. 172–7. Macmillan, New York.

Burton, J. C. (1977). In *Exploiting the legume–Rhizobium symbiosis in tropical agriculture* (ed. J. M. Vincent, A. S. Whitney and J. Bose) pp. 293–311. College of Tropical Agriculture, Misc. Publ. No. 145, University of Hawaii, Honolulu.

Burton, M. O. and Lochhead, A. G. (1952). *Can. J. Bot.* **30**, 521–4.

Cabezes de Herre, E. (1956). *Soils Fertils*, **19**, 422.

Caldwell, B. B. (1969). *Agron. J.* **61**, 813–15.

—— and Weber, D. F. (1970). *Agron. J.* **62**, 12–4.

Campêlo, A. B. and Döbereiner, J. (1969). *Pesquisa Agropecuaria Bras.* **4**, 67–72.

Camugli, E. N. and Gomez Etchart, O. E. G. (1974). *Rev. Fac. Agron.* (*Argentina*) **50**, 41–7.

Carg, F. C., Tauro, P., and Grover, R. K. (1973). *Ind. J. Microbiol.* **13**, 179–81.
Chandler, M. R. (1978). *J. exp. Bot.* **29**, 749–55.
Chatel, D. L. and Parker, C. A. (1972). *Soil Biol. Biochem.* **4**, 289–94.
Child, J. J. (1975). *Nature, Lond.* **253**, 350–1.
—— and Kurz, W. G. W. (1978). *Can. J. Microbiol.* **24**, 143–8.
Chonkar, P. K. and Subba Rao, N. S. (1966). *Can. J. Microbiol.* **12**, 1253–61.
Chowdhury, M. S. (1977). In *Exploiting the legume-*Rhizobium *symbiosis in tropical agriculture* (ed. J. M. Vincent, A. S. Whitney, and J. Bose) pp. 385–411. College of Tropical Agriculture, Misc. Publ. No. 145, University of Hawaii, Honolulu.
—— Marshall, K. C., and Parker, C. A. (1968). *Aust. J. agric. Res.* **19**, 919–25.
Cole, M. A. and Elkan, G. H. (1979). *Appl. environ. Microbiol.* **37**, 867–70.
Conn, H. J., Bottcher, E. J., and Randall, C. (1945). *J. Bacteriol.* **49**, 359–73.
Cook, F. D., Watson, D. W., and Allen, O. N. (1949). *Soc. Am. Bacteriol. 49th Gen. Meet., Abst. of Papers*, pp. 59–60.
Cowan, S. T. (1968). *A dictionary of microbial taxonomic useage.* Oliver and Boyd, Edinburgh.
Cowles, J. R., Evans, H. J., and Russell, S. A. (1969). *J. Bacteriol.* **97**, 1460–5.
Craig, A. S. and Williamson, K. I. (1972). *Arch. Mikrobiol.* **87**, 165–71.
—— Greenwood, R. M. and Williamson, K. I. (1973). *Arch. Mikrobiol.* **89**, 23–32.
Curley, R. L. and Burton, J. C. (1975). *Agron. J.* **67**, 807–8.
Dadarwal, K. R. and Sen, A. N. (1973). *Ind. J. agric. Sci.* **43**, 82–7.
Damirgi, S. M. and Johnson, H. W. (1966). *Agron. J.* **58**, 223–4.
—— Frederick, L. R., and Anderson, I. C. (1967). *Agron. J.* **59**, 10–12.
Daniel, R. M. and Appleby, C. A. (1972). *Biochim. biophys. Acta* **275**, 347–54.
Danso, S. K. A., Keya, S. O., and Alexander, M. (1975). *Can. J. Microbiol.* **21**, 884–95.
Dart, P. J. (1969). Rothamsted Exp. Sta. Rept. for 1968, Pt. 1, 89–90.
—— (1975). In *The development and function of roots* (ed. J. G. Torrey and D. T. Clarkson) pp. 467–506. Academic Press, London.
—— (1977). In *A treatise on dinitrogen fixation, Section III. Biology* (ed. R. W. F. Hardy and W. S. Silver) pp. 367–472. John Wiley, New York.
—— and Mercer, F. V. (1963). *Arch. Mikrobiol.* **46**, 382–401.
—— —— (1966). *J. Bacteriol.* **91**, 1314–9.
Date, R. A. (1972). *J. appl. Bacteriol.* **35**, 379–87.
—— (1974). *Proc. Ind. natn. Sci. Acad.* **40 B**, 700–12.
—— (1977). In *Exploiting the legume–*Rhizobium *symbiosis in tropical agriculture* (ed. J. M. Vincent, A. S. Whitney, and J. Bose) pp. 293–311. College of Tropical Agriculture, Misc. Publ. No. 145, University of Hawaii, Honolulu.
—— and Decker, A. M. (1965). *Can. J. Microbiol.* **11**, 1–8.
—— and Halliday, J. (1979). *Nature, Lond.* **277**, 62–4.
—— and Norris, D. O. (1979). *Aust. J. agric. Res.* **30**, 1–19.
—— Burt, R. L., and Williams, W. T. (1979). *Agro-Ecosystems* **5**, 57–67.
Davis, R. J. (1962). *J. Bacteriol.* **84**, 187–8.
DeLey, J. and Rassel, A. (1965). *J. gen. Microbiol.* **41**, 85–91.
—— Park, I. W., Tijlgat, R., and van Ermengem, J. (1966). *J. gen. Microbiol.* **42**, 43–56.
Demolon, A. (1952). *Rev. gen. Bot.* **58**, 489–519.
—— and Dunez, A. (1936). *Ann. Agron. N. S.* **6**, 434–54.
Diatloff, A. (1968). *Qd. J. agric. Sci.* **25**, 165–7.
—— (1970*a*). *Aust. J. exp. agric. Anim. Husb.* **10**, 562–7.
—— (1970*b*). *Qd. J. agric. An. Sci.* **27**, 279–93.
—— and Ferguson, J. E. (1970). *Trop. Grassl.* **4**, 223–8.
—— and Luck, P. E. (1972). *Trop. Grassl.* **6**, 33–8.
Dilworth, M. J. (1974). *A. Rev. Plant Physiol.* **25**, 81–114.

—— and Williams, D. C. (1967). *J. gen. Microbiol.* **48**, 31–6.

Döbereiner, J. (1977). In *Genetic engineering for nitrogen fixation* (ed. A. Hollaender) pp. 451–62. Plenum Press, New York.

—— and Day, J. M. (1976). In *Symposium on nitrogen fixation* (ed. W. E. Newton and C. J. Nyman) pp. 518–38. Washington State University Press, Pullman.

Drapeau, R., Fortin, J. A., and Gagnon, C. (1973). *Can. J. Bot.* **51**, 681–2.

Drozanska, D. (1966). *Acta Microbiol. Polon.* **15**, 323.

Dubé, A. J., Dodman, R. L., and Flentje, N. T. (1971). *Aust. J. biol. Sci.* **24**, 57–65.

Dudman, W. F. (1964). *J. Bacteriol.* **88**, 782–94.

—— (1968). *J. Bacteriol.* **95**, 1200–1.

—— (1977). In *A treatise on dinitrogen fixation, Section IV. Agronomy and Ecology* (ed. R. W. F. Hardy and A. H. Gibson) pp. 487–508. John Wiley, New York.

Egeraat, A. W. S. M. van (1975). *Pl. Soil* **42**, 381–6.

—— (1976). *Pl. Soil* **45**, 191–9.

Elkan, G. H. (1971). *Pl Soil*, Special Volume, pp. 85–104.

—— and Kwik, I. (1968). *J. appl. Bacteriol.* **31**, 399–404.

—— and Usanis, R. A. (1971). *Inst. J. Syst. Bact.* **21**, 295–8.

El Essawi, T. M. E. and Abdel-Ghaffar, A. S. (1967). *J. appl. Bacteriol.* **30**, 354–61.

Ertola, R. J., Mazza, L. A., Balatti, A. P., Cuevas, C. M., and Daguerre, R. (1969). *Soil Sci.* **108**, 373–80.

Evans, J., Barnet, Y. M., and Vincent, J. M. (1979a). *Can. J. Microbiol.* **25**, 974–8.

—— —— —— (1979b). *Can. J. Microbiol.* **25**, 968–73.

Faizah, A. W., John, C. K., and Broughton, W. J. (1976). In *Seed technology in the tropics*, p. 91–6, Denerbit Universiti Pertanian, Sereking.

Fletcher, W. W. and Alcorn, J. W. S. (1958). In *Nutrition of the legumes* (ed. E. G. Hallsworth) Butterworths Scientific Publ. London.

Fottrell, P. F., O'Connor, S., and Masterson, C. L. (1964). *Ir. J. Agric. Res.* **3**, 246–9.

Franco, A. A. and Vincent, J. M. (1976). *Pl. Soil* **45**, 27–48.

Fred, E. B., Baldwin, I. L., and McCoy, E. (1932). *Root nodule bacteria and leguminous plants.* University of Wisconsin Studies in Science.

Gaur, Y. D. and Sen, A. N. (1979). *New Phytol.* **83**, 745–54.

Gerson, T. and Patel, J. J. (1975). *Appl. Microbiol.* **30**, 193–8.

Gibbons, A. M. and Gregory, K. F. (1972). *J. Bacteriol.* **111**, 129–41.

Gibson, A. H. (1968). *Aust. J. agric. Res.* **19**, 907–18.

—— and Brockwell, J. (1968). *Aust. J. agric. Res.* **19**, 891–905.

—— Scowcroft, W. R., and Pagan, J. D. (1977). In *Recent developments in nitrogen fixation* (ed. W. Newton, J. R. Postgate, and C. Rodriguez-Barrueco) pp. 387–417. Academic Press, New York.

Gissmann, L. and Lotz, W. (1975). *J. gen. Virol.* **27**, 379–83.

Godfrey, C. A. (1972). *J. gen. Microbiol.* **72**, 399–402.

Gollobin, G. S. and Levin, R. A. (1974). *Arch. Microbiol.* **101**, 83–90.

Goodchild, D. J. and Bergersen, F. J. (1966). *J. Bacteriol.* **92**, 204–13.

Gordon, J. and Gordon, M. (1943). *J. Pathol. Bacteriol.* **55**, 63–8.

Goss, O. M. and Shipton, W. A. (1965). *J. Agric. Western Aust.* **6**, 659–61.

Graham, P. H. (1963a). *Aust. J. biol. Sci.* **16**, 557–9.

—— (1963b). *Antonie van Leeuwenhoek* **29**, 281–91.

—— (1963c). *J. gen. Microbiol.* **30**, 245–8.

—— (1964a). *Antonie van Leeuwenhoek* **30**, 68–72.

—— (1964b). *J. gen. Microbiol.* **35**, 511–7.

—— (1971). *Arch. Mikrobiol.* **78**, 70–5.

—— and Parker, C. A. (1964). *Pl. Soil* **20**, 283–96.

—— —— Oakley, A., Lange, R. T., and Sanderson, I. J. V. (1963). *J. Bacteriol.* **86**, 1353–4.

Gregory, K. F. and Allen, O. N. (1953). *Can. J. Bot.* **31**, 730–8.

Gresshoff, P. M. and Rolfe, B. G. (1978). *Planta* **142**, 329–33.

—— Skotnicki, M. L., Eadie, J. F., and Rolfe, B. G. (1977). *Plant Sci. Lett.* **10**, 299–304.

Gupta, V. K. (1973). *Sci. Cult.* **39**, 197–8.

Habte, M. and Alexander, M. (1977). *Arch. Microbiol.* **113**, 181–3.

—— —— (1978). *Soil Biol. Biochem.* **10**, 1–6.

Hagedorn, C. (1979). *Soil Sci. Soc. Am. J.* **43**, 921–5.

Hahn, N. J. (1966). *Can. J. Microbiol.* **12**, 725–33.

Hale, C. N. and Mathers, D. J. (1977). *N. Z. J. Agric. Res.* **20**, 69–73.

Ham, G. E., Frederick, L. R., and Anderson, I. C. (1971). *Soil Sci. Soc. Am. Proc.* **63**.

Hamdi, Y. A. (1968). *Arch. Mikrobiol.* **63**, 227–31.

—— (1969). *Pl. Soil* **31**, 111–21.

—— (1977). *Zbl. Bakt.* II, *Bd.* **132**, 350–60.

Harris, J. R. (1953). *Nature, Lond.* **172**, 507–8.

Hattingh, M. J. and Louw, H. A. (1966a). *S. Afr. J. agric. Sci.* **9**, 453–60.

—— —— (1966b). *S. Afr. J. agric. Sci.* **9**, 239–52.

Heberlein, G. T., DeLey, J., and Tijtgat, R. (1967). *J. Bacteriol.* **94**, 116–24.

Hendry, G. S. and Jordan, D. C. (1969). *Can. J. Microbiol.* **15**, 671–5.

Herridge, D. F. and Roughley, R. J. (1975a). *J. appl. Bacteriol.* **38**, 19–27.

—— —— (1975b). *Aust. J. exp. Agric. Anim. Husb.* **15**, 264–9.

Hill, L. R. (1966). *J. gen. Microbiol.* **44**, 419–37.

Hofer, A. W. (1941). *J. Bacteriol.* **41**, 193–224.

Holding, A. J. and Lowe, J. F. (1971). *Pl. Soil* Special Volume, pp. 153–66.

Holland, A. A. (1966). *Pl. Soil* **25**, 238–48.

—— (1970). *Pl. Soil* **32**, 293–302.

—— and Parker, C. A. (1966). *Pl. Soil* **25**, 329–40.

Hubbell, D. H., Morales, V. M., and Umali-Garcia, M. (1978). *Appl. environ. Microbiol.* **35**, 210–13.

Hughes, D. G. and Vincent, J. M. (1942). *Proc. Linn. Soc. N.S. W.* **67**, 142–52.

Humphrey, B. A. and Vincent, J. M. (1962). *J. gen. Microbiol.* **29**, 557–61.

—— —— (1965). *J. gen. Microbiol.* **41**, 109–18.

—— —— (1969). *J. gen. Microbiol.* **59**, 411–25.

Hussien, Y. A., Tewfik, M. S., and Hambdi, Y. A. (1974). *Soil Biol. Biochem.* **6**, 377–81.

Hutchinson, J. (1959). *The families of flowering plants.* Vol. 1. *Dicotyledons.* Clarendon Press, Oxford.

—— (1969). *Evolution and phylogeny of flowering plant.* Academic Press, London.

Imshenetskii, A. A. (1963). In *Evolutionary biochemistry* (ed. A. I. Oparin) pp. 139–48. Macmillan, New York.

Inukai, S., Sato, K., and Shimizu, S. (1977). *Agric. biol. Chem.* **41**, 2229–34.

Ireland, J. A. and Vincent, J. M. (1968). *Trans. 9th Int. Cong. Soil Sci.*, Vol. 2, pp. 85–93.

Isaac, I. and Heale, J. B. (1961). *Ann. appl. Biol.* **49**, 675.

Ishizawa, S. (1955). *J. Sci., Soil Manure, Japan.* **26**, 31–2.

Israilsky, V. P. and Rushkova, A. S. (1967). *Nauk USSR, Ser. Biol.* **4**, 567

Iswaran, V. (1975). *Zbl. Bakt. Abt.* II **130**, 365–6.

Jain, J. M. and Rewari, R. B. (1976). *Zbl. Bakt. Abt.* II, 131–69.

Jarvis, B. D. W., Dick, A. G., and Greenwood, R. M. (1980). *Int. J. Sys. Bacteriol.* **30**, 42–52.

—— McLean, T. S., Robertson, I. G. C., and Fanning, G. R. (1977). *N.Z. J. agric. Res.* **20**, 235–48.

Jensen, H. L. (1942). *Proc. Linn. Soc. N.S.W.* **67**, 98–108.

—— (1943). *Proc. Linn. Soc. N.S.W.* **68**, 207–20.

—— (1969). *Tidsskr. Planteavl.* **73**, 61–72.

—— and Schroder, M. J. (1965). *J. appl. Bacteriol.* **28**, 473–8.

Jimbo, T. (1930). *Bot. Mag.* **44**, 158–68.

Johnson, G. V., Evans, H. J., and Ching, T. M. (1966). *Plant Physiol.* **41**, 1330–6.

Johnson, H. W. and Means, U. M. (1964). *Agron. J.* **56**, 60–2.

Johnson, M. D. and Allen, O. N. (1952*a*). *Antonie van Leeuwenhoek* **18**, 1–12.

—— —— (1952*b*). *Antonie van Leeuwenhoek* **18**, 13–22.

Johnston, A. W. B. and Beringer, J. E. (1976). *Nature, Lond.* **263**, 502–4.

—— Beynon, J. L., Buchanan-Wollaston, A. V., Setchell, S. M., Hirsch, P. R., and Beringer, J. E. (1978). *Nature, Lond.* **276**, 634–6.

Jordan, D. C. (1952) *Can. J. Bot.* **30**, 125–30.

—— (1962). *Bact. Rev.* **26**, 119–41.

—— and Allen, O. N. (1974). In *Bergey's manual of determinative bacteriology* (ed. R. E. Buchanan and N. E. Gibbons) 8th edn, pp. 261–4. Williams and Wilkins, Baltimore.

—— and Coulter, W. H. (1965). *Can. J. Microbiol.* **11**, 709–20.

—— and Grinyer, I. (1965). *Can. J. Microbiol.* **11**, 721–5.

—— and San Clemente, C. L. (1955). Can. J. Microbiol. **1**, 659–67.

—— Yamamura, Y., and McKague, M. E. (1969). *Can. J. Microbiol.* **15**, 1005–12.

Kamata, E. (1962). *Proc. Crop Sci. Soc. Jap.* **31**, 78–82.

Kandasamy, D., Kesavan, R., Ramasamy, K., and Prasad, N. N. (1974). *Ind. J. Microbiol.* **14**, 25–30.

Kapoor, K. K., Singh, D. P., Khandelwal, K. C., and Mishra, M. M. (1977). *Pl. Soil* **47**, 249–52.

Karanth, N. G. K. and Vasantharajan, V. N. (1974). *Proc. Ind. natn. Sci. Acad. B* **40**, 576–85.

Kaszubiak, H. (1965). *Acta Microbiol. Pol.* **14**, 309–14.

Katznelson, H. (1955). *Nature, Lond.* **175**, 551–2.

—— and Zagallo, A. C. (1957). *Can. J. Microbiol.* **3**, 879–84.

Kecskés, M. and Manninger, E. (1962). *Can. J. Microbiol.* **8**, 157–9.

—— and Vincent, J. M. (1969). *Agrokem. Talajtan* **18**, 57–70.

Keister, D. L. and Evans, W. R. (1976). *J. Bacteriol.* **129**, 149–53.

Keith, D. D., Tortora, J. A., Ineichen, K., and Leimgruber, W. (1975). *Tetrahedron* **31**, 2633–6.

Keya, S. O. and Alexander, M. (1975). *Soil Biol. Biochem.* **7**, 231–7.

Keyser, H. H. and Munns, D. N. (1979*a*). *Soil Sci. Soc. Am. J.* **43**, 519–23.

—— —— (1979*b*). *Soil Sci. Soc. Am. J.* **43**, 500–503.

—— —— and Hohenberg, J. S. (1979). *Soil Sci. Soc. Am. J.* **43**, 719–22.

Khatri, A. A. and Choksey, M. (1973). *Sci. Cult.* **39**, 282–4.

Kidby, D. K. and Goodchild, D. J. (1966). *J. gen. Microbiol.* **45**, 147–52.

Kleczkowska, J. (1957). *Can. J. Microbiol.* **3**, 171–80.

—— Nutman, P. S., and Bond, G. (1944). *J. Bacteriol.* **48**, 673–5.

—— —— Skinner, F. A., and Vincent, J. M. (1968). In *Identification methods for micro-biologists* (ed. B. M. Gibbs and D. A. Shapton) pp. 51–65. Academic Press, New York.

Kleczkowski, A. and Thornton, H. G. (1944). *J. Bacteriol.* **48**, 661–72.

Koleshko, O. I. (1975). *Microbiology, Moscow* **44**, 291–3.

Kononova, M. M. (1966). *Soil organic matter.* Pergamon Press, New York.

Koontz, F. P. and Faber, J. E. (1961). *Soil Sci.* **91**, 228–32.

Kowalski, M. (1966). *Acta Microbiol. Pol.* **15**, 119–28.

—— Ham, G. E., Frederick, L. R., and Anderson, I. C. (1974). *Soil Sci.* **118**, 221–8.

Krasil'nikov, N. A. and Korenyako, A. I. (1944). *Mikrobiologiya* **13**, 39–44.

Kulkarni, J. H., Sardeshpande, J. S., and Bagyaraj, D. J. (1975). *Zbl. Bakt. Abt.* II **130**, 41–4.

Kurz, W. G. W. and LaRue, T. A. (1975). *Nature, Lond.* **256**, 407–8.

Kuykendall, L. D. and Elkan, G. H. (1976). *Appl. environ. Microbiol.* **32**, 511–9.

Labandera, C. A. and Vincent, J. M. (1975). *Pl. Soil* **42**, 327–47.

Lahiri, K. K. (1974). *Proc. Ind. natn. Sci. Acad. B. Biol. Sci.* **40**, 644–9.

Lakshmi-Kumari, M., Biswas, A., Vijayalakshmi, K., Narayana, H. S., and Subba Rao, N. S. (1974). *Proc. Ind. natn. Sci. Acad. B. Biol. Sci.* **40**, 528–34.

Lance, G. N. and Williams, W. T. (1967). *Aust. Computer J.* **1**, 15–20.

—— —— (1968). *Aust. Computer J.* **1**, 82–5.

Lange, R. T. (1960). *Rhizobium of south western Australia.* PhD Thesis, University of Western Australia.

—— (1961). *J. gen. Microbiol.* **26**, 351–9.

—— (1966). In *Symbiosis*, (ed. S. M. Henry) Vol. 1, pp. 99–170. Academic Press, New York.

Law, I. J. and Strijdom, B. W. (1974). *Phytophylactica* **6**, 221–8.

Leifson, E. and Erdman, L. W. (1958). *Antonie van Leeuwenhoek* **24**, 97–110.

Leuck, E. W. and Rice, E. L. (1976). *Bot. Gaz.* **137**, 160–4.

Levin, R. A. and Montgomery, M. P. (1974). *Pl. Soil* **41**, 669–76.

Libbenga, K. R. and Bogers, R. J. (1974). In *The biology of nitrogen fixation* (ed. A. Quispel) pp. 430–72. North·Holland, Amsterdam.

Lim, G. and Ng, H. L. (1977). *Pl. Soil* **46**, 317–27.

Lindemann, W. C., Schmidt, E. L., and Ham, G. E. (1974). *Soil Sci.* **118**, 274–9.

Loneragan, J. F. (1972). In *Use of isotopes for study of utilization by legume crops.* pp. 17–54. Tech. Rep., Int. Atomic Energy Agency, Vienna.

Longley, B. J., Berge, T. O., Van Lanen, J. M., and Baldwin, I. L. (1937). *J. Bacteriol.* **33**, 29–30.

Lotz, W. and Mayer, F. (1972). *J. Virol.* **9**, 160–73.

Lowe, R. H., Evans, H. J., and Ahmed, S. (1960). *Biochem. Biophys. Res. Commun.* **3**, 675

MacKenzie, C. R. and Jordan, D. C. (1970). *Biochem. Biophys. Res. Commun.* **40**, 1008–12.

—— Vail, W., and Jordan, D. (1973). *J. Bacteriol.* **113**, 387–93.

Mackenzie, S. L., Lapp, M. S., and Child, J. J. (1979). *Can. J. Microbiol.* **25**, 68–74.

McComb, J. A., Elliot, J., and Dilworth, M. J. (1975). *Nature, Lond.* **256**, 409–10.

Marques Pinto, C., Yao, P. Y., and Vincent, J. M. (1974). *Aust. J. agric. Res.* **25**, 317–29.

Marshall, K. C. (1956). *Nature, Lond.* **177**, 92.

Martinez-de Drets, G. and Arias, A. (1972). *J. Bacteriol.* **109**, 467–70.

—— Arias, A., and Rovira de Cutinella, M. (1974). *Can. J. Microbiol.* **20**, 605–9.

Masterson, C. L. (1968). *Int. Cong. Soil Sci., Trans. 9th (Adelaide)* **11**, 95–102.

—— and Sherwood, M. T. (1974). *Irish J. agric. Res.* **13**, 91–9.

—— and Turner, S. (1968). *An Foras. Taluntais Soils Div. Res. Report* pp. 71–75.

Means, U. M., Johnson, H. W., and Erdman, L. W. (1961). *Soil Sci. Soc. Am. Proc.* **25**, 105–8.

Méndez-Castro, F. A. and Alexander, M. (1976). *Rev. lat-amer. Microbiol.* **18**, 155–8.

Milthorpe, F. L. (1945). *J. Aust. Inst. Agric. Sci.* **11**, 89–92.

Mino, Y. (1970). *J. Jap. Soc. Grassl. Sci.* **60**, 16–21.

—— Mino, S., Hayaski, H., and Chiji, H., (1968). *Res. Bull. Obihiro Univ.* **5**, 453.

Mishra, K. C. and Gaur, A. C. (1975). *Zbl. Bakt. Abt.* II **130**, 598–602.

Moffett, M. L. and Colwell, R. R. (1968). *J. gen. Microbiol.* **51**, 245–66.

Mostafa, M. A. and Mahmoud, M. Z. (1951). *Nature, Lond.* **167**, 446–7.

Munns, D. N. (1977). In *Exploiting the legume-Rhizobium symbiosis in tropical agriculture* (ed. J. M. Vincent, A. S. Whitney, and J. Bose) pp. 211–38. College of Tropical Agriculture, Misc. Publ. No. 145, University of Hawaii, Honolulu.

Murthy, N. B. K. and Raghu, K. (1976). *Pl. Soil* **44**, 491–3.

Nalbandyan, A. D. (1976). *Dokl. Vses. Akad. Skh. Nauk* **8**, 14–5.

—— Ovsepyan, E. A., Ovsepyan, M. V., and Stepanyan, M. D. (1976). *Dokl. Akad.*

Nauk Armyanskoi **62**, 43–5.

Namdeo, K. N. and Dube, J. N. (1973). *Ind. J. exp. Biol.* **II**, 114–6.

Napoli, C. A. and Hubbell, D. H. (1976). *Abstr. Annu. Meet. Am. Soc. Microbiol.* **76**, 170.

Neal, O. R. and Walker, R. H. (1935). *J. Bacteriol.* **30**, 173–87.

Nicholas, D. B. (1971). *J. Aust. Inst. agric. Sci.* **37**, 69–70.

Norris, D. O. (1956). *Emp. J. exp. Agric.* **24**, 247–70.

—— (1958*a*). *Nature, Lond.* **182**, 734–5.

—— (1958*b*). *Aust. J. agric. Sci.* **9**, 629–32.

—— (1959). *Aust. J. agric. Res.* **10**, 651–98.

—— (1965). *Pl. Soil* **22**, 143–66.

—— and Date, R. A. (1976). In *Tropical pasture research: principles and methods* (ed. N. H. Shaw and W. W. Bryan) pp. 134–74. Commonw. Agric. Bur., Farnham Royal.

—— and 'tMannetje, L. (1964). *E. Afr. Agric. For. J.* **29**, 214–35.

Nuti, M. P., Ledeboer, A. M., Lepidi, A. A., and Schilperoort, R. A. (1977). *J. gen. Microbiol.* **100**, 241–8.

Oblisami, G., Balaraman, K., Venkataramanan, C. V., and Rangaswami, G. (1973). *Madras Agric. J.* **60**, 462–4.

Odeyemi, O. and Alexander, M. (1977*a*). *Soil Biol. Biochem.* **9**, 247–51.

—— —— (1977*b*) *Appl. environ. Microbiol.* **33**, 784–90.

Okafor, N. and Alexander, M. (1975). *Soil Biol. Biochem.* **7**, 405–6.

Ordogh, F. and Szende, K. (1961). *Acta microbiol. Acad. Sci. Hung.* **8**, 65–71.

Paau, A. S., Lee, D., and Cowles, J. R. (1977). *J. Bacteriol.* **129**, 1156–8.

Pagan, J. D., Child, J. J., Scowcroft, W. R., and Gibson, A. H. (1975). *Nature, Lond.* **256**, 406–7.

Page, W. J. (1978). *Can. J. Microbiol.* **24**, 209–14.

Pandher, M. S. and Kahlon, S. S. (1978). *Ind. J. Microbiol.* **18**, 81–4.

Pankhurst, C. E. (1977). *Can. J. Microbiol.* **23**, 1026–33.

—— and Craig, A. S. (1978). *J. gen. Microbiol.* **106**, 207–19.

Pariiskaya, A. N. and Gorelova, O. P. (1976). *Microbiology* **45**, 747–50.

Parker, C. A. (1957). *Nature, Lond.* **179**, 593–4.

—— (1968). In *Festskrift til Hans Laurits Jensen.* Gadgaard Nielsons Bogtrykkeri, Lemrig, Denmark.

—— and Graham, P. H. (1964). *Pl. Soil* **20**, 383–96.

—— and Grove, P. L. (1970). *J. appl. Bacteriol.* **33**, 253–5.

—— and Scutt, P. B. (1960). *Biochim. Biophys. Acta* **38**, 230–8.

—— Trinick, M. J., and Chatel, D. L. (1977). In *A treatise on dinitrogen fixation. Section IV. Agronomy and ecology* (ed. R. W. F. Hardy and A. H. Gibson) pp. 311–52. John Wiley, New York.

Paromenskaya, L. N. (1975). Int. Symp. Interaction of Herbicides, Micro-organisms, and Plants, Wroclaw, 1973.

Patel, J. J. (1974). *Pl. Soil* **41**, 395–402.

Patrick, Z. A. and Toussour, T. A. (1970). In *Ecology of soil-born plant pathogens, prelude of biological control* (ed. K. F. Baker and W. C. Snyder) pp. 440–59. University of California Press, Berkley.

Paul, E. A. and Newton, J. D. (1961). *Can. J. Microbiol.* **7**, 7–13.

Pechy, K. K. and Szende, K. (1974). *Agrartud. Kozl.* **33**, 65–7.

Pereira, A. L. G., Abrahao, J., and Oliveira, B. S. (1974). *O Biologico* **40**, 214–5.

Peterson, H. B. and Gooding, T. H. (1941). *Nebraska Agr. Exp. Sta. Bull.* **121**, 1–24.

Peterson, P. J., Greenwood, R. M., Belling, G. B., and Bathurst, N. O. (1971). *Pl. Soil* Special Volume, pp. 111–4.

Pillai, R. N. and Sen, A. (1970). *Ind. J. agric. Sci.* **40**, 1017–8.

Postgate, J. R. (1974). *Symp. Soc. gen. Microbiol.* **24**, 263–92.

Purchase, H. F. and Nutman, P. S. (1957). *Ann. Bot., N.S.* **21**, 439–54.

—— Vincent, J. M., and Ward, L. W. (1951). *Proc. Linn. Soc. N.S.W.* **76**, 1–6.

Purushothaman, D. and Balaraman, K. (1973). *Curr. Sci.* **42**, 507–8.

Rajagopalan, T. (1938). *Ind. J. agric. Sci.* **8**, 379–402.

Rao, N. R. and Dhala, S. A. (1978). *Ind. J. Microbiol.* **18**, 9–12.

Reijnder, L., Vesser, L., Aalbers, A., van Kammen, A., and Houwers, A. (1975). *Biochim. biophys. Acta* **414**, 206–16.

Rensburg, H. J. van and Strijdom, B. W. (1969). *Phytophylactica* **1**, 201–4.

—— Hahn, J. S., and Strijdom, B. W. (1973). *Phytophylactica* **5**, 119–22.

Roberts, G. P., Leps, W. T., Silver, L. E., and Brill, W. J. (1980). *Appl. environ. Microbiol.* **39**, 414–22.

Robinson, A. C. (1969*a*). *Aust. J. agric. Res.* **20**, 827–41.

—— (1969*b*). *Aust. J. agric. Res.* **20**, 1053–60.

Robison, R. S. (1945). *Proc. Soil Sci. Soc. Am.* **10**, 206–10.

Robson, A. D., Edwards, D. G., and Loneragan, J. F. (1970). *Aust. J. agric. Res.* **21**, 601–12.

Roponen, I. E., Valle, E., and Ettala, T. (1970). *Physiol. Plant.* **23**, 1198–205.

Roslycky, E. B. (1967). *Can. J. Microbiol.* **13**, 431.

—— Allen, O. N., and McCoy, E. (1962). *Can. J. Microbiol.* **8**, 71–8.

Roughley, R. J. (1968). *J. appl. Bacteriol.* **31**, 259–65.

—— (1976). In *Symbiotic nitrogen fixation in plants* (ed. P. S. Nutman) pp. 125–36. Cambridge University Press.

—— Blowes, W. M., and Herridge, D. F. (1976). *Soil Biol. Biochem.* **8**, 403–7.

Ruhloff, M. and Burton, J. C. (1951). *Soil Sci.* **72**, 283–90.

Ruiz-Argueso, T., Cabrera, E., and Santa Maria, J. (1977). *An. Ind./Ser. General.* **5**, 23–30.

Russell, P. E. and Jones, D. G. (1975). *Pl. Soil* **42**, 119–29.

Sabet, Y. S. (1946). *Nature, Lond.* **157**, 656–7.

Sardeshpande, J. S., Balasubramanya, R. H., Kulkarni, J. H. and Bogyaraj, D. J. (1977) *Pl. Soil* **47**, 75–80

Saubert, G. G. P. (1949). *Ann. Bot. Gdn. Buitenz.* **51**, 177–97.

Schaede, R. (1940). *Planta* **31**, 1–21.

—— (1951). *Planta* **39**, 154–70.

Scheffler, J. G. and Louw, H. A. (1967). *S. Afr. J. Agr. Sci.* **10**, 161–74.

Schmidt, E. L. (1978). In *Interactions between non-pathogenic soil micro-organisms and plants* (ed. Y. R. Dommergues and S. V. Krupa) pp. 269–303. Elsevier, Amsterdam.

Schreven, D. A. van (1964). *Pl. Soil* **21**, 283–302.

—— (1972). *Pl. Soil* **37**, 49–55.

Schwinghamer, E. A. (1964). *Can. J. Microbiol.* **10**, 221–33.

—— (1967). *Antonie van Leeuwenhoek* **33**, 121–36.

—— (1968). *Can. J. Microbiol.* **14**, 355–67.

—— (1970). *Aust. J. biol. Sci.* **23**, 1187–96.

—— (1971). *Soil Biol. Biochem.* **3**, 355–63.

—— (1975). *J. gen. Microbiol.* **91**, 403–13.

—— and Belkengren, R. P. (1968). *Arch. Mikrobiol.* **64**, 130–45.

—— and Brockwell, J. (1978). *Soil Biol. Biochem.* **10**, 383–7.

—— and Dudman, W. F. (1973). *J. appl. Bacteriol.* **36**, 263–72.

—— and Reinhardt, D. J. (1963). *Aust. J. biol. Sci.* **16**, 597–605.

—— Pankhurst, C. E., and Whitfeld, P. R. (1973). *Can. J. Microbiol.* **19**, 359–68.

Scowcroft, W. R. and Gibson, A. H. (1975). *Nature, Lond.* **253**, 351–2.

Sen, M., Pal, T. K., and Sen, S. P. (1969). *Antonie van Leeuwenhoek* **35**, 533–40.

Sen, H. A. (1938). *Bibl. Genet.* **12**, 175–345.

Sethi, R. P. and Subba Rao, N. S. (1972). *Symposium legume inoculants–inoculants–Sci. and*

Technol. New Delhi, India. 27.

Shemakhanova, N. M. and Oleinikov, R. R. (1971). *Dokl. Akad. Nauk SSSR* **200**, 1240–1.

Sherwood, M. T. (1972). *J. gen. Microbiol.* **71**, 351–8.

Singer, M., Holding, A. J., and King, J. (1964). *8th. Int. Cong. Soil Sci.* pp. 1021–5.

Singh, C. S., Lakshmi-Kumari, M., Biswas, A., and Subba Rao, N. S. (1973). *Ind. J. Microbiol.* **13**, 125–8.

Skinner, F. A., Roughley, R. J., and Chandler, M. H. (1977). *J. appl. Bacteriol.* **43**, 287–97.

Skotnicki, M. L. and Rolfe, B. G. (1977). *Microbios* **20**, 15–28.

—— —— (1978). *J. Bacteriol.* **133**, 518–26.

Škrdleta, V. (1965a). *Ved. Pr. Vysk Ust. Rostl. Vyroby, Prahe-Ruzyne* 177.

—— (1965b). *Pl. Soil* **23**, 43–8.

—— (1970). *Soil. Biol.* Biochem. **2**, 167–71.

—— and Karimová, J. (1969). *Arch. Mikrobiol.* **66**, 25–8.

Skyring, G. W., Quadling, C., and Rouatt, J. W. (1971). *Can. J. Microbiol.* **17**, 1299–311.

Smith, C. R., Funke, B. R, and Schulz, J. T. (1978). *Soil Biol. Biochem.* **10**, 463–6.

Smith, R. S. and Miller, R. H. (1974). *Agron. J.* **66**, 564–7.

Smyk, B. and Ettlinger, L. (1963). *Annls. Inst. Pasteur* **105**, 341–8.

Sneath, P. H. A. (1957). *J. gen. Microbiol.* **17**, 201–26.

Souto, S. M., Coser, A. C, and Döbereiner, J. (1970). In *Proc. V. Reunion Latinoamericana de Rhizobium*, Departmento Nacional de Pesquisa Agropecuária, Rio de Janeiro, Brazil.

—— —— —— (1972). *Pesqui. Agropecu. Bras. Ser. Zootec.* **7**, 1–5.

Spiess, L. D., Lippincott, B. B., and Lippincott, J. A. (1977). *Bot. Gaz.* **138**, 35–40.

Sporne, K. R. (1956). *Biol. Rev.* **31**, 1–29.

Staniewski, R. (1970). *Can. J. Microbiol.* **16**, 1003–9.

—— and Kowalski, M. (1965). *Acta microbiol. Pol.* **14**, 231–6.

—— —— and Lomanska, I. (1963). *Acta microbiol. Pol.* **12**, 187–91.

—— —— Gogaez, E., and Sokolowska, F. (1962). *Acta microbiol. Pol.* **11**, 245–54.

Staphorst, J. L. and Strijdom, B. W. (1975). *Phytophylactica* **7**, 95–6.

—— —— (1976). *Phytophlactica* **8**, 47–54.

Stebbins, G. L. (1950). *Variation and evolution in plants.* Columbia University Press, New York.

Steinborn, J. and Roughley, R. J. (1974). *J. appl. Bacteriol.* **37**, 93–9.

—— —— (1975a). *J. appl. Bacteriol.* **39**, 213–6.

—— —— (1975b). *J. appl. Bacteriol.* **39**, 133–8.

Stevens, J. W. (1923). *J. infect. Dis.* **33**, 557–66.

Stout, J. D. (1962). *J. gen. Microbiol.* **27**, 209–19.

Strijdom, B. W. and Allen, O. N. (1966). *Can. J. Microbiol.* **12**, 275–83.

—— —— (1969). *Phytophylactica* **1**, 147–52.

Subba Rao, N. S. and Vasantha, P. (1965). *Naturwiss.* **52**, 44.

—— Lakshmi-Kumari, M., Singh, C. S., and Biswas, A. (1974). *Proc. Ind. natn. Sci. Acad. B. Biol. Sci.* **40**, 544–7.

Sutton, W. (1974). *Biochim. biophys. Acta* **366**, 1–10.

—— and Robertson, J. G. (1974). In *Mechanisms of regulation of plant growth* (ed. R. L. Bielski, A. R. Ferguson, and M. Creswell) pp. 23–30. Roy. Soc. New Zealand.

—— Jepsen, N., and Shaw, B. (1977). *Plant Physiol.* **59**, 741–4.

Swamiappan, M. and Chandy, K. C. (1975). *Curr. Sci.* **44**, 558–9.

Takahashi, I. and Quadling, C. (1961). *Can. J. Microbiol.* **7**, 455–65.

Tan, I. K. P. and Broughton, W. J. (1981). *Soil Biol. Biochem.* **13**, 389–93.

Tewfik, M. S., Embaba, M. S., and Hamdi, Y. A. (1975). *Zbl. Bakt. Abt.* II **130**, 725–31.

Thompson, J. A. (1960). *Nature, Lond.* **187**, 619–20.

Thorne, D. W. and Walker, R. H. (1936). *Soil Sci.* **42**, 231–40.

Thornton, G. D., DeAlencar, J., and Smith, F. B. (1949). *Soil Sci. Soc. Am. Proc.* **14**, 188–91.

Tjepkema, J. and Evans, H. J. (1975). *Biochem. Biophys. Res. Commun.* **65**, 625–8.

'tMannetje, L. (1967). *Antonie van Leeuwenhoek* **33**, 477–91.

Trinick, M. J. (1965a). *Investigations into the rhizobia of tropical legumes with particular reference to Leucaena leucocephala (Lam.) de Wit.* MSc thesis, University of Sydney, N.S.W. Australia.

—— (1965b). *Aust. J. Sci.* **27**, 263–4.

—— (1968). *Expl. Agric.* **4**, 243–53.

—— (1970). *The ecology of Rhizobium — interactions between rhizobium strains and other soil microorganisms.* PhD Thesis University of Western Australia.

—— (1973). *Nature, Lond.* **244**, 459–60.

—— (1976). In *1st Int. Symp. nitrogen fixation* (eds. W. E. Newton and C. J. Nyman) pp. 507–17. Washington State University Press, Pullman.

—— (1979). *Can. J. Microbiol.* **25**, 565–78.

—— (1980a). *J. appl. Bacteriol.* **49**, 39–53.

—— (1980b). *New Phytol.* **85**, 37–45.

—— (1981). *Appl. environ. Microbiol.* In press.

—— and Galbraith, J. (1980). *New Phytol.* **86**, 17–26.

—— and Parker, C. A. (1981a). *Soil Biol. Biochem.* In press.

—— —— (1981b). *Soil biol. Biochem.* In press.

Truesdell, H. W. (1917). *Soil Sci.* **3**, 77–98.

Tsien, H. C., Cain, P. S., and Schmidt, E. L. (1977). *Appl. environ. Microbiol.* **34**, 854–6.

Tu, J. C. (1979). *Physiol. Plant Pathol.* **14**, 171–7.

Tutin, T. G. (1958). In *Nutrition of the legumes* (ed. E. G. Hallsworth) pp. 3–14. Butterworths Scientific Publ. London.

Upchurch, R. G. and Elkan, G. H. (1977). *Can. J. Microbiol.* **23**, 1118–22.

Van Den Berg, E. H. R. (1977). *Pl. Soil* **48**, 629–39.

Venkataraman, G. S., Roychaudhury, P., Henriksson, L. E., and Henriksson, E. (1975). *Curr. Sci.* **44**, 520–1.

Vest, G., Weber, D. F., and Sloger, C. (1973). In *Soybeans: improvement, production, and uses* (ed. B. E. Caldwell) pp. 353–90. Am. Soc. Agron., Monograph 16. Madison.

Vincent, J. M. (1941). *Proc. Linn. Soc. N.S.W.*, **66**, 145–54.

—— (1942). *Proc. Linn. Soc. N.S.W.*, **67**, 82–6.

—— (1954). *Proc. Linn. Soc. N.S.W.* **79**, 4–32.

—— (1958). In *Nutrition of the legumes* (ed. E. G. Hallsworth) pp. 108–23. Butterworths Scientific Publ. London.

—— (1962a). *Proc. Linn. Soc. N.S.W.* **87**, 8–38.

—— (1962b). *J. gen. Microbiol.* **28**, 653–63.

—— (1965). In *Soil nitrogen* (ed. M. V. Bartholomew and F. E. Clark) pp. 384–35. Am. Soc. Agron., Madison.

—— (1974). In *The biology of nitrogen fixation* (ed. A. Quispel) pp. 265–341. North Holland, Amsterdam.

—— (1977). In *A treatise on dinitrogen fixation. Section III. Biology* (ed. R. W. F. Hardy and W. S. Silver) pp. 277–366. John Wiley, New York.

—— and Colburn, J. R. (1961). *Aust. J. Sci.* **23**, 269–70.

—— and Humphrey, B. A. (1963). *Nature, Lond.* **199**, 149–51.

—— —— (1968). *J. gen. Microbiol.* **54**, 397–405.

—— —— (1970). *J. gen. Microbiol.* **63**, 379–82.

—— and Waters, L. M. (1953). *J. gen. Microbiol.* **9**, 357–70.

—— —— (1954). *Aust. J. agr. Res.* **5**, 61–76.

—— Humphrey, B. A., and Škrdleta, V. (1973). *Arch. Mikrobiol.* **89**, 79–82.

—— Thompson, J. A., and Donovan, K. O. (1962). *Aust. J. agric. Res.* **13**, 258–70.

Virtanen, A. I. and Linkola, H. (1948). *Suomen Kemistilehti* (B) **21**, 12–3.

Vyas, S. R. and Prasad, N. (1960). *Proc. Ind. Acad. Sci. B* **61**, 242.

Waksman, S. A. (1927). *Principles of soil microbiology.* Waverly Press, Baltimore.

Weber, D. F. and Miller, V. (1972). *Agron. J.* **64**, 796–8.

Werner, D. and Berghauser, K. (1976). *Arch. Microbiol.* **107**, 257–62.

Whitehead, D. C. (1964). *Nature, Lond.* **202**, 417–8.

Wilcockson, J. and Werner, D. (1978). *J. gen. Microbiol.* **108**, 151–60.

Wilkins, J. (1967). *Aust. J. agric. Res.* **18**, 299–304.

Wilson, D. O. and Reisenauer, H. M. (1970). *J. Bacteriol.* **102**, 729–32.

Wilson, J. K. (1944). *Soil Sci.* **58**, 61–9.

—— (1947). *Proc. Soil Sci. Soc. Am.* **12**, 215–6.

Wilson, P. W. (1940). *The biochemistry of symbiotic nitrogen fixation.* University of Wisconsin Press, Madison.

Winarno, R. and Lie, T. A. (1979). *Pl. Soil* **51**, 135–42.

Wolf, M. and Baldwin, I. L. (1940). *J. Bacteriol.* **39**, 344.

Yadav, N. K. and Vyas, S. R. (1971). *Ind. J. agric. Sci.* **41**, 1123–5.

—— —— (1973). *Folia Microbiol.* **18**, 242–7.

Yu, K. K. Y. and Jordan, D. C. (1971). *Can. J. Microbiol.* 1283–6.

Zablotowicz, R. M., Eskew, D. L., and Focht, D. D. (1978). *Can. J. Microbiol.* **24**, 757–60.

Ziemiecka, J. (1963). *Ann. Inst. natn. Agron.* (*Paris*) **1**, 65–72.

4 Carbohydrate metabolism

G. H. ELKAN AND L. D. KUYKENDALL

4.1 Introduction

Evans, Emerich, Ruiz-Argueso, Maier, and Albrecht calculated that a total of 28 mol of ATP are consumed in the reduction of 1 mol of nitrogen. Although this high energy requirement is readily supplied by catabolism of carbohydrates, photosynthesis, and hence carbohydrate supply has been cited as a major limiting factor in nitrogen fixation by the *Rhizobium*–legume symbiosis (Bethlenfalvay and Phillips 1977; Hardy 1977; Pate 1977; Ronson and Primrose 1979). Thus, for the full exploitation of dinitrogen fixation, elucidation of carbohydrate metabolism in *Rhizobium* is of basic importance.

The pioneering studies of Fred (1911–12) and Zipfel (1911) established the usefulness of glucose, mannitol, sucrose, and maltose as energy sources for the root-nodule bacteria. Thorne and Walker (1936) studied the nutrition of *Rhizobium* and concluded that no complex, unidentified substances were required for growth (Walker and Brown 1930). These studies were followed by numerous investigations to determine the nutritional availability of various carbon compounds. For example, Neal and Walker (1934) used oxygen consumption as a criterion for suitability of carbon compounds for *Rhizobium meliloti* and *Rhizobium japonicum*. They showed that whereas *R. meliloti* utilized a wide variety of carbon sources, *R. japonicum* was much more restricted.

Allen and Allen (1950) stated that 'few groups of bacteria have been so thoroughly studied as have the rhizobia'. Yet, much of the carbohydrate work had been directed toward nutritional studies designed for establishing growth conditions in culture or for taxonomic purposes. Even now, 30 years later, the information available on the pathways of carbohydrate metabolism is incomplete, fragmented, and sometimes contradictory. Recent investigations have employed mutant methodology. Duncan and Fraenkel (1979) stated that 'the general intermediary metabolism of *Rhizobium* has not been extensively studied'. Ronson and Primrose (1979) also concluded that relatively little is known about the pathways of central carbohydrate metabolism in rhizobia. Recognizing the relative neglect of this most important topic, the purpose of this review is to bring together that which is known about carbohydrate metabolism in rhizobia.

The genus *Rhizobium* has traditionally and informally been divided into two groups. According to Allen and Allen (1950), the term 'fast-growers' commonly designates the rhizobia associated with *Medicago, Trifolium, Phaseolus,* and *Pisum,* these grow much faster in culture (less than one-half the doubling time) than do the 'slow-growers' examplified by *Glycine, Vigna,* and *Lupinus* rhizobia. From an extensive study of acid production by *Rhizobium* involving 717 strains, Norris (1965) hypothesized that the ancient form of the symbiont is represented by the slow-growing strains which produce alkali and are commonly associated with legumes of tropical origin. Disputing this hypothesis, Graham (1964a) contended that the differences between the slow-growing and fast-growing rhizobia were too great to be based solely on evolutionary differentiation of root nodule bacteria from an organism similar to the present day slow-growing type. Carbohydrate nutritional differences between fast- and slow-growing strains have been shown. In their review, Allen and Allen (1950) suggested that the slow-growing rhizobia are more specific in their carbohydrate requirements in every respect. The great differences in carbohydrate utilization between the fast-growing and slow-growing root-nodule bacteria were also confirmed by Graham (1964a). He concluded that carbohydrate utilization tests are valid criteria for the subdivision of *Rhizobium.* For example, *R. trifolii* and *R. leguminosarum* can utilize 20 different carbohydrates and tri-carboxylic-acid-cycle intermediates for growth; *R. japonicum* can utilize only eight of the twenty.

The relative fastidiousness of the slow-growing isolates has been substantiated by newer studies (Graham 1964a, b; Elkan and Kwik 1968). As will be discussed in later sections, specific enzymes may differ in the two groups, Martinez-de Drets and Arias (1972) proposed an enzymatic basis for differentiation between these groups based on presence of an NADP-6-phosphogluconate dehydrogenase (EC 1.1.1.43) found in the fast-growing, but not in the slow-growing strains. While the major pathways seem to be similar, there is evidence that the preferred pathways may be different. However, most of the metabolic studies were done with only one or very few *Rhizobium* strains. Of the studies discussed in this chapter, considerable nutritional variability was reported between species (Fred 1911–1912; Gaur and Mareckova 1977; Mulongoy and Elkan 1977a, b) and similar nutritional variability was shown between 36 strains within the same species (Elkan and Kwik 1968).

The great similarities between the fast-growing species of rhizobia and the contrast between this group and the slow-growing ones (such as *R. japonicum*) led Graham (1964*b*) to propose that *R. trifolii*, *R. leguminosarum*, and *R. phaseoli* be consolidated into a single species and that a new genus, *Phytomyxa*, be created to contain strains of slow-growing *Rhizobium* (type species: *Phytomyxa japonicum*). This proposal merits our support because the numerous striking differences in carbon metabolism between these groups clearly justify their assignment to taxonomically distinct genera of bacteria. Vincent, Nutman, and Skinner (1979) reached the tentative conclusion that the slow-growing *Rhizobium* should be assigned to a separate genus based largely on antigenic distinctions and flagellar arrangements.

4.2 Disaccharides

A number of early investigators demonstrated that disaccharides were energy sources for some rhizobia. Sucrose and maltose were the two most frequently reported, Zipfel (1911); Fred (1911–12); Sarles and Reid (1935). Neal and Walker (1936) showed that *R. meliloti* could utilize maltose, lactose, or sucrose as acceptable energy sources and verified that the slow-growing *R. japonicum* utilized these substrates only very poorly, if at all. Georgi and Ettinger (1941) showed that three fast-growing species could grow well on cellobiose, whereas the slow-growing species grew poorly or not at all. These papers reinforced the conclusion that the fast-growing *Rhizobium* can utilize disaccharides as growth substrates, whereas the slow-growing isolates cannot.

Graham (1964*a*) examined the carbohydrate response of 95 strains of rhizobia in seven species. Maltose, sucrose, and lactose were included. While these disaccharides were utilized by 36 out of 40 fast-growing strains, they were utilized by only nine of 55 slow-growing strains. Elkan and Kwik (1968) examined the carbohydrate response of 36 *R. japonicum* strains. None of these could utilize sucrose; on lactose thirteen strains grew poorly, none grew well, and 23 did not grow at all. Six grew well on maltose, 23 did poorly, and seven could not utilize this substrate. Shmyreva and Plaksina (1972) examined the ability of three strains each of *R. lupini* and *R. japonicum* to grow on either sucrose, lactose, or maltose and found no growth of these slow-growing *Rhizobium* with these substrates. They thus corroborated the work of Elkan and Kwik (1968) and concluded that the inability to utilize disaccharides was 'one of the peculiarities of slow-growing nodule bacteria'.

Martinez-de Drets and Arias (1970) examined the growth on sucrose by 64 fast- and slow-growing *Rhizobium* strains, confirmed the difference in the abilities of these two groups to utilize the disaccharides maltose, lactose, and sucrose, and established the enzymatic basis of sucrose utilization by the fast-growing strains. For the metabolism of sucrose, an inducible β-glucosidase (EC 3.2.1.21) (invertase) was found in the fast-growing strains, but invertase was not present in the slow-growing strains. The invertase, found exclusively in the fast-growing species, was induced by either sucrose, lactose, or maltose, but was detected at only a low constitutive level in cells grown on either glucose or fructose. A glucosido-invertase was present in all extracts and, in addition, some strains had a β-D-fructofuranoside-invertase as well. No sucrose phosphorylase (EC 2.4.1.7) activity was found in either of the fast-growing or slow-growing strains.

The enzymatic basis for lactose metabolism in *R. meliloti* was reported by Niel, Guillaume and Bechet (1977). These workers reported two different β-galactosidase (EC 3.3.1.23) activities present in strain 2011, one inducible and one low level constitutive. The two enzymes were distinguished by differing concentrations of ammonium sulphate required for their precipitation, and were separable by gel electrophoresis. The inducible enzyme was missing in a mutant strain only able to hydrolyse lactose slowly. Ucker and Signer (1978) studied the effect of succinate on lactose metabolism in *R. meliloti*. They described a phenomenon similar to catabolite repression, since succinate was found to repress β-galactosidase synthesis in *R. meliloti*. A lactose negative (*lac*) mutant, with unaltered symbiotic properties, lacked β-galactosidase activity. Ucker and Signer also described the isolation of a pleiotropic mutant, unable to grow on a mixture of cellobiose and maltose, in which β-galactosidase activity was not inducible. This mutant made relatively few nodules and did not fix nitrogen, but revertants occurring at a high frequency regained both properties.

It is important to note that *Rhizobium* can be distinguished from *Agrobacterium* by the absence of 3-ketolactose production (*Bergey's Manual*). Gaur and Mareckova (1977) surveyed 54 strains of *R. phaseoli* for production of 3-ketolactose and found that none of these strains did, confirming the validity of the 3-ketolactose test in distinguishing *Agrobacterium* from *Rhizobium*.

4.3 Polyols

D-mannitol is the traditional carbon source used for the *in vitro* cultivation of *Rhizobium* bacteria (Georgi and Ettinger 1941; Bergersen 1960). This hexitol produces a high cell yield, equal to or greater than that obtained with D-glucose (Singh, Singh, and Sidhu 1967; Shmyreva, Krasnopol'skaya, and Plaksina 1969); D-mannitol is the standard source of energy for rhizobia and, in general, the one with which the widest range of *Rhizobium* strains give the best growth (Norris 1965). Yet, Elkan and Kwik (1968) found that only 21 out of 36 *R. japonicum* strains tested show a good growth response with D-mannitol as the carbon source; 11 out of 36 strains did not utilize mannitol.

Wilson (1937) showed that polyols were oxidized by rhizobia in an unusual manner. That is, the metabolic rate was initially low and increased with time. This was an early indication that polyol metabolic enzymes are inducible. Burris, Phelps, and Wilson (1942) first interpreted and described the growth of *Rhizobium* on polyhydric alcohols as due to adaptive or inducible enzymes. They found that the oxidation of polyols was much greater when the bacteria had been previously grown on the same substrate. Martinez-de Drets and Arias (1970) investigated polyol metabolism in *R. meliloti* and found that this species possessed two distinct enzymes, one for the metabolism of D-mannitol and D-arabitol, and one for D-sorbitol metabolism. The NAD-specific enzyme responsible for the dehydrogenation of D-mannitol and D-arabitol was a D-arabitol dehydrogenase (EC 1.1.1.11) whereas a sorbitol dehydrogenase (EC 1.1.1.14) was found to be responsible for the oxidation of D-sorbitol. They also found that *R. meliloti* has an adenosine triphosphate-linked hexokinase which acts on fructose (fructokinase) (EC 2.7.1.4) and a phosphohexose isomerase (EC 5.3.1.9) which acts on fructose-6-phosphate, but that a hexose isomerase capable of interconverting glucose and fructose was absent. Hornez, Courtois, and Derieux (1976) discerned that fructose and glucose were metabolized via different pathways in *R. meliloti*. Fructose is directly phosphorylated and utilized through the Embden–Meyerhof–Parnas (EMP) pathway.

We (Kuykendall and Elkan 1976) isolated derivatives of USDA strain 110 of *R. japonicum* that differed in ability to utilize D-mannitol as a carbon source; these also differed by about twenty times in nitrogen-fixation activity $(C_2H_2 \rightarrow C_2H_4)$ with soyabeans. The symbiotically competent strains I-110 and S-110 cannot utilize mannitol, whereas strains L1-110 and L2-110 can. Using the derivatives cloned from USDA strain 110, the enzymatic basis for D-mannitol utilization in *R. japonicum* was examined (Kuykendall and Elkan 1977). In strains

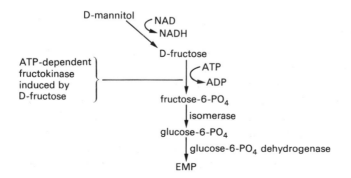

L1-110 and L2-110 the ability to utilize D-mannitol was determined by the presence of an inducible NAD-dependent D-mannitol dehydrogenase (EC 1.1.1.67) capable of using either D-mannitol or D-arabitol as substrate. The kinetics of the induction of D-mannitol dehydrogenase synthesis in *R. japonicum* was followed and, in the presence of D-mannitol, the specific activity of the enzyme increased linearly with time after 1 h, doubling within 2–2.5 h. After 5 h, the enzyme specific activity had increased approximately four-fold. D-mannitol also induces its own specific transport system in *R. japonicum* (Mulongoy and Elkan 1978). In *R. japonicum*, the enzyme substrate preference was for D-mannitol over D-arabitol since specific activities were two-fold higher with D-mannitol. In contrast, Martinez-de Drets and Arias (1970) found that the analogous enzyme in *R. meliloti*, a D-arabitol dehydrogenase, had three-to five-fold higher activities with D-arabitol as substrate than with D-mannitol as substrate. This enzymatic specificity difference between fast- and slow-growing rhizobia may reflect siginificant evolutionary differences separating these groups into distinct origins.

D-Mannitol has a profound effect on D-glucose metabolism in *R. japonicum*. Its presence in the growth-medium stimulates the synthesis of the NAD-linked 6-phosphogluconate dehydrogenase (EC 1.1.1.43) and it represses the glucose uptake system two- to three-fold (Mulongoy and Elkan 1978). Addition of D-mannitol to cell suspensions of either strain I-110 or strain L1-110 immediately results in a 50 per cent reduction in adenosine triphosphate (ATP) levels (Mulongoy and Elkan 1978). This is thought to be at least partly due to D-mannitol kinase activity. This activity is present in extracts of *R. japonicum* strains I-110 and L1-110 (Mulongoy and Elkan 1978). However it only may represent a vestige of the system found in most bacteria in which mannitol initially is transported into the cell by an inducible phosphoenol pyruvate (PEP)-dependent phosphotransferase (EC 2.7.1.40) system and is subsequently converted to fructose-6-phosphate by an inducible mannitol-1-phosphate dehydrogenase (EC 1.1.1.17). An unknown component (resistant to hydrolysis) of *R. japonicum* extracellular polysaccharides is implied to be a mannitol phosphate polymer (Keele, Wheat, and Elkan 1974) although no data has been presented. This conjecture warrants investigation to determine whether the mannitol kinase activity present in *R. japonicum* is responsible for the synthesis of precursors for complex and, as yet, unidentified polysaccharides.

As Martinez-de Drets and Arias (1970), in the study of *R. meliloti*, we did not detect mannitol-1-phosphate dehydrogenase in *R. japonicum* (Kuykendall and Elkan 1977). We conclude that in slow-growing as well as fast-growing *Rhizobium*, mannitol metabolism begins with a dehydrogenation producing D-fructose as it does in *Pseudomonas aeruginosa*, rather than by a phosphorylation of the free hexitol as it does in *Escherichia coli*. The fructose is then phosphorylated via a non-specific hexokinase to form fructose-6-phosphate which then enters the Embden–Meyerhof–Parnas pathway.

Five different polyol dehydrogenases have been described in *R. trifolii* by Primrose and Ronson (1980): inositol dehydrogenase (EC 1.1.1.18), specific for inositol; ribitol dehydrogenase (EC 1.1.1.56), specific for ribitol; D-arabitol dehydrogenase, which oxidizes D-arabitol, D-mannitol, and D-sorbitol; xylitol dehydrogenase, which oxidizes xylitol and D-sorbitol; dulcitol dehydrogenase (EC 1.1.1.16), which oxidizes dulcitol, ribitol, xylitol and sorbitol.

Ronson and Primrose (1979) corroborated the observation of Martinez-de Drets and Arias (1970) that D-glucose represses D-mannitol utilization by *R. meliloti*. Glucose was observed to repress the polyol dehydrogenase activities induced by mannitol or inositol only 20–40 per cent and, also almost totally repress activities induced by ribitol or dulcitol. For this repression to occur, glucose has to be metabolized to at least glucose-1-phosphate since D-mannitol dehydrogenase synthesis was not repressed by glucose in a glucokinase mutant (*glk*) of *R. meliloti* (Bergerson 1960). However, we found (Kuykendall and Elkan 1977) that levels of D-mannitol dehydrogenase were higher in *R. japonicum* strain L1-110 cells grown on medium containing both D-mannitol and D-glucose than on medium containing only D-mannitol alone. D-Glucose enhances the induction of a system for active D-mannitol uptake approximately two-fold in *R. japonicum*; active D-mannitol uptake is induced by the presence of either D-mannitol or D-fructose in the growth medium (Mulongoy and Elkan 1978). Although catabolite repression has been shown to occur in *R. meliloti* and *R. trifolii*, it has not been shown for *R. japonicum*. These data illustrate the need for continuing research on catabolic pathways using mutant methodology as an investigative tool.

4.4. Hexoses

The specific catabolic pathways for hexose utilization by *Rhizobium* are as yet only roughly defined. The first suggestion of a specific pathway for carbohydrate catabolism in *Rhizobium* was made by Jordan (1952) who presented evidence for oxidation of glucose via the Embden–Meyerhof–Parnas pathway. Katznelson (1955) reported that pyruvate and triose phosphate could be produced from 6-phosphogluconate by cell-free extracts prepared from glucose-grown cells of *R. phaseoli*, *R. meliloti*, *R. leguminosarum*, and *R. trifolii*. He further concluded that in the course of hexose-phosphate utilization pyruvate was formed via the

Entner–Doudoroff pathway. In a subsequent report, Katznelson and Zagallo (1957) demonstrated the presence not only of enzymes of the Entner–Doudoroff pathway, but also of enzymes of the Embden–Meyerhof–Parnas and pentose-phosphate pathways. Two key enzymes of the pentose-phosphate pathway found in extracts of the fast-growing *Rhizobium* species were NADP–glucose-6-phosphate dehydrogenase (EC 1.1.1.49) and NADP-6-phosphogluconate dehydrogenase (EC 1.1.1.43). Tuzimura and Meguro (1960) reported the oxidation of α-ketoglutarate, fumarate, succinate, and fructose-1,6-diphosphate by whole cells of *R. japonicum*, indicating the presence of the Embden–Meyerhof–Parnas pathway and tricarboxylic acid cycle (TCA).

Keele, Hamilton, and Elkan (1969) studied glucose catabolism in *R. japonicum*. Using the radiorespirometric method and by assaying for key enzymes of the known major energy-yielding pathways, the values obtained for $^{14}CO_2$ evolution from specifically labelled ^{14}C-glucose showed equality between C-1 and C-4 and between C-3 and C-6. This pattern was consistent with the glucose appearing as the carboxyl groups of pyruvate. Thus, the Entner–Doudoroff pathway appeared to account for 100 per cent of the catabolism of glucose in growing cells of *R. japonicum*. Since no 6-phosphogluconate dehydrogenase (NADP) activity was detected, an active pentose-phosphate pathway was apparently lacking. Glucose-6-phosphate dehydrogenase (EC 1.1.1.49), hexokinase (EC 2.7.1.1.), transketolase (EC 2.2.1.1), and an enzyme system (Entner–Doudoroff dehydrase and aldolase) which produced pyruvate from 6-phosphogluconate were found. The transketolase activity suggested that essential pentose phosphates could be synthesized from fructose-6-phosphate by transketolase and transaldolase reactions. The $^{14}CO_2$-evolution pattern from specifically labelled pyruvate indicated the presence of an active tricarboxylic acid cycle.

Both fast-growing and slow-growing *Rhizobium* species have been shown to possess the Entner–Doudoroff pathway (Katznelson 1955; Katznelson and Zagallo 1957; Kersters and De Ley 1968; Keele *et al.* 1969; Martinez-de Drets and Arias 1972; Mulongoy and Elkan 1977a, b). Jordan (1962) concluded that the Embden–Meyerhof–Parnas and pentose-phosphate pathways were present in *Rhizobium*. Siddiqui and Banerjee (1975) found the key enzyme of the Embden–Meyerhof–Parnas pathway, fructose 1,6-diphosphate aldolase (EC 4.1.2.13), in cell-free extracts of slow-growing and fast-growing species.

Keele, Hamilton, and Elkan (1970) followed their study of glucose catabolism with a study of D-gluconate catabolism in *R. japonicum*. In search of additional support for their earlier findings, showing the involvement of the Entner–Doudoroff pathway and tricarboxylic acid cycle as the sole catabolic pathways, they found a preferential release of $^{14}CO_2$ from C-1 and C-4, indicating degradation primarily via the

Entner–Doudoroff pathway, but discovered that inequalities between C-1 and C-4 (a preferential release of the C-1 carbon of gluconate) and between C-3 and C-6 indicated another pathway (C-1 decarboxylating). However, no NADP-6-phosphogluconate dehydrogenase activity was detected, thus eliminating the pentose-phosphate pathway. Having found gluconate-dehydrogenase activity present, they postulated that a nonphosphorylated ketogluconate pathway which enters the tricarboxylic acid cycle at α-ketoglutarate was an ancillary pathway for D-gluconate catabolism in *R. japonicum*. Radiorespirometric experiments used 2-ketogluconate-1-^{14}C, 2-ketogluconate-6-^{14}C, 2,5-diketogluconate-1-^{14}C, and 2,5-diketogluconate-6-^{14}C. Gluconate-grown cells rapidly oxidized these compounds without a lag period and the ratios of C-1 to C-6 obtained were the same as those for D-gluconate oxidation. These observations clearly indicated that the ^{14}C-labelled compounds were oxidized by way of same pathway as that for gluconate.

Martinez-de Drets and Arias (1972) extenced Keele *et al.* (1969) observations on the absence of NADP-6-phosphogluconate dehydrogenase in *R. japonicum*. From a comparative study of the fast- and slow-growing *Rhizobium* species, they concluded that the presence of this enzyme constituted an enzymatic basis for distinguishing the two groups, since only the fast-growing species possessed it. Then, Mulongoy and Elkan (1977*b*) found that *R. japonicum*, although lacking an NADP-specific 6-phosphogluconate dehydrogenase, does have an NAD-linked enzyme acting on 6-phosphogluconate. Chromatographic study of the reaction mixtures with partially purified preparations of this enzyme from strain L1-110 indicated that a phosphorylated ketohexonic compound was produced. Thus the enzyme possibly could initiate a new pathway, distinct from either the pentose-phosphate pathway and/or the hexose cycle and could be characterized by a preferential release of the C-1 and C-6 carbons of glucose. However, the data of Martinez-de Drets, Gardiol, and Arias (1977) appeared to show that the NAD-6-phosphogluconate dehydrogenase of *R. japonicum* is a decarboxylating enzyme(s). They found that partially purified preparations of this enzyme from strain 5006, when presented with 6-phosphogluconate-1-^{14}C in the presence of NAD, gave a stoichiometric production of $^{14}CO_2$. Although different strains were used for these independent investigations, it seems unlikely that the two strains studied have different pathways leading from 6-phosphogluconate. This enzyme system clearly deserves further study, particularly in order to resolve these differences. This enzyme may initiate a new pathway for the metabolism of hexoses in *R. japonicum*; the other possibility is that of participation of the pentose-phosphate pathway in hexose utilization.

Several studies have examined naturally occurring variation in symbiotic nitrogen fixation among wild-type isolates of *Rhizobium* for a relationship between carbohydrate utilization and efficiency in symbiotic nitrogen fixation. Georgi and Ettinger (1941) concluded that no differentiation of efficient and inefficient strains of *Rhizobium* was possible on the basis of utilization of various carbohydrates. Katznelson and Zagallo (1957) compared the abilities of effective and ineffective strains of *R. meliloti*, *R. leguminosarum*, and *R. phaseoli* to metabolize 6-phosphogluconate to pyruvate, to oxidize glucose-6-phosphate, and to carry out aldolase (EC 4.1.2.7) and phosphohexokinase (EC 2.7.1.11) reactions. They found no distinct relationship between these metabolic properties and symbiotic competence. Symbiotically-competent strains did, however, oxidize succinate more rapidly than did non-competent strains. The main limitation was that few strains were used. Gupta and Sen (1965) conducted a study in which the nitrogen fixation efficiency of a *Rhizobium* isolate from legumes of each of four species was compared to the extent of glucose utilization by those strains *in vitro*. They examined 40 isolates each from *Pisum sativum* (pea) *Trigonella foenumgraecum* (fenugreek), *Phaseolus mungo* (black gram), and *Dolichos lablab* (lablab) for differences in glucose consumption capacity and in phosphate utilization and found that nitrogen-fixing efficiency (nitrogen content of plants when grown in a nitrogen-deficient medium and inoculated with a particular isolate) was positively correlated with both asymbiotic metabolic parameters, but particularly with the extent of glucose utilization. Magu and Sen (1969) studied the respiration rate of 15 strains of *R. trifolii* and of ten strains of *R. leguminosarum* when growing on glucose, maltose, and mannitol and compared these results with those of plant inoculation experiments where nitrogen fixation efficiency was again estimated by the increase in nitrogen content over the controls. They did not find any correlation between efficiency and respiratory rate on these carbon sources. They noted, however, that there was a stimulatory effect by the amino acid glycine on respiratory rate during glucose utilization, and that this stimulation was greater in efficient strains than in inefficient strains. Petrova, Chermenskaya, and Kretovich (1974) reported higher levels of glucose 6-phosphate dehydrogenase in nodules formed by effective strains of *R. lupini* than those formed by ineffective strains. This was correlated with a lower concentration of free sugar in bacteroids.

The closely-related derivatives of strain 311b110 of *R. japonicum* differing in nitrogen-fixing efficiency isolated by Kuykendall and Elkan (1976) also differed in glucose utilization. Whereas strain I-110 grew faster on D-glucose than strain L1-110, strain I-110 grew more slowly on D-fructose than strain L1-110. This shows that glucose and fructose are metabolized differently and suggests a relationship between efficient glucose utilization and efficient symbiotic nitrogen fixation. These facts led to a reinvestigation of D-glucose catabolism in *R. japonicum*, with an emphasis on comparing the mechanisms utilized by strains I-110 and L1-110. Radiorespirometric and enzymatic analyses revealed that glucose-grown cells of both strains I-110 and L1-100 possess an active tricarboxylic acid cycle and that they both metabolize glucose by simultaneous operation of the Embden–Meyerhof–Parnas and Entner–Doudoroff pathways (Mulongoy and Elkan, 1977*a*). This study revealed that the enzyme system of *R. japonicum* strains I-110 and L1-110 includes all the Embden–Meyerhof–Parnas and Entner–Doudoroff pathway enzymes. The radiorespirometric patterns for glucose utilization (C3,4>C1>C2>C6) repeated the earlier results obtained by Keele *et al.* (1969) with ATCC 10324. The cumulative yield of $^{14}CO_2$ from glucose C3 was about two-thirds that from C4; thus the pattern for $^{14}CO_2$ evolution from glucose was C4>C3>C1>C2>C6 for strain I-110 and C4>C1>C3>C2>C6 for strain L1-110. The importance of the Embden–Meyerhof–Parnas pathway is shown by the extensive conversion of C3 and C4 of glucose to $^{14}CO_2$, since glycolytic sequences result in the conversion of glucose into two molecules of pyruvate in which the carboxyl groups derive from the C3 and C4. The carboxyl carbons of pyruvate are then first released in the tricarboxylic acid cycle giving extensive $^{14}CO_2$ release when glucose is labelled in the C3 or C4 position. Strain I-110, characterized by shorter generation times on glucose and greater nitrogen-fixing efficiency (Kuykendall and Elkan 1976), oxidizes glucose more extensively than strain L1-110 (Mulongoy and Elkan 1977*a*). The ratio of cumulative carbon dioxide yields from C1 and C3 of glucose was used to approximate the relative participation of the Embden–Meyerhof–Parnas and Entner–Doudoroff pathways. The C1/C3 ratio provides an estimation of the relative use of the two pathways. Strain I-110, as indicated by a C1/C2 < 1, utilizes the Embden–Meyerhof–Parnas pathway preferentially whereas the Entner–Doudoroff pathway apparently predominates in strain L1-110 (Mulongoy and Elkan 1977*a*) since C1/C2 ≥ 1. Enolase levels are more than four-fold lower in strain L1-110 than in strain I-110; this enzyme probably controls the flow of carbon through the Embden–Meyerhof–Parnas pathway. Kuykendall and Elkan (1976) hypothesized that the higher nitrogen fixation efficiency of strain I-110 may be due to its higher efficiency in glucose utilization. Yet, these two strains differ by the same order of magnitude in ability to synthesize nitrogenase *in vitro* as they do in association with soyabeans when grown on carbon sources such as D-gluconate, which they metabolize equally (Upchurch and Elkan 1977). The levels of *in vitro* nitrogenase expression are much less than one per cent of those found in symbiotic cells in nodules and conclusions based on *in vitro* expression may not be valid. Bergerson (1960), however, proposed that hexoses may be the main substrates used by bacteroids *in vivo*.

Ronson and Primrose (1979) utilized a mutant methodology approach to determine carbohydrate utilization in *R. trifolii* and the carbon source supplied to the microsymbiont by the clover plant. The carbohydrate-negative mutants of *R. trifolii* isolated and characterized included mutants defective in glucokinase (*glk*; EC 2.7.1.2), fructose transport (*fup*), the Entner–Doudoroff pathway and pyruvate carboxylase (*pyc*; EC 6.4.1.1.). Significantly, all the mutants, including one double mutant (*glk fup*), formed effective symbioses on *T. pratense*, showing that neither glucose nor fructose are supplied to the microsymbiont by the host plant. In addition, the inability to demonstrate phosphofructo-kinase (EC 2.7.1.11) activity in cell-free extracts of *R. trifolii* and the detection of only low activities of fructose-1,6-diphosphate aldolase (EC 4.1.2.13) indicated that the Embden–Meyerhof–Parnas pathway was not physiologically important in the strain studied. Operation of the Entner–Doudoroff pathway was shown by the production of pyruvate from 6-phosphogluconate. Pyruvate carboxylase (EC 6.4.1.1) mutants were unable to grow on any carbon source except succinate. Thus, as for *Agrobacterium* sp and *Pseudomonas aeruginosa*, pyruvate carboxylase is required by *R. trifolii* for growth on hexoses, pentoses, and trioses.

Carbohydrate-negative mutants have also been isolated and studied in *R. meliloti*. Arias, Cervenansky, Gardiol, and Martinez-de Drets isolated a phosphoglucoseisomerase-negative (*pgi*) mutant of *R. meliloti* which gave little nitrogen fixation in association with the *M. sativa* host plant. This mutant strain was isolated following N-methyl-*N'*-nitro-*N*-nitrosoguanidine (NTG) mutagenesis and screening glycerol-grown cells for lack of ability to utilize mannitol. A mutant was isolated that did not grow on mannitol, sorbitol, fructose, mannose, ribose, arabitol, or xylose, but grew on glucose, maltose, gluconate, L-arabinose, and other carbohydrates. This mutant strain accumulated high levels of fructose-6-phosphate when fructose was present, resulting in a toxic accumulation that prevented growth on available carbohydrates such as L-arabinose. A revertant selected for fructose utilization regained phosphoglucose-isomerase activity, wild-type phenotype growth on all carbohydrates, and symbiotic nitrogen-fixing ability. Duncan and Fraenkel (1979) selected a mutant of *R. meliloti* unable to grow on L-arabinose. This mutant was also unable to grow on acetate or pyruvate; it lacked α-ketoglutarate dehydrogenase (succinyl CoA synthetase, EC 6.2.1.5.) activity. In *R. meliloti*, fructose and glucose are metabolized differently (Hornez *et al.* 1976); 2-ketogluconate accumu-lates in cells growing on glucose, and fructose is directly phosphorylated and utilized via the Embden–Meyerhof–Parnas glycolytic sequence. *R. meliloti* cells growing on fructose or mannitol produce abundant extracellular polysaccharide, but not when growing on glucose or galactose (Hornez *et al.* 1976). The pH of cultures in glucose or galactose medium drops to pH 4.5, but pH of cells growing on (and producing polysaccharide from) fructose and mannitol does not drop.

The results of Arias *et al.* (1979) suggest that *R. meliloti* does not have a functional Embden–Meyerhof–Parnas glycolytic sequence and that glucose and gluconate are metabolized primarily via the Entner–Doudoroff pathway and partly by the pentose-phosphate pathway. The inability of the *pgi* mutant to grow on ribose, xylose, and arabitol suggests that the metabolism of these pentoses occurs via the non-oxidative pentose-phosphate pathway leading to fructose-6-phosphate. Growth occurs on L-arabinose, however, since this pentose is utilized by a non-phosphorylated pathway leading to α-ketoglutarate (Duncan and Fraenkel 1979). The pleiotropic effects of the *pgi* mutation in *R. meliloti* on carbohydrate utilization were recognized by Arias *et al.* (1979) as being essentially identical to the pleiotrophy shown by a *pgi* mutant in *Pseudomonas aeruginosa* as described by Phibbs, McCowen, Feary, and Blevins (1978). The results of these studies on *R. meliloti* and *P. aeruginosa*, carbohydrate-negative mutants, clearly indicates great similarities in the possible pathways of carbohydrate metabolism.

Generally, fast-growing *Rhizobium* are said to be 'acid-producing' and slow-growing *Rhizobium* are known as 'alkali-producing' or, in some instances, 'non-acid-producing'. Walker and Brown (1930) first cautioned the use of this generalization. In a study of 23 strains of *R. meliloti* and 12 strains of *R. japonicum* pH changes occurring in glucose and galactose media, they found that although the reactions produced by most *R. meliloti* strains were acid, this was not true in all cases and certain *R. japonicum* strains produced more acid than some *R. meliloti* strains. There was considerable variation in the fermentation characteristics of different strains of the same species. The ranges between the two species overlapped, thus pH changes in glucose or galactose media was not a valid distinct differentiation between the two species.

4.5 Pentoses

The preferred carbon source for the slow-growing species represented by *R. japonicum* is L-arabinose (Allen and Allen 1940; Neal and Walker 1934). Neal and Walker stated that 'arabinose was distinctly superior to the other carbonaceous compounds studied, as a source of energy for *R. japonicum*'. Pedrosa and Zancan (1974) studied L-arabinose catabolism in *R. japonicum* and found a pathway operating similarly to one previously described in a pseudomonad. In this pathway, L-arabinose is first converted to 2-keto-3-deoxy-L-arabonate in three steps: (1) dehydrogenation of L-arabinose to form L-arabonolactone, (2) hydrolysis of L-arabonolactone to L-arabonate, and (3) dehydration of L-arabonate to form 2-keto-3-deoxy-L-arabonate. The 2-keto-3-deoxy-L-arabonate is then cleaved by a specific aldolase to yield glycoaldehyde and pyruvate.

Their finding of an active NAD/NADP-dependent glycoaldehyde dehydrogenase suggested a significant role for glyoxylate cycle enzymes (Johnson, Evans, and Ching 1966).

L-Arabinose metabolism was studied in *R. meliloti* by Duncan and Fraenkel (1979), who isolated an L-arabinose-negative mutant which also failed to grow on either acetate or pyruvate and grew more slowly than the parent strain on other carbon sources. The strain, described as ineffective on *M. sativa*, lacked α-ketoglutarate dehydrogenase (*kgd*) and produced revertants that had the enzyme, had wild-type carbohydrate utilization, and effectively fixed nitrogen on *Medicago*. The pathway for L-arabinose metabolism in *R. meliloti* was found by Duncan and Fraenkel (1979) differed from that found by Pedrosa and Zancan (1974) for *R. japonicum*; it led to α-ketoglutarate rather than pyruvate and glycoaldehyde. *R. meliloti* lacked 2-keto-3-deoxy-L-arabonate aldolase, but possessed α-ketoglutarate semialdehyde dehydrogenase. Their conclusion that *R. meliloti* has the L-arabinose pathway which leads to α-ketoglutarate accounts for the inability of an α-ketoglutarate dehydrogenase-negative mutant to utilize L-arabinose.

4.6 Tricarboxylic acid cycle intermediates

Organic acids that are intermediates in the tricarboxylic acid (TCA) cycle are utilized by many micro-organisms as carbon and energy sources. However, some microbes which possess a complete TCA cycle fail to actively transport these compounds, and consequently they cannot metabolize them. Thus, *R. japonicum* has an active TCA cycle (Keele *et al.* 1969, 1970; Mulongoy and Elkan 1977*a*) but only two strains out of 36 examined (Elkan and Kwik 1968) grew well *in vitro* with succinate as a carbon source, and none of the 36 utilized either citrate or malate. Earlier, Graham studied the utilization of carbohydrates of diverse isolates of different species of *Rhizobium* and observed that the fast-growing species gave excellent growth on most carbon sources tested including TCA intermediates, whereas the slow-growing *Rhizobium* were much more restricted in their choice of metabolizable substrates (Graham 1964*a*; Gupta and Sen 1965). Specifically, *R. trifolii* and *R. leguminosarum* could utilize 20 different carbohydrates and TCA cycle intermediates, whereas *R. japonicum* could utilize only eight of the 20. Proctor (1963) found that there was relatively weak oxidation of TCA intermediates by *Rhizobium* compared to other carbohydrates. Using manometric techniques, he surveyed the oxidation of more than 50 compounds by four distinct strains of *Rhizobium* from *Trifolium pratenase*, *Lotus uliginosus*, and *Galega officinalis*. Except for oxalosuccinate, TCA intermediates were only weakly oxidized, aconitate was extensively oxidized by one of the strains. Proctor also found very rapid oxidation of proline, leading him to suggest that this amino acid may be a 'true substrate for the rhizobia in nodules'. Tuzimura and Meguro (1960) studied the oxidation of various carbohydrates and TCA-cycle intermediates by *Rhizobium* extracted from *G. max* nodules compared with *in vitro* cultured *R. japonicum* grown on either glucose or succinate. Cells derived from nodules actively oxidized TCA-cycle intermediates, but did not metabolize glucose or other hexoses (or sucrose or mannitol) except for fructose 1,6-diphosphate. Succinate-grown cultured cells showed the same pattern; only glucose-grown cells actively metabolized glucose. They concluded: 'These facts suggest that the energy sources of *Rhizobium* in symbiotic state are organic acids but not carbohydrates'. Katznelson and Zagallo (1957) examined the substrate-specific metabolic activity of fast-growing species, comparing effective and ineffective strains; effective strains of *Rhizobium* oxidized succinate more rapidly than ineffective strains. Working with a strain of *R. meliloti* that has been useful in genetic studies and which grows most rapidly with succinate as the carbon source, Ucker and Signer (1978) demonstrated a catabolite-repression-like phenomenon in which succinate produced an immediate cessation of β-galactosidase synthesis when added to cells growing on lactose. They produced a *lac* mutant with less than one-fortieth of the wild-type β-galactosidase activity.

The extensive mutant methodological study conducted by Ronson and Primrose (1979) clearly addresses the question of plant carbohydrate supply to the microsymbiont. They isolated carbohydrate-negative mutants in specific enzymes such as glucokinase (*glk*, EC 2.7.1.2) fructose uptake (*fup*), the Entner–Doudoroff pathway, and pyruvate carboxylase (*pyc*, EC 6.4.1.1) and then found that all of these mutants formed an effective symbiosis on *T. pratense*. Their conclusion was that the plant supplies the microsymbiotic cell with tricarboxylic acid intermediates as their primary source of energy.

The enzymes of the glyoxylate cycle were examined in nodules and in free-living *Rhizobium* by Johnson *et al.* (1966). This study followed from the observation that *G. max* nodules and bacteria extracted from soyabean nodules contain relatively large amounts of fatty acids, as well as by the facts that (1) poly-β-hydroxybutyrate is an important energy-storage form in *Rhizobium* and (2) that utilization of poly β-hydroxybutyrate (pBHB) would be expected to result in formation of acetoacetate and acetyl-CoA. They surveyed *Rhizobium* isolates from *Phaseolus*, *Vigna*, *Lupinus*, *Glycine*, *Medicago*, *T. pratense* and *Pisum*. Interestingly, the activity of malate synthetase was high in nodules of plant species in symbiosis with slow-growing species, but was barely detectable in nodules of plant species hosting the fast-growers. Significant isocitrate lyase (EC 4.1.3.1) activity was lacking in bacteria derived from any type of nodule, indicating that the glyoxylate cycle does not operate in *Rhizobium* symbiotic metabolism. They therefore postulated a structural role for the fatty acids found in nodules. When oleate was utilized as the carbon source for *in vitro* cultivation of *Rhizobium*, significant levels of isocitrate lyase were found (Johnson *et al.* 1966). Duncan and Fraenkel (1979) confirmed and extended these experiments. They assayed the glyoxylate cycle enzymes in a strain of *R. meliloti* and in an α-ketoglutarate dehydrogenase mutant of this strain. Malate synthetase (EC 4.1.3.2) activity was present in cultures grown on all carbon sources but isocitrate lyase activity was present only in cells grown on acetate; this shows that the glyoxylate cycle is probably used only in acetate growth, but not for either pyruvate or L-arabinose metabolism (Duncan and Fraenkel, 1979).

Stovall and Cole (1978) found that *R. japonicum* cells extracted from soyabean root nodules rapidly oxidized ^{14}C-labelled succinate, pyruvate, and acetate, releasing $^{14}CO_2$ in a manner consistent with the operation of the tricarboxylic acid cycle and a partial glyoxylate cycle. They found incorporation of ^{14}C into macromolecular cell components indicating that microsymbiont cells can utilize the tricarboxylic acid cycle both for energy production and as a source for carbon compounds used in biosynthesis. Their data suggest that the anapleurotic enzymes, phosphoenolpyruvate or pyruvate carboxylases and malate synthetase, are functional in *R. japonicum* symbiotic cells and that acetate carbon is used for malate formation (to replenish this intermediate of the tricarboxylic acid cycle).

4.7 Triose metabolism

In the nutritional survey of 36 strains of *R. japonicum* reported by Elkan and Kwik (1968) several trioses were included. Twenty-nine of the strains utilized glycerol; only six strains used pyruvate as the sole energy and carbon source. Neither lactate nor propionate were utilized by any of the strains. Since whole cells were used in these studies, transport problems might be a factor in the inability to metabolize the latter compounds. Graham (1964a) found that pyruvate, the only triose included in his study, could serve as sole carbon source for 53 of 95 strains representing seven *Rhizobium* species. Representatives of the fast-growing species of *Rhizobium* as well as *R. japonicum* were shown by Johnson *et al.* (1966) to oxidize acetate and pyruvate and yet to be unable to utilize these compounds for growth. They suggested that this was due to an inability of the organism to produce four-carbon intermediates from acetate because of the absence of isocitrate lyase (EC 4.1.3.1).

Arias and Martinez-de Drets (1976) examined glycerol metabolism in four strains of *R. japonicum* and in one strain of *R. trifolii*, all of which could grow on glycerol. Cell-free extracts of glycerol-grown cells of both slow-growing and fast-growing rhizobia contained a glycerol kinase (EC 2.7.1.30). This enzyme, specifically induced by glycerol, catalyses the phosphorylation of glycerol to glycerolphosphate and is located in the soluble fraction of the cell extract and required ATP. No phosphoenol pyruvate phosphotransferase activity was found in the different fractions of the extracts. A glycerolphosphate dehydrogenase catalysing the oxidation of glycerolphosphate to dihydroxyacetone phosphate was detected in the five *Rhizobium* extracts. This enzyme was found in the particulate fraction. No glycerol dehydrogenase (EC 1.1.1.6) activity was detected. These data show that in *Rhizobium*, glycerol is metabolized through the phosphorylated pathway similar to *E. coli* (Koch, Hayashi, and Lin 1964; Cozzarelli, Freedberg, and Lin 1968), *Aerobacter aerogenes* (Rush, Karibian, Karnovsky, and Magasanik 1957), and *Rhodopseudomonas capsulata* (Lueking, Tokuhisa, and Sojka 1973). However, *Rhizobium* cannot grow on glycerol-phosphate, whereas *E. coli* can. The pathway for glycerol utilization does not appear to differ in fast- and slow-growing organisms, but the authors are careful to state that more strains and species need to be examined before a definitive conclusion is reached.

Since rhizobia have pyruvate carboxylase as well as an active tricarboxylic acid cycle, pyruvate is utilized via the tricarboxylic acid cycle. An alternate role for pyruvate has been described by Trinchant and Rigaud (1974). Cell-free extracts of *R. meliloti* contained a soluble lactate dehydrogenase (EC 1.1.1.27). This enzyme is found to catalyse the reduction of pyruvate to lactate with NADH. In addition the enzyme reduces indole-3-pyruvic acid to indole-3-lactic acid. Earlier Rigaud and Trinchant (1973) found an alcohol dehydrogenase (EC 1.1.1.1) that carries out the reduction, using NADH, of indole-3-acetaldehyde to form tryptopol. The role of these enzymes and pyruvate in indole metabolism is proposed.

De Hertogh, Mayeux, and Evans (1964) demonstrated that propionate is oxidized by bacteroids from soyabean nodules and by cells of *R. japonicum* and *R. meliloti*. They demonstrated the presence of enzymes required for the metabolism of propionate by conversion to succinate via methylmalonate and then oxidation of succinate via the tricarboxylic acid cycle. Cell-free extracts of *R. meliloti* and *R. japonicum* and soyabean bacteroids had the capacity to catalyse the activation of propionate to propionyl-CoA when supplied with propionate, ATP, and coenzyme A. The carboxylation of propionyl CoA to form methylmalonyl-CoA via propionyl CoA carboxylase was also demonstrated. Radiorespirometric experiments with specifically labelled propionate yielded patterns consistent with propionate utilization by a series of reactions resulting in the formation of succinate, which is then oxidized via the tricarboxylic acid.

4.8 Conclusions

The mechanisms of carbohydrate metabolism in *Rhizobium* are particularly deserving of research, since the efficiency of energy derivation from carbon-source utilization is probably a determining factor in the efficiency of symbiotic nitrogen fixation. Our knowledge of these systems is incomplete and fragmented. The possible use of the pentose-phosphate pathway in hexose metabolism by *R. japonicum* is an example of much research being required before a meaningful understanding is achieved.

The taxonomic status of *Rhizobium*, as proposed by Graham (1964*b*), is well addressed by comparing the metabolism of carbohydrates between the fast-growing and slow-growing isolates. The differences (Table 4.1) found help to justify their assignment to taxonomically-distinct genera of bacteria.

TABLE 4.1

Summary of known aspects of carbohydrate metabolism in Rhizobium meliloti *(fast-growing) and* Rhizobium japonicum *(slow-growing)*

Aspect of carbohydrate metabolism	*Rhizobium* species	
	R. meliloti	*R. japonicum*
(1) The nutritional availability of various carbon compounds	Very versatile, for example disaccharides such as sucrose are readily used as are TCA cycle intermediates	More restricted in choice of substrates for growth, for example, non-utilization of disaccharides and often of TCA cycle intermediates
(2) Substrate preference of D-Mannitol/D-Arabitol dehydrogenase	D-Arabitol	D-Mannitol
(3) Catabolite repression	Observed	Not yet observed
(4) Presence of Embden–Meyerhof–Parnas and Entner–Doudoroff enzymes	Demonstrated, but the EMP pathway may not be physiologically significant	Demonstrated, strains have been shown to operate both pathways simultaneously
(5) Pentose-phosphate pathway	Demonstrated NADP-6-phosphogluconate dehydrogenase activity present	Unclear (?), no NADP-6-phosphogluconate dehydrogenase, but the NAD-linked enzyme may/may not be decarboxylating
(6) L-Arabinose catabolism	Pathway leads to α-ketoglutarate	Different pathway leading to glycoaldehyde and pyruvate

References

Allen, E. K. and Allen, O. N. (1950). *Bacteriol Rev.* **14,** 273–330.

Arias, A. and Martinez-de Drets, G. (1976). *Can. J. Microbiol.* **22,**, 150–3.

—— Cervenansky, C., Gardiol, A., and Martinez-de Drets, G. (1979). *J. Bacteriol.* **137,** 409–14.

Bergersen, F. J. (1960). *Aust. J. Biol. Sci.* **14,** 349–60.

Bethlenfalvay, G. J. and Phillips, D. A. (1977). In *Genetic engineering for nitrogen fixation*, (ed, A. Hollaender) pp. 401–8. Plenum Press, New York.

Buchanan, R. E. and Gibbons, N. E. (eds) (1974). *Bergey's manual of determinative bacteriology*, 8th edn. Williams and Wilkins, Baltimore.

Burris, R. H., Phelps, A. S., and Wilson, J. B. (1942). *Proc. Soil Sci. Soc. Am.* **7,** 272–5.

Cozzarelli, N. R., Freedberg, W. B. and Lin, E. C. C. (1968). *J. molec. Biol.* **31,** 371–87.

De Hertogh, A. A., Mayeux, P. A., and Evans, H. J. (1964). *J. biol. Chem.* **239,** 2446–53.

Duncan, M. J. and Fraenkel, D. G. (1979). *J. Bacteriol.* **37,** 415–19.

Elkan, G. H. and Kwik, I. (1968). *J. appl. Bacteriol* **31,** 399–404.

Evans, H. J., Emerich, D. W., Ruiz-Argueso, T., Maier, R. J., and Albrecht, S. L. (1980). In *Nitrogen fixation*, (ed. W. E. Newton and W. H. Orme-Johnson) Vol. II, pp. 69–86. University Park Press, Baltimore.

Fred, E. B. (1911–1912). Va. Agric. Sta. Ann. Rept. 145–73.

Gaur, Y. D., and Mareckova, H. (1977). *Folia Microbiol.* **22,** 311–12.

Georgi, C. E. and Ettinger, J. M. (1941). *J. Bacteriol.* **41,** 323–40.

Graham, P. H. (1964a). *Antonie van Leeuwenhoek* **30,** 68–72.

—— (1964b). *J. gen. Microbiol.* **35,** 511–17.

Gupta, K. G. and Sen, A. N. (1965). *Ind. J. Sci.* **35,** 39–42.

Hardy, R. W. F. (1977). In *Genetic engineering for nitrogen fixation*, (ed. H. Hollaender) pp. 369–97. Plenum Press, New York.

Hornez, J., Courtois, B., and Derieux, J. (1976). *C. R. Acad. Sci. Paris*, **283,** 1559–62.

Johnson, G. V., Evans H. J., and Ching, T. (1966). *Plant Physiol.* **41,** 1330–6.

Jordan, D. C. (1952). *Can. J. Bot.* **30,** 693–700.

—— (1962). *Bacteriol. Rev.* **26,** 119–41.

Katznelson, H. (1955). *Nature, Lond.* **175,** 551–2.

—— and Zagallo, A. C. (1957). *Can. J. Microbiol.* **3,** 879–84.

Keele, B. B., Hamilton, P. B., and Elkan, G. H. (1969). *J. Bacteriol.* **97,** 1184–91.

—— —— —— (1970). *J. Bacteriol.* **101,** 698–704.

—— Wheat, R. W., and Elkan, G. H. (1974). *J. gen. appl. Microbiol.* **20,** 187–96.

Kersters, K. and De Ley, J. (1968). *Antonie van Leeuwenhoek* **34,** 393–408.

Koch, J. P., Hayashi, S., and Lin, E. C. C. (1964). *J. biol. Chem.* **239,** 3106–8.

Kuykendall, L. D. and Elkan, G. H. (1977). *J. gen. Microbiol.* **98,** 291–5.

—— —— (1976). *Appl. Environ. Microbiol.* **32,** 511–19.

Lueking, D., Tokuhisa, D., and Sojka, G. (1973). *J. Bacteriol.* **115,** 897–903.

Magu, S. P. and Sen, A. N. (1969). *Arch. Mikrobiol.* **68,** 355–61.

Martinez-de Drets, G. and Arias, A. (1970). *J. Bacteriol.* **103,** 97–103.

—— —— (1972). *J. Bacteriol.* **109,** 467–70.

—— —— and Rovira de Cutinella, M. (1974). *Can. J. Microbiol.* **20,** 605–9.

—— A. Gardiol, and Arias, A. (1977). *J. Bacteriol.* **103,** 1139–43.

Mulongoy, K. and Elkan, G. (1977a). *J. Bacteriol.* **131**(1), 179–87.

—— —— (1977b). *Can. J. Microbiol.* **23,** 1293–8.

—— —— (1978). *Current Microbiol.* **1,** 335–40.

Neal, O. R. and Walker, R. H. (1934). *J. Bacteriol.* **30,** 173–87.

Niel, C., Guillaume, J. B., and Bechet, M. (1977). *Can. J. Microbiol.* **23,** 1178–81.

Norris, D. O. (1965). *Pl. Soil* **22,** 143–66.

Pate, J. S. (1977). In *A treatise on dinitrogen fixation*, Section III, *Biology* (ed. R. W. F. Hardy and W. S. Silver) pp. 473–518. Wiley-Interscience, New York.

Pedrosa, F. O. and Zancan, G. T. (1974). *J. Bacteriol.* **119,** 336–7.

Petrova, A. N., Chermenskaya, I. E. and Kretovich, V. L. (1974). *Dokl. Biochem.* (Engl. Transl. *Dokl. Akad. Nauk. SSSR*) **217,** 479–80.

Phibbs, P. V., Jr., McCowen, S. M., Feary, T. W., and Blevins, W. T. (1978). *J. Bacteriol.* **133,** 717–28.

Primrose, S. B. and Ronson, C. W. (1980). *J. Bacteriol.* **141,** 1109–14.

Proctor, M. H. (1963). *N.Z. J. Sci.* **6,** 17–26.

Rigaud, J. and Trinchant, J. (1973). *Physiol. Plant* **28,** 160–5.

Ronson, C. W. and Primrose, S. B. (1979). *J. Bacteriol.* **139,** 1075–8.

—— —— (1979). *J. gen. Microbiol.* **112,** 77–88.

Rush, D., Karibian, D., Karnovsky, M., and Magasanik, B. (1957). *J. biol. Chem.* **226,** 891–9.

Sarles, W. B. and Reid, J. (1935). *J. Bacteriol.* **30,** 651.

Shmyreva, T. V., Krasnopol'skaya, V. S., and Plaksina, T. B. (1969) *Prikladnaya Biokhimiya i Mikrobiologiya* **5**(5), 567–72.

—— and Plaksina, T. B. (1972). *Prikl. Biochem. Mikrobiol.* **8,** 26–9.

Siddiqui, K. A. and Banerjee, A. K. (1975). *Folia Microbiol.* **20,** 412–17.

Singh, R., Singh, N., and Sidhu, G. S. (1967). *Indian J. Microbiol.* **7,** 143–50.

Stovall, I. and Cole, M. (1978). *Plant Physiol.* **61,** 787–90.

Thorne, D. W. and Walker, R. H. (1936). *Soil Sci.* **42,** 231–40.

Trinchant, J. and Rigaud, J. (1974). *Physiol. Plant* **32,** 394–9.

Tuzimura, K. and Meguro, H. (1960). *J. Biochem.* **47,** 391–7.

Ucker, D. S. and Signer, E. R. (1978). *J. Bacteriol.* **136,** 1197–200.

Upchurch, R. G. and Elkan, G. H. (1977). *Can. J. Microbiol.* **23,** 1118–22.

Vincent, J. M., Nutman, P. S., and Skinner, F. A. (1979). In *Identification methods for microbiologists* (ed. F. A. Skinner) 2nd edn., pp. 49–69. Soc. Appl. Bact. Tech. Ser. 14. Academic Press, London.

Virtanen, A. I., Nordlund M., and Hollo, E. (1934). *Biochem. J.* **28,** 796–802.

Walker, R. H. and Brown, P. E. (1930). *Soil. Sci.* **30,** 219–29.

Wilson, P. W. (1937). *J. Bacteriol.* **35,** 601–23.

Zipfel, H. (1911). *Z. Bakt. Parasitenk.* II, **32,** 97–137.

5. Genetics

J. E. BERINGER, N. J. BREWIN, AND A. W. B. JOHNSTON

5.1 Introduction

During the last few years there has been a rapid increase in our knowledge of *Rhizobium* genetics and hence in our ability to manipulate these bacteria genetically. The progress of this research has been well documented, both in original publications and in reviews (see reviews by Dénarié, Truchet, and Bergeron 1976; Dénarié and Truchet 1976; Beringer 1976, 1980; Schwinghamer 1977; Beringer, Brewin, Johnston, Schulman, and Hopwood 1979). In this chapter we have attempted to review recent developments in *Rhizobium* genetics for the non-specialist, and to provide sufficient references for those who wish to study the literature on this subject in more detail.

5.2 Mutation

Genetic studies depend upon mutants to provide variability. Natural variation occurs and specific mutants can be found by screening very large populations for strains with altered characteristics (such as drug-resistant or symbiotically-defective mutants) or more readily by selecting directly for specific mutations. Either screening procedure can be facilitated by mutagenizing the rhizobia to increase the proportion of mutants in the population.

Most of the mutagens described for other micro-organisms have been reported to be capable of inducing mutations in the fast-growing rhizobia. However, there appear to be significant differences in the frequencies of mutation observed for different mutagen–strain combinations (see Cunningham 1980). In slow-growing strains, spontaneous drug-resistant mutants are readily available, but very few auxotrophic mutants have been reported. This may reflect difficulty in mutating such bacteria or may be due to the tendency of these bacteria to clump, so that single colonies seldom arise from individual bacteria. By taking care to dissociate clumps of *R. japonicum* after mutagenesis, D. Kuykendall (personal communication) was able to isolate a number of auxotrophic mutants.

Techniques for obtaining spontaneous, chemical- or radiation-induced mutants of *Rhizobium* have been described many times (Schwinghamer 1962; Heumann 1968; Beringer 1974; Dénarié *et al.* 1976) and will not be discussed here. However, it is worth noting that most media used for culturing rhizobia contain large amounts of glucose or mannitol, both of which give rise to the formation of large slimy colonies. This limits the number of colonies that can be handled per Petri dish and makes replica plating and other manipulations less efficient.

Transposon mutagenesis has recently been described for fast-growing *Rhizobium* species (Beringer, Beynon, Buchanan-Wollaston, and Johnston 1978*a*). Transposons are specific sequences of DNA which replicate only when they are integrated into a plasmid, or bacterial, or virus chromosome. For some transposons the site of integration is almost random; when it occurs within a gene it inactivates the gene function (Kleckner, Roth, and Botstein 1977). The various advantages of transposon mutagenesis have been discussed by Kleckner *et al.* (1977). Transposons are particularly useful for inducing mutants with a phenotype that is not easily scored. An obvious example concerns the isolation of symbiotically-defective mutants in *Rhizobium*. Thus mapping and the selection for such genes can be done on Petri dishes and only a limited number of plant tests are needed to confirm that the drug resistance and symbiotic defect are always closely linked. Even though transposon mutagenesis has only been used for generating mutants in *Rhizobium* for about two years, the majority of symbiotically-defective mutants currently available for genetic analysis have been generated by this method, using the transposon Tn5.

Transposons can integrate into a plasmid and convert it into a drug resistance plasmid which can readily be selected in crosses. Johnston, Beynon, Buchanan-Wollaston, Setchell, Hirsch, and Beringer (1978*c*) and Brewin, Beringer, Buchanan-Wollaston, Johnston and Hirsch (1980) took advantage of this to demonstrate the presence and transfer of indigenous *Rhizobium* plasmids; for example a plasmid originating in *R. leguminosarum* was found to be transmissible to other *Rhizobium* species and carried genes involved in host range (see below).

5.3 Gene transfer

In order to map genes and to construct new strains, gene transfer between bacteria is required. Four main methods of gene transfer occur in bacteria; conjugation, transduction, transformation, and (protoplast) fusion. In conjugation, donor and recipient cultures of bacteria are mixed, pair formation occurs and relatively large segments of deoxy-ribonucleic acid (DNA) are transferred from the donor to the recipient. Transduction involves the transfer of bacterial DNA by bacteriophages. This usually results from a small proportion of bacteriophages packaging host DNA instead of phage DNA. When the DNA is injected into a new host it can become integrated by replacing corresponding segments of the recipient chromosome. Transformation is the process whereby DNA extracted from donor bacteria is taken up by recipient bacteria and thereafter becomes integrated. Protoplast fusion has not yet been reported for rhizobia. In this process, cell walls are enzymatically digested, leaving bacteria surrounded by membranes; fusion between these membrane-bound bacteria can be very efficient under the influence of polyethylene glycol and high frequencies of recombination are obtainable (see Hopwood, Wright, Bibb, and Cohen 1977).

Conjugation

This is the method of choice for constructing linkage maps of bacteria. Large fragments of DNA are transferred and it is relatively easy to detect linkage between genes, even if they are not close to one another on the chromosome. In most of the well-documented systems of conjugation in bacteria, self-transmissible extra-chromosomal DNA circles (plasmids) have been shown to be necessary to make bacteria donors of DNA. The classical example is the sex plasmid F which promotes chromosome transfer in *Escherichia coli*. With the exception of the work of Heumann and his colleagues (Heumann 1968; Heumann, Pühler, and Wagner 1971) using a non-nodulating derivative of *R. lupini*, all the chromosome mapping by conjugation in *Rhizobium* has been by R plasmid-mediated conjugation using drug resistance plasmids of the P1 incompatability group which were originally isolated in *Pseudomonas aeruginosa* (see Table 5.1). This is because indigenous sex plasmids had not been detected in *Rhizobium* at the time when genetic mapping started. Thus donors were made by introducing plasmids with known chromosome donor properties from other bacterial genera.

Two R plasmids, RP4 and R68.45, have been used for chromosomal mapping. Both confer tetracycline and kanamycin resistance in *Rhizobium* species. Data for transfer frequencies and chromosome mobilization properties are given in Table 5.1, which shows that R68.45 was more efficient at promoting recombination in *R. leguminosarum* strain 300 than was RP4. R68.45 was selected by Haas and Holloway (1976) for having better chromosome mobilization ability in *P. aeruginosa* than its parent, R68, which has similar properties to RP4. Fortunately R68.45 also showed these good donor properties in *Rhizobium*. However, loss of

TABLE 5.1
Reports of plasmid-mediated gene transfer in Rhizobium

| *Rhizobium* species | Plasmid | Approximate frequency | | Reference[c] |
		Plasmid transfer[a]	Recombination[b]	
R. leguminosarum	RP4	10^{-2}	10^{-7}	1
	R68.45	10^{-2}	10^{-4}	2
R. meliloti	R68.45	10^{-1}	10^{-4}	3
R. meliloti	RP4	10^{-1}	10^{-4}	4
R. meliloti	R68.45			5
R. phaseoli	R68.45	10^{-2}	10^{-4}	
R. trifolii	R68.45	10^{-2}	10^{-4}	6
R. leguminosarum	pRL1JI			
	pRL3JI	10^{-2}	10^{-6}	7
	pRL4JI			

[a]   Plasmid transfer frequency estimated per recipient.

[b]   Reconstruction frequencies estimated per plasmid transferred.

[c]   1, Beringer and Hopwood (1976); 2, Beringer, Hoggan, and Johnston (1978b); 3, Kondorosi *et al.* (1977a); 4, Meade and Signer (1977); 5, Casadesus and Olivares (1979b); 6, Johnston and Beringer (1977); 7, Hirsch (1979).

chromosome donor ability by derivatives carrying R68.45 is quite frequent, and comparisons between different plasmids should take this into consideration.

Also included in Table 5.1 are data for plasmids pRL1JI, pRL3JI, and pRL4JI. These are conjugative plasmids (which transfer at about 10^{-2}) and carry genes for bacteriocin production and which were first detected in three field isolates of *R. leguminosarum,* strains 248, 306, and 309 respectively. While they are not particularly efficient sex plasmids they do show that indigenous *Rhizobium* plasmids are able to promote conjugation.

Fairly detailed circular linkage maps for *R. meliloti* and *R. leguminosarum* have been published and data for interspecific crosses between *R. leguminosarum,* and *R. trifolii,* and *R. phaseoli* indicate that these three species have genetically indistinguishable chromosomes (Johnston and Beringer, 1977). Unfortunately, as with other aspects of genetic research, we know much less about the linkage maps of slow-growing rhizobia. R plasmid transfer has been demonstrated in *R. japonicum* (Kuykendall, 1979), but a linkage map of the *R. japonicum* chromosome will probably not be available in the near future.

Gene transfer in crosses between *R. leguminosarum* and *R. meliloti* does occur, though it appears that recombination of the transferred DNA into the host chromosome is a rare event (Johnston, Setchell, and Beringer 1978b). Johnston *et al.* (1978b) took advantage of this apparent lack of good DNA homology to select R plasmids carrying *R. meliloti* chromosomal DNA (R-primes). Because R68.45 was used to construct the R-primes, they were transmissible to *E. coli* and *P. aeruginosa*; it was found that *R. meliloti* tryptophan genes were able to function in *P. aeruginosa*, but not in *E. coli* (Johnston, Bibb, and Beringer 1978a). A range of R-primes carrying *Rhizobium* DNA is now available (A. W. B. Johnston and A. Kondorosi, personal communication) and these may be used to compare mutations in different species. Such comparisons can be used to 'map' auxotrophic mutations, since all those mutants whose function is restored by the DNA carried on an R-prime must be defective in genes from the corresponding region on the chromosome.

Transformation

Transformation is the most widely reported method of gene transfer for *Rhizobium* species (see reviews by Balassa 1963; Dénarié and Truchet 1976). Surprisingly, therefore, none of the groups involved in constructing the recent genetic maps of fast-growing *Rhizobium* species has utilized transformation for linkage studies. Because transformation has not been utilized with well-studied rhizobia its impact on *Rhizobium* genetic studies has been limited, though its potential for the manipulation of plasmids suggests that it will have an important role in the future.

Transduction

Until 1979 the only reports of generalized transduction in *Rhizobium* were from the laboratory of Kowalski (1967, 1970) or by workers using his phages for studies with *R. meliloti* (Kowalski and Dénarié 1972). Specialized transduction of a cysteine marker by phage 16–3 in *R. meliloti* had also been reported by Svab, Kondorosi, and Orosz (1978). The lack of reports for other species is surprising in view of the report by Kowalski (1970) that many *R. meliloti* phages were capable of generalized transduction.

Recently Buchanan-Wollaston (1979) described generalized transduction in *R. leguminosarum* and reported that the phage could also be used to transduce strains of *R. trifolii*. Casadesus and Olivares (1979a, b) have reported generalized transduction in *R. meliloti* using a newly isolated phage. Both these recently isolated phages have been used for fine-structure mapping of their respective host species. Plasmid-borne genes have also been transferred in *R. leguminosarum* by transduction (Johnston et al. 1978c), and linkage between symbiotically defective mutations and known plasmid-borne genes has been demonstrated (Buchanan-Wollaston, Beringer, Brewin, Hirsch, and Johnston 1980). Thus, unlike transformation, transduction has served a useful role in extending the range of genetic techniques presently available for the analysis of rhizobia.

This discussion of transduction would not be complete without a mention of the genetic analysis of a restricted transducing phage 16–3 of *R. meliloti*, since this was the first report of fine-structure mapping in *Rhizobium*. A partial genetic map of the phage was first reported by Orosz and Sik (1970) and since then this has been extended and there is now a physical-restriction map (Dallmann, Orosz, and Sain 1979).

5.4 *Rhizobium* **plasmids**

There is a certain irony in the fact that, just at the time when methods of gene mobilization by conjugation and transduction were being developed for the *Rhizobium* chromosome, the emphasis of *Rhizobium* genetic research began to switch from chromosomal to extrachromosomal genes.

It has been apparent for some time that strains of a number of *Rhizobium* species contain plasmids. Earlier methods of plasmid isolation, using variations on the 'cleared lysis' procedure, succeeded in isolating plasmids of molecular weight up to 70×10^6 (Sutton 1974; Klein, Jemison, Haak, and Matthysse 1975; Tshitenge, Luyundula, Lurquin, and Ledoux 1975; Zurkowski and Lorkiewicz 1976). More recently, interest has focused on the very large plasmids (molecular weights $> 100 \times 10^6$) which are found in *R. leguminosarum* and *R. trifolii* (Nuti, Ledeboer, Lepidi, and Schilperoort 1977), *R. meliloti* (Casse, Boucher, Julliot, Michel, and Dénarié 1979), *R. japonicum* (Gross, Vidavar, and Klucas 1979) and *R. phaseoli* (J. L. Beynon, personal communication). There have been a number of experimental results which indicate that plasmids play a crucial role as determinants of the symbiotic ability of *Rhizobium*.

A non-nodulating (Nod⁻) derivative of a *R. leguminosarum* strain was found to have lost one of the plasmids present in the Nod⁺ parent (Casse *et al.* 1979). Similarly, loss of nodulating ability in *R. trifolii* was associated with the loss of a large plasmid (Zurkowski and Lorkiewicz 1979). In both these cases the Nod⁻ derivatives were isolated by exposing *Rhizobium* to elevated temperature, a procedure known to select for the loss of certain plasmids in *Agrobacterium tumefaciens* (van Larabeke, Engler, Holsters, van den Ebacker, Zaenen, Schilperoort, and Schell 1974). A non-reverting Nod⁻ mutant of the genetically well-characterized *R. leguminosarum* strain 300 has been isolated. This strain, 6015, has the same number of plasmid bands (three) as its Nod⁺ parent, but the largest plasmid of the Nod⁺ strains is replaced by a substantially smaller one indicating the deletion of plasmid DNA in strain 6015 (P. R. Hirsch, personal communication).

At least some of the genes coding for nitrogenase (*nif* genes) appear to be plasmid-linked in *R. leguminosarum* and *R. trifolii*. Nuti, Lepidi, Prakash, Schilperoort, and Cannon (1979) and Ruvkun and Ausubel (1980) constructed *in vitro* plasmids containing various sections of the *nif* region of *Klebsiella pneumoniae*. These plasmids were radioactively labelled for use as probes and some were found to hybridize specifically with *Rhizobium* plasmid DNA. Interestingly, only some of the structural genes for nitrogenase showed sufficient homology to hybridize; the remaining *K. pneumoniae nif* genes gave no detectable hybridization with *Rhizobium* DNA.

Earlier genetic studies had also pointed to the presence of *Rhizobium nif* genes on plasmids. There have been reports that the ability to fix nitrogen could be transferred by conjugation from *R. trifolii* to other bacterial genera including *Klebsiella*, *Escherichia*, and *Agrobacterium* (Dunican and Tierney 1974; Skotnicki and Rolfe 1978; Stanley and Dunican 1979). The relatively high frequencies of transfer reported are consistent with the *nif* genes being present on a transmissible plasmid. Bishop, Dazzo, Appelbaum, Maier, and Brill (1977) found that DNA isolated from *R. japonicum* could be used to transform *nif* mutants of *Azotobacter vinelandeii* to a Nif$^+$ phenotype. Given the fact that these two genera are not closely related, and hence chromosomal recombination between them might be severely reduced, one way of explaining this result is to postulate that a *Rhizobium* plasmid carrying *nif* genes was transferred.

Genetic studies have also shown that symbiotically important properties can be transferred by conjugation at relatively high frequencies between *Rhizobium* strains, indicating that the determinants of these characters are carried on transmissible plasmids. Higashi (1967) found that the ability to infect clover could be transferred by conjugation from *R. trifolii* (the species that normally nodulates clover) to *R. phaseoli* (which nodulates *Phaseolus* beans but not clover) at frequencies greater than would be expected for the transfer of chromosomal genes.

Some strains of *R. leguminosarum* possess transmissible plasmids. As has been mentioned, three field isolates of this species, all of which produce bacteriocin of medium molecular weight, can transfer bacteriocinogenicity at frequencies of about 1 per cent by conjugation to other *R. leguminosarum* strains (Hirsch 1979). One of these plasmids, pRL1JI, is of particular interest because it appears to carry genes that determine the ability of *Rhizobium* to induce the development of functional nodules on the roots of the host plant.

When pRL1JI was transferred from the field isolate (strain 248) in which it was found to derivatives of *R. leguminosarum* strain 300, it was seen on agarose gels that the transconjugants had acquired an extra band (molecular weight *c.* 120×10^6) which corresponded to one of the bands present in strain 248 (P. R. Hirsch, personal communication). To facilitate the selection of its transfer, the transposon Tn5 was introduced into pRL1JI by the method of Beringer *et al.* (1978*a*) to form the kanamycin-resistant derivative pJB5JI. When this plasmid was transferred to the Nod⁻ strain 6015 (see above) all the transconjugants induced nitrogen-fixing nodules (Johnston *et al.* 1978*c*). In addition, Tn5-marked derivatives of pRL1JI have been transferred to a number of mutants of *R. leguminosarum* strain 300 which induced nodules that did not fix nitrogen (Fix⁻). Five such mutants were used as recipients, and for three of them it was found that all the transconjugants were Fix⁺ (Brewin *et al.* 1980). Thus pRL1JI appears to carry genes that govern nodulation and nitrogen fixation in peas.

In contrast, when the other two bacteriocenogenic plasmids, pRL3JI and pRL4JI, were transferred to strain 6015 or the Fix⁻ mutants that were suppressible by pRL1JI, none of the transconjugants was restored to a Nod⁺ Fix⁻ phenotype. However, in crosses in which strain 6015 was the recipient and the donors were derivatives of strain 300 carrying pRL3JI or pRL4JI, Nod⁺ could be transferred at frequencies of about 10^{-3} per bacteriocinogenic plasmid transfer (Brewin *et al.* 1980). It appears that pRL3JI and pRL4JI can mobilize Nod⁺ genes from strain 300 at relatively high frequency, but they do not themselves carry such genes capable of suppressing the Nod⁻ phenotype of strain 6015.

We have also transferred the Tn5-marked derivative (pJB5JI) of pRL1JI to strains of *R. trifolii* and *R. phaseoli*, two species that are closely related to *R. leguminosarum* (Graham 1964). All the pJB5JI transconjugants tested acquired the ability to nodulate peas and in most cases retained the ability to nodulate clover or *Phaseolus* as appropriate (Johnston *et al.* 1978*c*). However, the majority of interspecific transconjugants did not nodulate peas as effectively as *R. leguminosarum*, nor was the nodulation of clover or *Phaseolus* as good as with *R. trifolii* or *R. phaseoli* respectively.

The nature of the interspecific transconjugants of *R. phaseoli* has been investigated in some detail (J. L. Beynon, unpublished observations). It was found that 98 per cent of the *R. phaseoli* transconjugants were of the type described above (i.e. rather poor at nodulating peas and *Phaseolus*), but that a small minority nodulated peas as effectively as did *R. leguminosarum* and essentially failed to nodulate *Phaseolus*. It was found that the *R. phaseoli* parent contained two large molecular weight ($c.\ 200 \times 10^6$) plasmids. All of the majority class of transconjugants examined differed from the *R. phaseoli* parent by the presence of an extra plasmid band corresponding to pJB5JI. The minority class also acquired the pJB5JI band but had lost the smaller of the resident *R. phaseoli* plasmids. This suggests that this plasmid specifies the ability to nodulate *Phaseolus* and also that when this plasmid and pJB5JI are present in the same cell, the joint presence of two 'host-range' plasmids reduces nodulation ability on both hosts.

A connection between *Phaseolus* nodulation and another phenotype, namely the ability to produce a dark pigment (probably melanin), was also noted. Pigment production has been found in all strains of *R. phaseoli* which have been examined, but in none of our strains of *R. leguminosarum* or *R. trifolii*. In the crosses described above the majority class transconjugants retained the ability to make melanin but the minority class failed to do so. It would appear, therefore, that genes for melanin production are also located on the smaller of the two plasmids in *R. phaseoli* strain 1233.

From the examples discussed in this section, it is clear that *Rhizobium* plasmids are very important as determinants of symbiotic properties and in the near future it may become apparent that other symbiotic functions are located on plasmids. It would, however, be premature to suggest that all symbiotically important genes are plasmid linked in all strains or that all of the large amount of plasmid DNA in *Rhizobium* strains is committed to determining symbiotic functions. In some strains, as much as 20 per cent of the total DNA may be in the form of plasmids. If even the majority of this potential information determines the symbiotic properties of *Rhizobium* this would indeed be a massive investment in the process of nodulation and nitrogen fixation.

5.5 *Rhizobium* genetics and nodule biochemistry

What role can *Rhizobium* genetics play in the understanding and improvement of symbiotic nitrogen fixation? First it can be used as an analytical tool to identify the important biochemical steps involved in the development and functioning of the root nodules. Secondly, it can be used as a breeding method for the construction of new strains of *Rhizobium* of potential agricultural value. It should be emphasized that *Rhizobium* genetics is still in its infancy, and that the current achievements of the approach are modest compared to the potential insights to be gained when the appropriate mutants become available.

Details of the biochemistry and morphology of nodule development are presented in Chapter 10, and we wish here simply to highlight the areas where genetic studies have made or could make a contribution, starting with a discussion of the specificity of infection.

Considerable evidence has accumulated that lectins from *Trifolium* and *Glycine* (which, though normally recovered from the seed, are also to be found on the root hairs of these plants) can bind specifically to strains of *Rhizobium* that nodulate these particular hosts (Bhuvaneswari, Pueppke, and Bauer 1977; Dazzo, Yanke, and Brill 1978). Most of these studies have involved comparisons of lectin-binding ability among field isolates of *Rhizobium* which must differ in many characteristics, and it is disappointing that a detailed genetic analysis of mutant strains that fail to nodulate and/or fail to bind lectins has not been carried out in order to establish the importance of lectin binding for more-or-less isogenic lines. In the case of the exopolysaccharide-deficient (slimeless) *R. leguminosarum* mutant (Sanders, Carlson, and Albersheim 1977) that does not nodulate *Pisum*, the biochemical pertubations induced may have caused a pleiotropic loss of phenotypic characters: perhaps nodulating ability might have been lost because a slimy matrix is required within the infection thread, rather than because exopolysaccharide is required for lectin-mediated binding to the surface of the root hair cell. This example shows that a distinction between causality and correlation may be a fine one as far as the lectin-recognition hypothesis is concerned.

Many auxotrophic mutants have been isolated because of the value of these mutations in linkage analysis and genetic crosses. Generally, auxotrophic mutants still produce effective (Fix$^+$) root nodules, implying that the relevant growth factor can be provided by the host plant for the growth of the infecting rhizobia. However, there are a number of exceptions to this rule. Purine auxotrophs of *R. meliloti* (Scherrer and Dénarié 1971) and *R. leguminosarum* (Pain 1979) are symbiotically defective; in the case of *R. leguminosarum*, the mutants fail to induce nodule formation unless exogenous adenine is supplied, in which case non-fixing nodules are formed.

Leucine auxotrophs of *R. meliloti* also form ineffective (Fix$^-$) nodules unless supplementary leucine is added to the roots (Dénarié *et al.* 1976). These mutants form nodules that differ from normal nodules in that bacteria are not released from infection threads (Truchet, Michel, and Dénarié 1979), a phenotype which has also been observed in a Fix$^-$ (but not auxotrophic) mutant of *R. leguminosarum* (Beringer, Johnston, and Wells 1977). The existence of nodules of this type suggests that nodule morphogenesis does not require the complete differentiation of the bacteroid form. In other words it is likely that, even when the rhizobia are confined within infection threads, they are capable of inducing and maintaining the normal meristematic activity and differentiation of neighbouring plant cells within the nodule, presumably by releasing a diffusible morphogen (Dénarié and Truchet 1979). The most interesting feature observed in the ineffective nodules induced by leucine auxotrophs of *R. meliloti* is cytogenetic (Truchet *et al.* 1980): the infected plant cells of the nodule are diploid unlike the polyploid cells that are normally found in effective nodules. Moreover, when exogenous leucine is supplied to these ineffective nodules, the bacteria are released from infection threads, and the infected plant cells become polyploid. Thus the endoreduplication of nodule cells apparently requires the intracellular presence of developing bacteroids and not merely the presence of rhizobia within an infection thread.

Many *Rhizobium* mutants resistant to antibiotics have been isolated and sometimes these mutations have been shown to affect symbiotic performance (reviewed in Dénarié *et al.* 1976). More recently a class of drug-resistant mutants of *R. trifolii* have been isolated, which were uncoupled in oxidative phosphorylation (Unc⁻). Following a neomycin-resistant selection technique which had been developed in *E. coli*, Skotnicki and Rolfe (1979) obtained several such mutants but, perhaps surprisingly, none of them was symbiotically defective. Thus it is possible that an entirely different adenosine triphosphatase complex is present in bacteroids from that present in the free-living bacterium and defective in the Unc⁻ mutants.

Several workers have attempted to use *Rhizobium* mutants to analyse carbohydrate oxidation pathways, both in free-living bacteria and indirectly within the nodule. Mutants of *R. meliloti* altered in β-galactosidase (Niel, Guillaume, and Bechet 1977), and phospho-glucoisomerase (Arias, Cervenansky, Gardiol, and Martinez-de Drets 1979) and oxo-2-glutarate dehydrogenase (Duncan and Fraenkel 1979) have been isolated and studied recently. Ronson and Primrose (1979) obtained mutants of *R. trifolii* unable to utilize glucose, fructose, or sucrose, but all were symbiotically proficient; they concluded that tricarboxylic acids are the probable carbon source for bacteroids, but this is an over-interpretation of the data. If a number of alternative carbon sources or assimilatory pathways can be utilized by bacteroids, no single mutation in any of them would be expected to result in a Fix⁻ phenotype.

In addition to the supply of photosynthate to nitrogen-fixing bacteroids, other aspects of energy metabolism have also been studied by genetic means. A riboflavin-requiring auxotroph of *R. trifolii* has been obtained (Pankhurst, Schwinghamer, Thorne, and Bergersen 1974) and shown to form Fix⁻ nodules on *Trifolium pratense* and two cultivars of *T. subterraneum*, unless supplementary riboflavin or flavine nucleotides were supplied to the roots. Fix⁺ nodules were found to contain twenty times as much flavine, and Fix⁻ nodules four times as much flavine, as non-nodulated root tissue, mainly in the form of flavine adenine dinucleotide or flavine adenine mononucleotide.

Up to 25 per cent of the energy flux through nitrogenase can be dissipated in the evolution of hydrogen gas (see Evans, Lepo, and Emerich, Vol. 3): some strains of *Rhizobium* possess an uptake hydrogenase that reoxidizes this hydrogen, thereby recovering some of the lost energy and contributing to the overall efficiency of the nitrogen fixation reaction. Recent debate on the importance of a hydrogen uptake (Hup) system has been assisted by the selection of a *hup* mutant of *R. japonicum* (Maier, Postgate, and Evans 1978). Similarly the identification of a mutant defective in nitrogenase component II (Maier and Brill 1976) has shown that induction of synthesis of nodule leghaemoglobin does not require the presence of a functional nitrogenase. Another membrane-bound enzyme that is frequently present in bacteroids is a respiratory nitrate reductase. This enzyme is apparently able to supply adenosine triphosphate for nitrogen fixation (Rigaud, Bergersen, Turner, and Daniel 1973), although its role cannot be essential because chlorate-resistant *Rhizobium* mutants (which lack a functional nitrate reductase) form nodules that fix nitrogen at normal rates (Gibson and Pagan 1977; Kiss, Vincze, Kálmán, Forrai, and Kondorosi 1979).

The initial product of nitrogenase is undoubtedly ammonia, but the subsequent interconversions of fixed nitrogen within nodules have been difficult to determine because a number of possible catabolic pathways exist both within the plant and within the nodules; thus the bulk of the fixed nitrogen obtained from the sap of a wide variety of legumes is in the form of amides, ureides, and organic acids (Pate 1977; Atkins, Herridge, and Pate 1978). The principle pathways for the assimilation of ammonia both in plants and bacteria are as follows (Brill 1975).

$$\text{I.} \quad \text{Oxo-2-glutarate} + NH^+_4 + NAD(P)H \xrightarrow{\text{GDH}} \text{glutamate} + NAD(P)^+ + H_2O$$

$$\text{II.} \quad \text{Glutamate} + NH_3 + ATP \xrightarrow{\text{GS}} \text{Glutamine} + ADP + Pi$$

$$\text{Glutamine} + \text{Oxo-2-glutarate} + NAD(P)H + H^+ \xrightarrow{\text{GOGAT}} 2\,\text{Glutamate} + NAD(P)^+$$

In free-living *Rhizobium* only route II, via glutamine synthetase (GS) and glutamate synthase (GOGAT), is involved in ammonium assimilation, and glutamine dehydrogenase (GDH) is thought to have a catabolic role (Kondorosi, Svab, Kiss, and Dixon 1977*b*; Ludwig 1978). Within the bacteroids of the root nodule, the activity of glutamate dehydrogenase is undetectable, and even the activity of the GS/GOGAT pathway becomes very much reduced (Brown and Dilworth 1975; Planqué, de Vries, and Kijne 1978), while the corresponding activities in the plant-cell fraction become enhanced during nodule development. These observations suggest that ammonia is excreted by bacteroids and subsequent assimilation proceeds within the host plant. Interestingly, when slow-growing strains of *Rhizobium* are induced to fix nitrogen *ex-planta*, about 95 per cent of the fixed nitrogen is excreted into the medium (O'Gara and Shanmugam 1976).

Genetic studies have indicated that the assimilation of nitrogen may be more complicated than the known biochemistry would suggest. The fact that mutants auxotrophic for glutamine form ineffective nodules suggests a possible role for glutamine synthetase in the control of nitrogen metabolism (Kondorosi, Kiss, Forrai, Vincze, and Banfalvi 1977*b*; Ludwig and Signer 1977). It is possible that the supply of glutamate from the plant cell to the bacteroid may exert a controlling influence on bacteroid metabolism (Shanmugam, O'Gara, Andersen, and Valentine 1978). Certainly the plant is capable of supplying glutamate for protein synthesis by *Rhizobium* within the nodule, because glutamate-requiring auxotrophs form effective nodules (Kondorosi *et al.* 1977*b*) and apparently the development of *ex-planta* nitrogen fixation requires a supply of fixed nitrogen, preferably in the form of glutamate (Ludwig 1978).

5.6 Concluding remarks

Studies on the genetics and molecular biology of *Rhizobium* have now reached a stage where many of the tools available for the genetic dissection of the bacterial contribution to the symbiosis have been assembled. There are methods for long-range and fine-scale chromosome mapping in a number of species. The importance of plasmids has been repeatedly demonstrated and methods of mapping these are available. Workers in a number of laboratories are accumulating symbiotically-defective mutants of various types and the analysis of these, combining genetic, biochemical, and morphological approaches, should tell us much about the various changes that *Rhizobium* induces both in the plant and in itself during its transformation from a humble soil microbe to its exalted position in the legume root nodules.

References

Arias, A., Cervenansky, C., Gardiol, A., and Martinez-de Drets, G. (1979). *J. Bacteriol.* **137,** 409.

Atkins, C. A., Herridge, D. F., and Pate, J. S. (1978). In *Isotopes in biological dinitrogen fixation*, p. 211. International Atomic Energy Agency, Vienna.

Balassa, G. (1963). *Bacteriol. Rev.* **27,** 228.

Beringer, J. E. (1974). *J. gen. Microbiol.* **84,** 188.

—— (1976). In *Proceedings of the first international symposium on nitrogen fixation*, Vol. 2. (ed. W. E. Newton and C. J. Nyman). p. 358. Washington State University Press, Pullman.

—— (1980). *J. gen. Microbiol.* **116,** 1.

—— and Hopwood, D. A. (1976). *Nature, Lond.* **264,** 291.

—— Beynon, J. L., Buchanan-Wollaston, A. V., and Johnston, A. W. B. (1978*a*). *Nature, Lond.* **276,** 633.

—— Brewin, N., Johnston, A. W. B., Schulman, H. M., and Hopwood, D. A. (1979). *Proc. R. Soc.,* **B204,** 219.

180   Genetics

—— Hoggan, S. A., and Johnston, A. W. B. (1978*b*). *J. gen. Microbiol.* **104,** 201.

—— Johnston, A. W. B., and Wells. B. (1977). *J. gen. Microbiol.* **98,** 339.

Bhuvaneswari, T. V., Pueppke, S. G., and Bauer, W. D. (1977). *Pl. Physiol.* **60,** 486.

Bishop, P. E., Dazzo, F. B., Appelbaum, E. R., Maier, R. J., and Brill, W. J. (1976). *Science N.Y.* **198,** 938.

Brewin, N. J., Beringer, J. E., Buchanan-Wollaston, A. V., Johnston, A. W. B., and Hirsch, P. R. (1980). *J. gen. Microbiol.* **116,** 261.

Brill, W. J. (1975). *Annu. Rev. Microbiol.* **29,** 109.

Brown, C. M. and Dilworth, M. J. (1975). *J. gen. Microbiol.* **86,** 39.

Buchanan-Wollaston, V. (1979). *J. gen. Microbiol.* **112,** 135.

—— Beringer, J. E., Brewin, N., Hirsch, P. R., and Johnston, A. W. B. (1980). *Molec. gen. Genet.* **178,** 185–90.

Casadesus, J. and Olivares, J. (1979*a*). *J. Bacteriol.* **139,** 316.

—— —— (1979*b*). *Molec. gen. Genet.* **174,** 203.

Casse, F., Boucher, C., Julliot, J. S., Michel, M., and Dénarié, J. (1979). *J. gen. Microbiol.* **113,** 229.

Cunningham, D. A. (1980). Genetic studies on the nitrogen fixing bacterium *Rhizobium trifolii*. Ph.D. Thesis. University of Edinburgh.

Dallmann, G., Orosz, L., and Sain, B. (1979). *Molec. gen. Genet.* **176,** 439.

Dazzo, F. B., Yanke, W. E., and Brill, W. J. (1978). *Biochim. biophys. Acta* **539,** 276.

Dénarié, J. and Truchet, G. (1976). In *Proceedings of the first international symposium on nitrogen fixation*, Vol. 2. (ed. W. E. Newton and C. J. Nyman), p. 343. Washington State University Press, Pullman.

—— —— (1979). *Physiol. Vég.* **17,** 643.

—— —— and Bergeron, B. (1976). In *Symbiotic nitrogen fixation in plants* (ed. P. S. Nutman), p. 77. Cambridge University Press.

Duncan, M. J. and Fraenkel, D. G. (1979). *J. Bacteriol.* **137,** 415.

Dunican, L. K. and Tierney, A. B. (1974). *Biochem. Biophys. Res. Commun.* **57,** 62.

Gibson, A. H. and Pagan, J. D. (1977). *Planta* **134,** 17.

Graham, P. H. (1964). *J. gen. Microbiol.* **35,** 511.

Gross, D. C., Vidavar, A. K., and Klucas, R. V. (1979). *J. gen. Microbiol.* **114,** 257.

Haas, D. and Holloway, B. W. (1976). *Molec. gen. Genet.* **144,** 234.

Heumann, W. (1968). *Molec. gen. Genet.* **102,** 132.

—— Pühler, A., and Wagner, E. (1971). *Molec. gen. Genet.* **113,** 308.

Higashi, S. (1967). *J. appl. Microbiol.* **13,** 391.

Hirsch, P. R. (1979). *J. gen. Microbiol.* **113,** 219.

Hopwood, D. A., Wright, H. M., Bibb, M. J., and Cohen, S. N. (1977). *Nature, Lond.* **268,** 171.

Johnston, A. W. B. and Beringer, J. E. (1977). *Nature, Lond.* **267,** 611.

—— Bibb, M. J., and Beringer, J. E. (1978*a*). *Molec. gen. Genet.* **165,** 323.

—— Setchell, S. M., and Beringer, J. E. (1978*b*). *J. gen. Microbiol.* **104,** 209.

—— Beynon, J. L., Buchanan-Wollaston, A. V., Setchell, S. M., Hirsch, P. R., and Beringer, J. E. (1978*c*). *Nature, Lond.* **276,** 634.

Kiss, G. B., Vincze, E., Kálmán, Z., Forrai, T., and Kondorosi, A. (1979). *J. gen. Microbiol.* **113,** 105.

Kleckner, N., Roth, J., and Botstein, D. (1977). *J. Molec. Biol.* **116,** 125.

Klein, G. E., Jemison, P., Haak, R. A., and Matthysse, A. G. (1975). *Experientia* **31,** 532.

Kondorosi, A., Kiss, G. B., Forrai, T., Vincze, E., and Banfalvi, Z. (1977*a*). *Nature, Lond.* **268,** 525.

—— Svab, Z., Kiss, G. B., and Dixon, R. A. (1977*b*). *Molec. gen. Genet.* **151,** 221.

Kowalski, M. (1967). *Acta. microbiol. polon.* **16,** 7.

—— (1970). *Acta microbiol. polon., ser. A* **2,** 109.

—— and Dénarié, J. (1972). *C. R. Acad. Sci. Paris, Série D* **275,** 141.

Kuykendall, L. D. (1979). *Appl. environ. Microbiol.* **37,** 862.

van Larabeke, N., Engler, G., Holsters, M., van den Ebacker, S., Zaenen, I., Schilperoort, R. A., and Schell, J. (1974). *Nature, Lond.* **252, 169.**

Ludwig, R. A. (1978). *J. Bacteriol.* **135,** 114–23.

—— and Signer, E. R. (1977). *Nature, Lond.* **267,** 245.

Maier, R. J. and Brill, W. J. (1976). *J. Bacteriol.* **127,** 763.

Maier, R. S., Postgate, J. R., and Evans, H. S. (1978). *Nature, Lond.* **276,** 494.

Meade, H. M. and Signer, E. R. (1977). *Proc. natn. Acad. Sci. U.S.A.* **74,** 2076.

Niel, C., Guillaume, J. B., and Bechet, M. (1977). *Can. J. Microbiol.* **23,** 1178.

Nuti, M. P., Ledeboer, A. M., Lepidi, A. A., and Schilperoort, R. A. (1977). *J. gen. Microbiol.* **100,** 241.

—— Lepidi, A. A., Prakash, R. K., Schilperoort, R. A., and Cannon, F. C. (1979). *Nature, Lond.* **282,** 533.

O'Gara, F. and Shanmugam, K. T. (1976). *Biochim. biophys. Acta* **437,** 313.

Orosz, L. and Sik, T. (1970). *Acta microbiol. Acad. Sci. Hung.* **17,** 185.

Pain, A. N. (1979). *J. appl. Bacteriol.* **47,** 53.

Pankhurst, C. E., Schwinghamer, E. A., Thorne, S. W., and Bergersen, F. J. (1974). *Plant Physiol.* **53,** 198.

Pate, J. S. (1977). In *A treatise on dinitrogen fixation*, Section III (ed. R. W. F. Hardy and W. S. Silver), p. 473. John Wiley, New York.

Planqué, K., De Vries, G. E., and Kijne, J. W. (1978). *J. gen. Microbiol.* **106,** 173.

Rigaud, J., Bergersen, F. J., Turner, G. L., and Daniel, R. M. (1973). *J. gen. Microbiol.* **77,** 137.

Ronson, C. W. and Primrose, S. B. (1979). *J. gen. Microbiol.* **112,** 77.

Ruvkin, G. B. and Ausubel, F. M. (1980). *Proc. natn. Acac. Sci. U.S.A.* **77,** 191.

Sanders, R. E., Carlson, R. W., and Albersheim, P. (1977). *Nature, Lond.* **271, 241.**

Scherrer, A. and Dénarié, J. (1971). *Pl. Soil* Special Volume 39.

Schwinghamer, E. A. (1962). *Am. J. Bot.* **49,** 269.

—— (1977). In *A treatise on dinitrogen fixation*, Section III (ed. R. W. F. Hardy and W. S. Silver), p. 577. John Wiley, New York.

Shanmugam, K. T., O'Gara, F., Andersen, K., and Valentine, R. C. (1978). *A. Rev. Plant Physiol.* **29,** 263.

Skotnicki, M. L. and Rolfe, B. G. (1978). *J. Bacteriol.* **133,** 518.

—— —— (1979). *Aust. J. biol. Sci.* **32,** 501.

Stanley, J. and Dunican, L. K. (1979). *Molec. gen. Genet.* **174,** 211.

Sutton, W. D. (1974). *Biochim. biophys. Acta* **366,** 1.

Svab, A., Kondorosi, A., and Orosz, L. (1978). *J. gen. Microbiol.* **106,** 321.

Truchet, G., Michel, M., and Dénarié, J. (1980). *Differentiation.* **16,** 163–72.

Tshitenge, G., Luyundula, N., Lurquin, P. F., and Ledoux, L. (1975). *Biochim. biophys. Acta* **414,** 357.

Zurkowski, W. and Lorkiewicz, Z. (1976). *J. Bacteriol.* **128,** 481.

—— —— (1979). *Arch. Microbiol.* **123,** 195.

6 Ultrastructure of the free-living cell

H.-C. TSIEN

6.1 Introduction

In the past twenty years, a considerable body of information on the ultrastructure of rhizobia has been published. Most of this information, however, is related to the study of rhizobial bacteroids and root nodule morphogenesis. To the author's knowledge, no general review on the ultrastructure of free-living rhizobial cells exists. It is therefore, appropriate to review the current knowledge on this topic. A survey of published information, together with unpublished results from our laboratories, are included in this chapter.

6.2 Methods for studying ultrastructure

In recent years, both electron optics and electron microscopic techniques for examining the cell structure and its constituent macromolecules have improved greatly. The chemistry involved in each step of specimen preparation is now well understood. Though structural artifacts have not been totally eliminated, critical interpretation of results is possible. During the preparation of biological specimens, fixation by chemicals is most critical and great care must be exercised.

Detailed methodology for the ultrastructural study of prokaryotes can be found in many papers and monographs (Hayat, 1970, 1972, 1977). Some frequently used techniques are listed below, together with their application to the study of *Rhizobium*.

1. Thin sectioning: This technique is most often used in examining fine structure and organization of cell components (Vincent, Humphrey, and North 1962; Tsien and Schmidt 1977).

2. Freeze fracturing and freeze etching: This is an invaluable technique employing cryofixation rather than chemical fixatives. With the application of freeze fracturing and etching in combination, structures were revealed for the first time; knowledge of the molecular architecture of bilayered membranes and cell-surface structures never before

observed are the result of utilizing this technique (Mackenzie, Vail, and Jordan 1973; Tsien and Schmidt 1977).

3. Negative staining: This is an extremely simple technique which gives very high resolution. It is most suitable for studying isolated subcellular components, such as flagella and fimbriae. A novel type of flagella from *Rhizobium lupini* was observed by using this technique (Schmitt, Bamberger, Acker and Mayer, 1977).

4. Shadow casting: This technique is very useful in demonstrating certain cell-surface structures and extracellular components, such as cellulose microfibrils produced by *Rhizobium trifolii* (Napoli, Dazzo, and Hubbell 1975).

5. Scanning electron microscopy: Thus far, scanning electron microscopy has been underutilized in *Rhizobium* studies. It should be of great value in demonstrating how the rhizobial cell adheres and attaches itself to the root-hair surface.

6. Immunoelectron microscopy: This is a very useful technique in studying the molecular architecture and the structural–functional relationships of the cell surface. It has been used for the demonstration of *Glycine max* lectin-binding receptors on the surface of *Rhizobium japonicum* (Calvert, LaLonde, Bhuvaneswari, and Bauer 1978; Tsien and Schmidt 1980).

6.3 General morphology

Free-living cells of all *Rhizobium* species are typically rod-shaped, Gram-negative bacteria of 0.5–0.8 μm in diameter and 1.2–4 μm in length. The size and shape of cells vary slightly depending on cultural conditions and the stage of growth. Long cells are frequently observed in a stationary phase culture as are some pleomorphic forms. Pleomorphic cells are either Y-shaped or swollen at one end, and are similar to the rhizobial bacteroids found in legume root-nodules infected with fast-growing *Rhizobium* species. This cellular distortion is frequently observed where (1) free-living rhizobia are cultured in a medium designed for the induction of nitrogen-fixing cells, (2) the oxygen supply is limited, or (3) where there is a high concentration of yeast extract in the medium (Jordan and Coulter 1965; Skinner, Roughley, and Chandler, 1977; Pankhurst and Craig 1978; Urban 1979).

Most cells in the exponential growth phase are motile. Subpolar or peritrichous arrangements of flagella have been reported (Leifson and Erdman 1958; De Ley and Rassel, 1965; Abdel-Ghaffar and Jensen 1966).

Reserve polymers identified as poly-β-hydroxybutyrate (PHB) and polyphosphate were reported in cultured rhizobial cells (Tsien and

Schmidt 1977). Granules of PHB, with their high reflective index and its lipophilic nature, are now easily recognized in a phase microscope, but were, at one time, erroneously identified as endospores. It is clear that no endospore formation occurs in rhizobia at any stage or condition of growth. Metachromatic granules which have been identified as polyphosphate have been observed in many species.

The production of extracellular polysaccharides (ECPS) in the form of capsule or slime is widely spread among rhizobial species (Dudman 1968). The participation of ECPS in the plant–microbial interaction is a most interesting topic in the structural–functional study of *Rhizobium* (Schmidt 1979).

6.4 Cell envelope

As Gram-negative bacteria, the cell envelope of rhizobia consists of two bilayered membranes, the outer membrane, or the cell wall, and the inner cytoplasmic membrane (Fig. 6.1). A periplasmic zone is located between the two membranes. In addition, a peptidoglycan–murin layer located immediately inside the outer membrane is a part of the cell envelope.

Outer membrane

The outer membrane of rhizobia as observed in thin sections is composed of a double-track structure made up of two electron-dense layers, which are separated by a light zone (Fig. 6.1). The thickness of this membrane as reported by Vincent *et al.* (1962) in a study of *R. trifolii* is 9 nm, consisting of two 3 nm dark lines separated by a 3 nm space. The outer membrane of *R. japonicum*, fixed with glutaraldehyde and osmium tetroxide, measures 12–15 nm in thickness and is composed of two unequal layers: a thin outer layer and a thick inner layer. The thick inner layer results from the close association of the inner layer of the bilayered outer membrane and a layer of peptidoglycan–murin complex (Fig. 6.1; Tsien and Higgins 1974). Observation of a separate peptidoglycan–murin layer from the outer membrane was partially achieved by treating cells with EDTA solution (Tsien and Schmidt, unpublished observations).

The cross-fractured outer membrane revealed by freeze fracturing also shows a two-track structure measuring approximately 15 nm across (Fig. 6.2(A), arrowhead). This agrees with findings produced by thin-section preparation. Occasionally, a convex fracture face of the outer membrane is observed (Fig. 6.2(B), arrowhead). In *R. japonicum*, the convex fracture face appears to be smooth and without detailed fine structure, although the same structure has been reported to be rough in *R. meliloti* (MacKenzie *et al.* 1973). The formation of a convex fracture

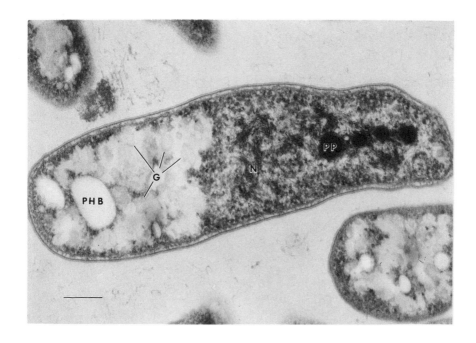

FIG. 6.1. A thin-sectioned cell of *Rhizobium japonicum* strain USDA 138. Cells from a late exponential phase culture was fixed with glutaraldehyde and osmium tetroxide and embedded in Epon 812. Thin sections were stained with uranyl acetate and lead citrate. The cell envelope is composed of two bilayered membrane: outer membrane and cytoplasmic membrane. Fibrous nuclear DNA, poly-β-hydroxybutyrate, glycogen, and polyphosphate are indicated by letters N, PHB, G, and PP respectively. Bars = 0.2 μm.

face as a small shelf was also observed in a cowpea *Rhizobium*, but no detailed structural feature was given (Van Brussel, Costerton and Child 1979).

The concave fracture face of the outer membrane has not been obtained in freeze-fractured whole cells, probably due to its being fractured away from the sample block. Nevertheless, a concave fracture face, appearing to be rough and particulate, has been obtained in an isolated outer membrane fragment (Fig. 6.2(C), arrowhead; Tsien and Schmidt, unpublished observation). The rough and particulate appearance of the concave fracture face of rhizobial outer membrane is in agreement with the outer membrane structure observed in other Gram-negative bacteria (Nanninga 1970). The surface of rhizobia

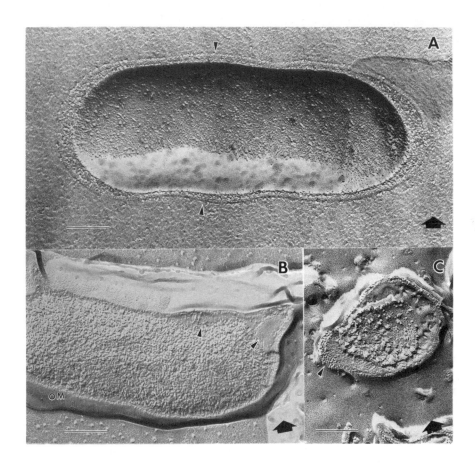

FIG. 6.2. Freeze fractures of *R. japonicum* cells revealing the fine structure of its outer membrane. (A) A cross fractured cell, illustrating a bilayered outer membrane (arrowhead). (B) A surface fractured cell illustrating a portion of convex fracture face of the outer membrane (arrowhead). OM indicates the surface of outer membrane. (C) A fractured outer membrane fragment shows the particulate nature of concave fracture face (arrowhead). Arrows in lower right-hand corner of all freeze fractured electron micrographs, and the shadow casted micrograph, (Fig. 8A) indicate the direction of shadowing. Bars = 0.2 μm.

revealed by etching shows no fine structural detail—only a smooth face (Figs. 6.2 and 3, OM).

The difficulty in obtaining a fracture face through the outer membrane is probably related to its chemical composition and molecular architecture. In Gram-negative bacteria, the hydrocarbon moieties make up only 25–40 per cent of phospholipids in the outer membrane, within which, the molecules of lipopolysaccharides are distributed exclusively in the outer layer (Costerton, Ingram, and Cheng 1974; Braun 1978). The distribution of structural proteins may also contribute to obtaining a fracture plane through the outer membrane. Outer membrane proteins in Gram-negative bacteria are known to be distributed largely in the outer layer (Braun, 1978). The structural analysis of the rhizobial outer membrane is so far incomplete, and considerable additional study will be needed in order to achieve a satisfactory understanding of how this membrane is built.

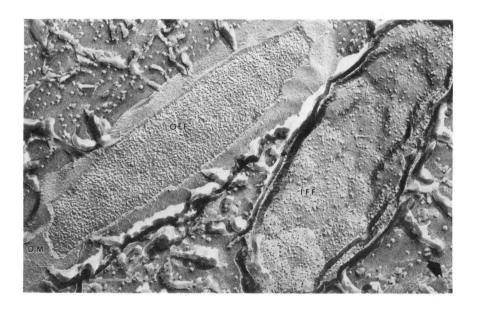

Fig. 6.3. Freeze fractured *R. japonicum* cells showing fracture faces of cytoplasmic membrane. Dense membrane particles are evenly distributed on the outer fracture face (OFF, or convex fracture face). A few particles are also located on the inner fracture face (IFF, or concave fracture face). OM indicates the surface of outer membrane. Bars = 0.2μm.

Cytoplasmic membrane

The cytoplasmic membrane of rhizobia, as revealed by axial sectioning of cells, is a tripartite structure consisting of two dense layers 3 nm in width and 2–3 nm apart (Fig. 6.1). The inner layer appears to be closely associated with cytoplasmic materials. The bilayer nature of the cytoplasmic membrane can be easily demonstrated by examining plasmolysed cells, physically damaged cells, or membrane fragments.

By using the technique of freeze fracturing, two major membrane fracture faces, the outer fracture face of the inner layer (OFF or convex fracture face) and the inner fracture face of outer layer (IFF or concave fracture face), are revealed (Fig. 6.3; Tsien and Schmidt, 1977). On OFF (Fig. 6.3), a fairly homogeneous distribution of particles ranging in diameter from 7 to 13 nm, was observed. These particles have been demonstrated in other bacteria to be proteinaceous in nature (Tsien and Higgins 1974). No report has been published on the exact dimension of rhizobial membrane particles or the size distribution of these particles. MacKenzie *et al.* (1973), in their study on the free-living and bacteroid forms of *R. meliloti*, reported that the membrane particles of bacteroids were smaller than those of free-living cells and did not completely cover the OFF; however, no dimensions were given for the particles of free-living rod form. Particles were also observed sparsely on IFF (Fig. 6.3). So far, no complimentary freeze fracture study has been done for rhizobial cells, so the information concerning the partition of membrane particles and the complementary properties of the two fracture faces is not available.

Periplasmic zone

The space between the outer membrane and the cytoplasmic membrane is commonly known as the periplasmic zone, in which numerous enzymes are located. These enzymes are largely hydrolytic in nature. Their periplasmic localization provides these enzymes with easy access to substrates which are too large to penetrate through the cytoplasmic membrane, and also retains the enzymes within the cell boundaries.

In a thin-sectioned rhizobial cell, the periplasmic zone is readily revealed as the space between two membranes (Fig. 6.1). Materials with medium electron density, which appear to be proteinaceous, are attached to either the cytoplasmic or outer membranes, or form bridges cross-linking two membranes. This zone is sometimes enlarged at one or both ends of the cell (Vincent *et al.* 1962). The periplasmic zone can also be observed in a freeze-fractured cell of *R. japonicum* (Fig. 6.2). Polar enlargement at both ends of the cell is apparent. No fracture plane along the periplasmic zone, however, has been observed.

The occurrence of periplasmic enzymes has been well documented in other Gram-negative bacteria (Costerton *et al.* 1974). In *R. leguminosarum*, Glenn and Dilworth (1979) demonstrated fourteen periplasmic proteins, among which alkaline phosphatase, cyclic phosphodiesterase, and inorganic pyrophosphatase were found. The role of these hydrolytic enzymes in the rhizobial infection process is a topic worthy of further detailed study.

6.5 Cytoplasm and inclusion polymers

The nuclear (DNA) material, which has been known as a chromatin body due to its Feulgen-positive reaction in light microscopy, is located in the central region of the cell. It is revealed in thin-sectioned rhizobia (Fig. 6.1) as fine fibrils clustered together to form fibrous bundles which are orderly arranged. The fibrils measured between 2 to 3 nm in diameter. In axially sectioned cells several bundles have been observed to aggregate either in one or two regions, possibly reflecting cells in different dividing stages. Condensed nucleoid has also been observed in an unfixed, freeze-fractured cell (Fig. 6.4, N) in which nuclear DNA also appears to be fibrous.

Ribosomes, revealed in the thin sectioning as 12–15 nm electron-dense particles, are located primarily in the periphery of the cytoplasm surrounding the nuclear fibrils (Fig. 6.1). A few of these particles have also been observed near the cytoplasmic membrane in the half of the cell with low electron density.

A polarity in the distribution of cytoplasmic materials has been reported in cultured *R. japonicum* cells (Tsien and Schmidt 1977). The nucleoplasm with its surrounding ribosomes is located at one end of the cell, while some granular materials are located at the other end (Figs. 6.1 and 6.4). This cytoplasmic polarity has not been reported for other rhizobial species.

In thin-sectioned cells (Fig. 6.1), the material of low electron density at one end of the cell consists of two morphologically distinct granules: numerous granules of 50–80 nm in diameter and a few granules of different sizes. The 50–80 nm granules, which appear to have low electron density, have been identified by cytochemical-staining technique as glycogen (G) (Tsien and Schmidt 1977). Glycogen granules in the same size-range have been observed in freeze-fractured cells (Fig. 6.4). The granules of irregular size can be easily identified as poly-β-hydroxybutyrate (PHB) by virtue of their characteristic morphological appearance in both thin-sectioned and freeze-fractured preparations. In thin sectioning (Fig. 6.1), PHB granules appear to be electron transparent and surrounded by a single layered membrane, while in freeze fracturing (Fig. 6.4) these granules form conical shaped bundles of elastic fibres stretching out from the fracture plane and heavily

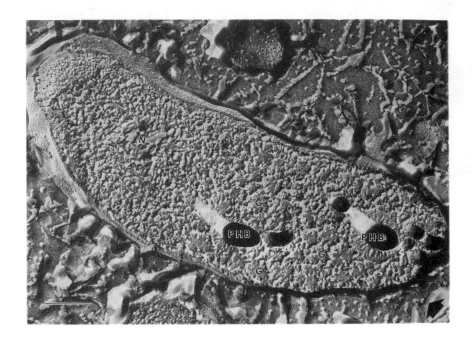

Fig. 6.4. A cross fractured *Rhizobium japonicum* cell. Inclusion polymers, including the granules of poly-β-hydroxybutyrate (PHB) and glycogen (G), are polarly distributed. Fibrous nuclear plasm (N) with its surrounding ribosomes is located at one end of the cell. Bars = 0.2 μm.

shadowed at the tip (Tsien and Schmidt 1977). The single membrane surrounding the PHB granule can also be seen in the freeze fractured preparations. Besides PHB and glycogen, several polyphosphate (PP) granules located in the nucleoplasm and darkly stained, can be observed with regularity in rhizobia (Fig. 6.1).

The occurrence of three inclusion polymers has been reported for both cultured cells and bacteroids (Craig and Williamson 1972). PHB granules are observed consistently in rhizobial bacteroids, however, they are found in free-living cells only at the late exponential growth phase and the stationary phase. Both polyphosphate and glycogen granules are present throughout entire growth phases. Aside from a probable role as a food reserve, it seems that there is little direct connection between their existence and the nitrogen fixation capability of rhizobial cells (Wong and Evans 1971).

6.6 Surface associated structures

Capsule and slime

The term capsule or microcapsule refers to a material external to the outermost part of the cell surface, made up of polysaccharides, proteins, or both, and which is characterized by a definite boundary. The term slime is generally used to refer to exopolysaccharide (ECPS) materials, which upon their synthesis, either diffuse away from the cell surface or remain to be attached to the cell surface, without a defined shape and boundary. The molecules of capsular ECPS are usually very large and those of slimy ECPS are smaller and heterogeneous in size. Both terms have been used to describe the ECPS.

By light microscopy, wet-film india ink methods, Dudman (1968), in a study of the capsulation of *Rhizobium* species, reported that strains of *R. trifolii*, *R. leguminosarum*, *R. lupini*, and *R. phaseoli* produce various amounts of capsulated cells, whereas strains of *R. japonicum*, cowpea *Rhizobium*, and *R. meliloti* produced no capsulated cell, but mucoid, gummy cultures. It has been demonstrated that the accumulation of the capsular ECPS depends upon the growth phase of the culture, the medium composition, and the strain used (Dudman 1964; Tsien and Schmidt 1980). Cultures of *R. japonicum* strain USDA 138 and *R. phaseoli* strain Nitragin 1038 became very viscous in the late exponential growth phase and cells were often clumped together in large aggregates. The encapsulation of cells in these cultures can be easily demonstrated by negative-staining electron-microscopy as low electron-dense amorphous materials surrounding cells (Fig. 6.8(B) and (C)). The cells of *R. japonicum* are partially covered by these materials and the cells of *R. phaseoli* are completely surrounded by materials of fibrous appearance, extending several microns from the cell surface.

Due to the extremely unstable nature of capsular ECPS, its visualization by thin-section electron microscopy is very difficult. Therefore, it is necessary to protect the polymer by complexing it with a cationic reagent, ruthenium red. Without fixing with ruthenium red, the ECPS, consisting of 99 per cent water (Sutherland 1977), is easily washed away or deformed upon dehydration during preparation for thin sectioning. Cells of *R. japonicum* fixed with ruthenium red show many electron-dense materials deposited on the cell surface (Fig. 6.5(A)). The capsule of *R. trifolii* was also demonstrated by complexing with ruthenium red resulting in cells evenly covered with ruthenium red positive materials of about 1 μm in thickness (Dazzo and Brill 1979).

Functionally, the rhizobial ECPS binds specifically with lectin isolated from specific legume host plant (Calvert *et al.* 1978; Dazzo and Brill 1979; Kamberger 1979; Tsien and Schmidt 1981). The interaction of the lectin and the rhizobial ECPS has been proposed as a basis for the

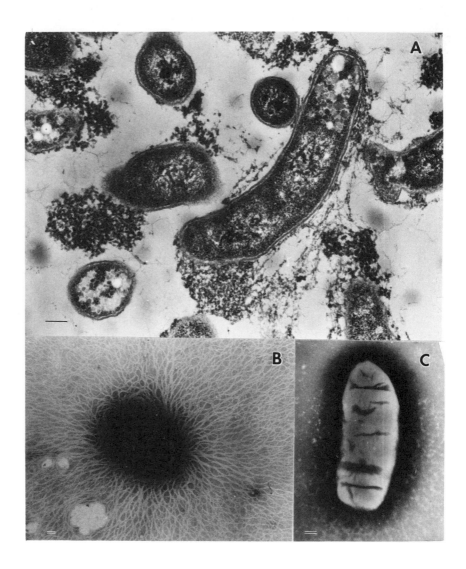

FIG. 6.5. (A) A ruthenium red stained cell of *Rhizobium japonicum* showing extracellular polysaccharide (ECPS). Ruthenium-red-positive materials are either attached at one end of the cell or distributed randomly about the cell. (B) and (C) Negatively stained rhizobial cells showing surrounding ECPS. (B) A cell of *Rhizobium phaseoli* is surrounded by fibrous ECPS. (C) A cell of *R. japonicum* is partially surrounded by ECPS. Bars = 0.2 μm.

specificity of *Rhizobium*–legume symbiotic association (Bohlool and Schmidt 1974; Bhuvaneswari, Pueppke, and Bauer 1977; Dazzo and Brill 1979; Schmidt 1979). Though rhizobial lipopolysaccharides and a cell-wall polysaccharide have also been reported to be involved in lectin binding (Wolpert and Albersheim 1976; Planqué and Klijne 1977; Kato, Maruyama, and Nakamura 1979), morphological evidence favours the fact that ECPS is responsible for the binding (Calvert *et al*. 1978; Tsien and Schmidt 1981). Immunoelectron microscopy was used to localize the binding sites of soyabean lectin. When cells of *R. japonicum* were labelled with soyabean lectin conjugated with ferritin and prepared for electron microscopy as shown in Fig. 6.6, ferritin particles were seen exclusively to be associated with the ECPS and not located on

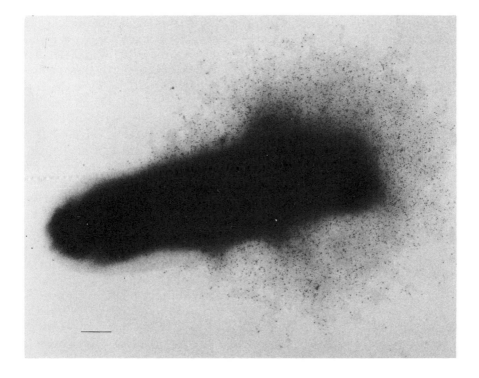

FIG. 6.6. Direct examination of unstained *Rhizobium japonicum* cells after interacting with ferritin-conjugated soyabean lectin. Ferritin particles were seen to be associated with amorphous materials of low electron density. No particle was seen to be associated with the outer membrane. Bars = 0.2 μm.

the outer membrane. This result indicates that ECPS but not LPS is the site of lectin binding.

Numerous compositional studies on rhizobial ECPS have been reported. The results indicate that the ECPS from fast growing rhizobia are compositionally homogeneous, whereas those from slow-growing rhizobia are heterogeneous. An ECPS component which binds soyabean lectin has been isolated from several strains of *R. japonicum* and a cowpea *Rhizobium*. Structural analysis of these polysaccharides indicates that they are compositionally identical (Tsien, Anderson and Schmidt 1980). The uniformity in the sugar composition of the lectin-binding ECPS component is an important feature with respect to the lectin-binding hypothesis as a prelude to the root-hair infection by rhizobia.

Flagella

Two types of flagellation, based on the arrangement, the number of flagella and flaggellar wavelength, have been reported in rhizobia (Leifson and Erdman 1958; De Ley and Rassel 1965; Abdel-Ghaffar and Jensen 1966). Peritrichous flagellation, which comprises two to ten flagella dispersed all over the cell, is frequently observed among fast-growing rhizobial species. The subpolar flagellation, which bears a single flagellum subpolarly attached to the cell at a right angle to the cell axis, is mainly associated with slow-growing rhizobia.

Flagella are made up of three morphologically distinct substructures: the filament, the hook, and the basal structure. Figure 6.7(A) shows a subpolar flagellated *R. japonicum* with a filament and a hook. The filament, which is made up of a protein-subunit component, flagellin, and arranged in a helix, extends 10–20 μm from the cell surface. The arrangement of flagellin subunits in a helical pattern is obvious in a partially purified flagella from *R. japonicum* (Fig. 6.7(B), arrowhead).

Schmitt *et al.* (1974) isolated a novel type of complex flagella from *R. lupini*, which has a characteristic 'zig-zag' patterned filament. The filament, which is made up of the flagellin subunits of molecular weight 43 000 daltons measured 16 nm in diameter and consisted of a cylindrical core surrounded by three bands of a helical sheath. The hook, 15 nm in diameter and 60–80 nm in length, was located at the base of the flagellar filament and was composed of 41 000 dalton protein subunits. The filament and the hook of these complex flagella have subsequently purified and chemically characterized (Maruyama, Lodderstaedt and Schmitt 1978).

No report on the fine structure of the basal structure of rhizobial flagella has appeared. For other Gram-negative bacteria, the basal structure consists of four ring-like structures which are fixed to a central rod and bound respectively to two layers of the cytoplasmic membrane, a peptidoglycan layer, and the outer membrane.

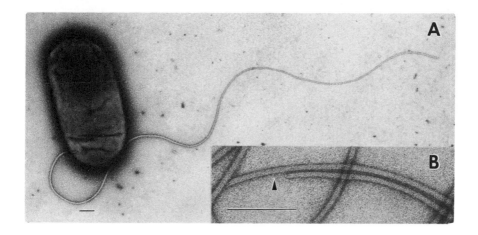

FIG. 6.7. (A) A negatively-stained cell of *R. japonicum* showing a sub-polar flagellum. A long filament is subpolarly attached to the cell by a hook. (B) A portion of isolated flagellar filaments at higher magnification. Helical arrangement of the filament subunits is clear (arrowhead). Bars = 0.2 μm.

Fimbriae and pili

Nonflagellar filamentous appendages occur widely among various types of bacteria and are generally considered to be fimbriae. The term pili or sex pili is often used to specify the non-flagellar structures which are functionally related to bacterial conjugation.

The occurrence of fimbriae in rhizobial species has been reported for *R. trifolii*, *R. leguminosarum* (De Ley and Rassel 1965), and *R. lupini* (Mayer 1969). The fimbriae of *R. lupini* strain 1/50, sta⁻, a non-starforming mutant, are tube-like structures measuring 3 nm in outer diameter, 0.8–1 nm in inner diameter and are singly or peritrichously inserted. In stationary growth-phase culture, fimbriae often stick together as a result of nonspecific cohesion forces. Fimbriated cells attach nonspecifically to a substrate or to a solid surface. In a star forming *R. lupini*, cells are clumped together by fimbriae to form star-shaped aggregates within which conjugation occurs between competent cells (Heumann, 1968).

The fimbriae have also been observed in *R. japonicum*, cowpea *Rhizobium*, and *R. phaseoli* (Tsien and Schmidt, unpublished results). As an example, Fig. 6.8(A) shows a fimbriated cell of *R. japonicum* strain USDA 71. Fimbriae of about 5 nm in diameter are of polar origin. Detached fimbriae also may be observed in a culture of *R. phaseoli* (Fig.

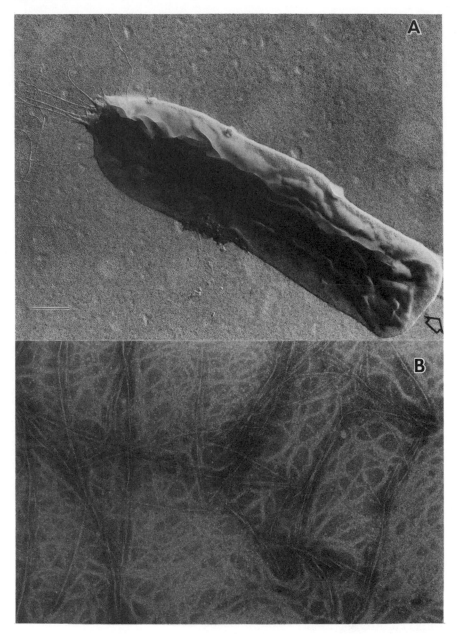

FIG. 6.8. (A) A shadow casted cell of *R. japonicum* USDA 71 showing polarly inserted fimbriae. Cells were air dried on a formvar–carbon coated grid and shadowed with platinum–carbon at an angle of 45°. (B) Some detached fimbriae from *R. phaseoli* negatively stained. Fibrous materials of low electron density in the background are extracellular polysaccharide. Bars = 0.2 μm.

6.8(B)). Since there is only very limited information concerning the occurrence, structure, and function of these organelles, more studies are needed, particularly with respect to a possible role in the rhizobial infection process.

The extracellular polar body of (EPB) *R. japonicum* described by Tsien and Schmidt (1977) is possibly partly fimbriae. The fine structure of EPB has been described to be globular materials surrounded by microfibrils. Through the EPB, *R. japonicum* cells aggregate to form rosettes or attach to the solid surface. The nature of extracellular polar body microfibrils (fimbriae?) and their role in attachment are worthy of further investigation.

Acknowledgements

The author would like to thank Dr. Edwin L. Schmidt for his encouragement and for reading the manuscript, and Ms. Lola J. Fredrickson for her assistance with English. The published and unpublished studies of the author cited here were supported by National Science Foundation grant DEB 7710172.

References

Abdel-Ghaffar, A. S. and Jensen, H. L. (1966). *Arch. Mikrobiol.* **54,** 393–405.
Amarger, N., Obaton, M., and Blachère, H. (1967). *Can. J. Microbiol.* **13,** 99–105.
Bailly, R. W., Greenwood, R. M., and Craig, A. (1971). *J. gen. Microbiol.* **65,** 315–24.
Bhuvaneswari, T. V., Pueppke, S. G., and Bauer, W. D. (1977). *Plant Physiol.* **60,** 486–91.
Bohlool, B. B. and Schmidt, E. L. (1974). *Science, N.Y.* **185,** 269–71.
Braun, V. (1978). In *Relations between structure and function in the prokaryotic cell* (ed. R. Y. Stanier, H. J. Roger, and J. B. Ward), Vol. 28. Sym. Soc. gen. Microbiol.
Calvert, H. E., LaLonde, M., Bhuvaneswari, T. V., and Bauer, W. D. (1978). *Can. J. Microbiol.* **24,** 785–93.
Costerton, J. W., Ingram, J. M., and Cheng, K. J. (1974). *Bacterial Rev.* **38,** 87–110.
Craig, A. S. and Williamson, K. I. (1972). *Archs Microbiol.* **87,** 165–71.
Dazzo, F. B. and Brill, W. J. (1979). *J. Bacteriol.* **137,** 1362–73.
De Ley, J. and Rassel, A. (1965). *J. gen. Microbiol.* **41,** 85–91.
Dudman, W. F. (1964). *J. Bacteriol.* **88,** 640–5.
—— (1968). *J. Bacteriol.* **95,** 1200–1.
—— (1976). *Carbohydr. Res.* **46,** 97–110.
—— (1978). *Carboydrate Res.* **66,** 9–23.
Glenn, A. R. and Dilworth, M. J. (1979). *J. gen. Microbiol.* **112,** 405–9.
Hayat, M. A. (1970, 1972, 1977). *Principles and techniques of electron microscopy.* Vols. I, II, and III. *Biological applications.* Van Nostrand Reinhold, New York.
Heumann, W. (1968). *Molec. gen. Genet.* **102,** 132–44.
Jordan, D. C. and Coulter, W. H. (1965). *Can. J. Microbiol.* **11,** 709–20.
Kamberger, W. (1979). *Archs Microbial.* **121,** 83–90.
Kato, G., Maruyama, Y., and Nakamura, M. (1979). *Agric. biol. Chem.* **43,** 1085–92.
Leifson, E. and Erdman L. W. (1958). *Antonie V. Leeuwenhoek J. Microbiol. Serol.* **24,** 97–110.
MacKenzie, C. R., Vail, W. J., and Jordan, D. C. (1973). *J. Bacteriol.* **113,** 387–93.

Maruyama, M., Lodderstaedt G., and Schmitt, R. (1978). *Biochim. biophys. Acta* **535**, 110–24.

Mayer, F. (1969). *Arch. Mikrobiol.* **68**, 179–86.

Nanninga, N. (1970). *J. Bacteriol.* **101**, 297–303.

Napoli, C., Dazzo, F., and Hubbell, D. (1975). *Appl. Microbiol.* **30**, 123–31.

Pankhurst, C. E. and Craig, A. S. (1978). *J. gen. Microbiol.* **106**, 207–19.

Planqué, K. and Klijne, J. W. (1977). *FEBS Lett.* **73**, 64–6.

Schmidt, E. L. (1979). *Ann. Rev. Microbiol.* **33**, 355–76.

Schmitt, R., Bamberger, I., Acker, G., and Mayer, F. (1974). *Archs Microbiol.* **100**, 145–62.

Skinner, F. A., Roughley, R. J., and Chandler, M. R. (1977). *J. appl. Bacteriol.* **43**, 287–97.

Sutherland, I. W. (1977). In *Surface carbohydrates of the prokaryotic cell* (ed. I. W. Sutherland). pp. 27–96. Academic Press, London.

Tsien, H. C., Anderson J. S., and Schmidt, E. L. (1980). *Am. Soc. Microbial. Abstr. Annu. Meet.* **1980**, 155.

—— and Higgins, M. L. (1974). *J. Bacteriol.* **118**, 725–34.

—— and Schmidt, E. L. (1977). *Can. J. Microbiol.* **23**, 1274–84.

—— —— (1980a) *Appl. environ. Microbiol.* 239, 1100–4.

—— —— (1981). *J. Bacteriol.* **145**, 1063–74.

Urban, J. E. (1979). *Appl. Environ. Microbiol.* **38**, 1173–8.

Van Brussel, A. A. N., Costerton, J. W., and Child, J. J. (1979). *Can. J. Microbiol.* **25**, 352–61.

Vincent, J. M., Humphrey, B., and North, R. J. (1962). *J. gen. Microbiol.* **29**, 551–5.

Wolpert, J. S. and Albersheim, P. (1976). *Biochem. Biophys. Res. Commun.* **70**, 729–37.

Wong, P. P. and Evans, H. J. (1971). *Plant Physiol.* **47**, 750–5.

7 Surface chemistry

R. W. CARLSON

7.1 Introduction

The surface carbohydrates of *Rhizobium* have been the subject of investigations directed toward determining the molecular basis by which a species of *Rhizobium* recognizes its legume host. The majority of these studies have concentrated on the sugar composition and sugar linkages of *Rhizobium* extracellular polysaccharides (EPCS), since it was hypothesized by Fåhraeus and Ljunggren (1959) that these molecules determine symbiotic specificity. It has only been within recent years that the structural sequence of several *Rhizobium* ECPS have been determined. These studies have shown that *Rhizobium* ECPS can be very complex structures containing a repeating oligosaccharide of up to eight sugars. *Rhizobium* lipopolysaccharides (LPS) have not been investigated as extensively as the ECPS, but these investigations have revealed that *Rhizobium* LPS are more structurally diverse than the ECPS. As yet there are no reports in the literature as to the exact structure of any *Rhizobium* LPS. The *Rhizobium* cell surface also contains flagella. Only recently have reports been published regarding the role of flagella; these suggest that flagella are not involved in symbiosis.

Studies directed at determining the role of *Rhizobium* ECPS and LPS in symbiosis have continued because of the large body of evidence which shows that the ECPS and LPS from other Gram-negative bacteria determine the specificity of their association with other organisms. The LPS of *Salmonella* and *Escherichia coli* are the strain-specific antigenic determinants of these bacteria and are known to be the determining factor in the specific infection of a host (Lüderitz, Staub, and Westphal 1966; Lüderitz, Jann, and Wheat 1968; Sharon 1975; Wilkinson 1977; Jann and Jann 1977; Ørskov, Ørskov, Jann, and Jann 1977). For certain types of *E. coli* the ECPS has been found to be the factor which determines infection (Ørskov *et al.* 1977; Jann and Jann 1977).

Reports in the literature strongly suggest that the ECPS and LPS of Gram-negative bacterial plant-pathogens play a role in determining the specificity of infection. For example, the presence of ECPS has been shown to be a requirement for the successful infection of rosaceous plants by *Erwinia amylovora* and the infection of potato by *Pseudomonas solanacearum* (Ayers, Ayers, and Goodman 1979; Sequeira and Graham 1977). In addition, LPS preparations from strains of *P. solanacearum* and *Agrobacterium tumefaciens* have been shown to specifically inhibit the infection of their respective hosts (Graham, Sequeira, and Huang 1977; Whatley, Bodwin, Lippincott, and Lippincott 1976; Whatley and Spiess 1977; Whatley, Hunter, Cantrell, Hendrick, Keegstra, and Sequeira 1980). Because of these examples it seems likely that the ECPS and LPS of *Rhizobium* also play a role in determining the specificity of symbiotic infection.

In this chapter the available literature regarding the purification and characterization of *Rhizobium* ECPS and LPS will be discussed, with regard to the role of these molecules in the symbiotic process.

7.2 *Rhizobium* **extracellular polysaccharides**

General characteristics of Rhizobium *EPCS*

There are primarily two types of ECPS from Gram-negative bacteria, homopolysaccharides and heteropolysaccharides (Sutherland 1977). Heteropolysaccharides are often acidic due to the presence of glucuronic acid, galacturonic acid, and pyruvic acid groups (Sutherland 1977). Homopolysaccharides are neutral, and are, in many cases, glucans (Sutherland 1977). *Rhizobium* bacteria have typical Gram-negative ECPS in that *Rhizobium* have both glucan homopolysaccharides and acidic heteropolysaccharides.

The acidic ECPS of Gram-negative bacteria are typically composed of repeating oligosaccharides which contain D-glucose, D-galactose, D-glucuronic acid, and D-galacturonic acid (Sutherland 1977). In some cases N-acetyl-D-glucosamine, N-acetyl-D-galactosamine are present and with less frequency L-fucose and L-rhamnose are found (Sutherland 1977). Acetyl- and pyruvic-acid groups are frequently found in the acidic heteropolysaccharide ECPS (Sutherland 1977).

The acidic ECPS of *Rhizobium* also contain these components. However, the acidic ECPS of several species of *Rhizobium* also contain unusual sugar components such as 4-O-methyl-β-D-glucuronic acid (Dudman 1978), 3-O-methyl-D-ribose, 6-O-methyl-D-galactose, 3-O-methyl-D-glucose and 6-dexoy-L-talose (Bailey, Greenwood, and Craig 1971; Kennedy and Bailey 1976; Dudman 1976; Kennedy 1978).

Purification of Rhizobium ECPS

The isolation of *Rhizobium* acidic ECPS is a relatively easy process because the acidic ECPS are often present in large amounts in the culture media. The culture media is centrifuged in order to remove the bacteria. The initial procedure consists of precipitating the ECPS from the now bacteria-free culture media by addition of two or three volumes of ethanol (Bailey *et al*. 1971). Occasionally the alcoholic media solution is allowed to stand overnight at 2 °C to ensure complete precipitation of the ECPS (Bailey *et al*. 1971). The precipitate is harvested by centrifugation, dissolved in water, dialysed against water and lyophilized (Bailey *et al*. 1971). For some slow-growing *Rhizobium* species the ECPS and the bacteria form a viscous clot (Bailey *et al*. 1971). In these cases the ECPS and the bacteria have been separated by making the culture media 0.5 N sodium hydroxide for 30 min prior to centrifugation (Bailey *et al*. 1971). After centrifugation the supernatant is acidified immediately with acetic acid and treated as described above (Bailey *et al*. 1971). Of course, this alkaline treatment could cause some chemical modification of the polysaccharide, such as the removal of any ester groups. This may be important since almost all *Rhizobium* ECPS contain acetyl groups.

The acidic ECPS can be further purified from the ethanol precipitated ECPS. The ethanol precipitated ECPS is dissolved in water and reprecipitated with a 5 per cent solution of cetyltrimethylammonium bromide (Zevenhuizen 1973; Sømme 1974). Cetyltrimethylammonium bromide is a cationic detergent and is used to precipitate acidic polymers; e.g. it is also used in the purification of nucleic acids. The precipitated acidic ECPS is dissolved in water, dialysed against water and lyophilized (Zevenhuizen 1973; Sømme 1974).

The acidic ECPS may also be present in the form of a capsule surrounding the bacteria. In this form the ECPS require more severe conditions for their isolation. Capsular ECPS can be isolated by extraction of the bacteria with 45 per cent aqueous phenol (Ørskov, Ørskov, Jann, and Jann 1977; Westphal and Jann 1965). The capsular polysaccharide is found in the water phase together with the LPS. The LPS sediment during ultracentrifugation, while the capsular ECPS remain in the supernatant. The capsular ECPS are further purified by precipitation with cetyltrimethylammonium bromide as described above. A 1 per cent phenol solution has also been used to remove *Rhizobium* ECPS capsules (Dazzo and Brill 1977, 1979).

To date the homopolysaccharides of *Rhizobium* have been neutral glucan polymers. Separation techniques have taken advantage of the difference in charge between the homopolysaccharides and the acidic heteropolysaccharides. The *Rhizobium* homopolysaccharide glucans have been purified from culture filtrates by selective precipitation with ethanol which removes primarily the acidic polysaccharide while the supernatant is enriched in the neutral glucan (York, McNeil, Darvill, and Albersheim 1980). The glucan is then separated from any remaining acidic polysaccharide by preparing a slurry in DEAE–cellulose which specifically absorbs the acidic polysaccharide. The purified glucan elutes as a partially included symmetrical peak on a Sephadex G-50 sizing column (York *et al.* 1980). The glucans prepared in this manner from one strain each of *R. leguminosarum*, *R. phaseoli*, *R. trifolii*, and *Agrobacterium tumefaciens* proved to be linear β-1,2-glucans (York *et al.* 1980). These β-1,2-glucans have also been isolated from *Rhizobium* using the hot phenol-water extraction procedure described for the isolation of capsular ECPS (see above) and LPS (see below) (Zevenhuizen and Scholten-Koerselman 1979). The water phase from the hot phenol–water extraction is extracted with diethyl ether to remove the phenol. Dialysis cannot be used to remove the phenol as the low molecular weight β-1,2-glucan will diffuse through the dialysis tubing. The acidic ECPS can be selectively precipitated in 0.2 M NaCl by 1 per cent cetyltrimethylammonium bromide. The neutral glucan and the slightly acidic LPS remain in solution. The LPS is separated from the glucan by fractionation using gel-filtration chromatography, since the LPS has a high molecular weight and the glucan has a relatively low molecular weight. Zevenhuizen and Scholten-Koerselman (1979) also found that the glucan could be partially purified by fractional precipitation with ethanol as previously described, and that the glucan was present in the culture filtrates but in small amounts.

Analytical procedures for Rhizobium *ECPS*

Standard methods of analysis of carbohydrates have been used in characterizing the *Rhizobium* ECPS. Compositions of *Rhizobium* ECPS have been determined by acid hydrolysis of the polysaccharide, followed by separation and characterization of the monosaccharide components. The hexoses produced by acid hydrolysis of the poly-saccharide have been separated and identified using paper chroma-tography (Sømme 1974; Zevenhuizen 1971; Humphrey and Vincent 1959). The hexoses have been quantified using the ferricyanide test (Sømme 1974) and also through the use of specific enzymes e.g. glucose oxidase and galactose oxidase (Zevenhuizen 1971). Colourimetric assays are also available for the determination of pyruvic acid and acetyl groups (Zevenhuizen 1971; Hestrin 1949; Sutherland 1969). Uronic acids have been identified by paper chromatography or paper electro-phoresis (Sømme 1974) and quantified colourimetrically with carbazole (Bitter and Muir 1962) or more recently with *m*-hydroxybiphenyl (Blumenkrantz and Asboe-Hansen 1973). The ECPS compositions have also been determined by gas chromatography of the alditol acetates. In this procedure the hexoses produced by acid hydrolysis of the ECPS are converted to their corresponding alditol acetates which are separated and quantified by gas chromatography. The various alditol acetates are identified by comparison of their gas chromatographic retention times to the retention times of standards (Albersheim, Nevens, English, and Karr 1967). When authentic standards are not available identification has been made by combined gas chromatography–mass spectrometry (Björndal, Hellerquist, Lindberg, and Svensson 1970; Lindberg and Lönngren 1978). The gas chromatographic techniques are more sensitive than paper chromatography and have been used in the more recent analyses of *Rhizobium* ECPS (Zevenhuizen 1973; Kennedy 1978; Zevenhuizen and Scholten-Koerselman 1979; York *et al.* 1980).

The techniques used to determine the structure of complex carbohydrates have been recently reviewed (Lindberg and Lönngren 1978; Svensson 1978). The polysaccharides are usually methylated, and subjected to partial acid hydrolysis. This produces a variety of oligosaccharides which contain overlapping hexose sequences. The reducing end of each oligosaccharide can be marked by reduction to the alditol using sodium borodeuteride. Any free hydroxyl groups on the oligosaccharides are alkylated with trideuteromethyl iodide or ethyl iodide. Thus the position of a trideuteromethyl or an ethyl group marks where each oligosaccharide was attached to another sugar prior to partial acid hydrolysis. These alkylated oligosaccharides can be separated and their sequences determined. In some instances the oligosaccharides can be separated and identified directly by combined gas chromatography–mass spectrometry (Lindberg and Lönngren 1978). A recent technique has been developed in which the alkylated oligosaccharides are purified by high-pressure liquid chromatography. The individual oligosaccharides are acid hydrolysed, reduced, acetylated, and analysed by combined gas chromatography–mass spectrometry (Valent, Darvill, McNeil, Robertsen, and Albersheim 1980). Using this method the sequence of various disaccharides, trisaccharides, and, in some cases, tetrasaccharides with overlapping structures can be determined. This method has been used to determine the structure of the ECPS from several strains of *R. trifolii* and *R. leguminosarum* (Albersheim, personal communication).

The structures of several *Rhizobium* ECPS have also been determined by using various techniques for the specific degradation of the polysaccharides. These techniques are both chemical and enzymatic. The ECPS from *R. meliloti* has been shown to have an octasaccharide repeating unit (see Table 7.3) by sequential degradation of the polysaccharide (Jansson, Kenne, Lindberg, Ljunggren, Lönngren, Ruden, and Svensson 1977). In this method the *R. meliloti* polysaccharide is first

TABLE 7.1

Percentage relative sugar compositions of slow-growing Rhizobium[a] *acidic ECPS*

Rhizobium strains	Glucose	Galactose	Mannose	Acetate	Pyruvic acid	Glucuronic acid	Galacturonic acid	4-O-Methylgalactose	3,6-Anhydrosugar	Xylose	Rhamnose	4-O-Methyl-glucuronic acid	6-O-Methylgalactose	6-Deoxytalose	3-O-Methylribose	References
R. japonicum																
1809	42	7.1	18	4.0	0	0	16	12	0	0	0	0	0	0	0	Dudman 1976
123	41	6.5	19	7.6	0	0	13	13	0	0	0	0	0	0	0	Dudman 1976
127	40	10	18	6.5	0	0	12	13	0	0	0	0	0	0	0	Dudman 1976
110	36	24	18	4.1	0	0	12	5.3	9.7	0	0	0	0	0	0	Dudman 1976
709	41	11	19	6.7	0	0	16	6.3	0	0	0	0	0	0	0	Dudman 1976
129	6.7	48	5.0	0	1.8	0	4.6	0	29	4.4	0	0	0	0	0	Dudman 1976
711	43	16	33	0	7.7	0	0	0	0	0	0	0	0	0	0	Dudman 1976
311b	0	0	0	0	0	0	0	0	0	0	71	29	0	0	0	Dudman 1978
CC708	0	0	0	0	0	0	0	0	0	0	70	30	0	0	0	Dudman 1978
CB1795	0	0	0	0	0	0	0	0	0	0	70	30	0	0	0	Dudman 1978
CIb110	40	9.0	17	5.4	0	0	21	7.8	0	0	0	0	0	0	0	Mort *et al.* 1979
CIb138	37	5.3	17	6.1	0	0	22	12	0	0	0	0	0	0	0	Mort *et al.* 1979
WH11																Zevenhuizen, personal communication
WH5R	36–41	6–8	13–16	11–13	0	0	12–15	8–10	0	0	0	0	0	0	0	
WH6R																
Cowpea strain																
CB756	37	9	20	0	0	0	20	0	0	0	Tr	1	0	9.0	4.0	Kennedy 1978

[a] The ECPS compositions of several other slow-growing *Rhizobium* are not reported here since several components of these ECPS remain unidentifed. However, the unreported data also show that slow-growing *Rhizobium* are highly variable in their compositions (Bailey *et al.* 1971). Tr = Trace amounts present.

methylated and free hydroxyl groups are produced by removing the pyruvate groups. The polysaccharide can then be sequentially degraded by repeated oxidation and β-elimination steps (Jansson *et al.* 1977; Svensson 1978). This method has also been used to determine the sequence of an *R. trifolii* ECPS (Jansson, Lindberg, and Ljunggren 1979).

The ECPS structures from *R. trifolii, R. phaseoli, R. leguminosarum,* and *R. meliloti* have also been examined by using several endoglucanase enzymes specific for different linkages (Anderson and Stone 1978). The products of the various enzyme digestions are characterized by paper chromatography and gel-filtration chromatography.

Many *Rhizobium* ECPS contain uronic acid (see Table 7.1), and a common feature of all these methods for structural analysis is the requirement that the uronic-acid residues be reduced. A commonly used method of reducing the uronic-acid carboxyl groups is to react the polysaccharide with a water-soluble carbodiimide which forms a lactone. The lactone can then be reduced to the alcohol by sodium borohydride (Taylor and Conrad 1972). Reduction is usually done with sodium borodeuteride in order to distinguish the hexose derived from the uronic acid from the hexoses which were present originally in the polysaccharide.

Rhizobium *ECPS chemistry*

Rhizobium bacteria have been divided into groups based on their growth rate in culture, the fast-growing organisms; e.g. *R. leguminosarum, R. trifolii, R. phaseoli, R. meliloti*; and the slow-growing ones; e.g. *R. japonicum* and *R. lupini* (Vincent 1974). Initial composition studies revealed that the ECPS from the fast-growing *Rhizobium* species largely consists of glucose, galactose, and uronic acid while the ECPS from slow-growing *Rhizobium* were highly variable in their composition, often containing unusual sugars (Bailey *et al.* 1971).

A compilation of the ECPS compositions for several species of *Rhizobium* is presented in Tables 7.1 and 7.2. These data verified the early data and show that the composition of the ECPS from the fast-growing isolates is the same from species to species of *Rhizobium*, and is widely variable even among strains of a single slow-growing *Rhizobium* species. One fast-growing species was unique, *R. meliloti. Rhizobium meliloti* ECPS contain either none or only low amounts of uronic acid.

Methylation studies revealed that ECPS from strains of *R. leguminosarum, R. trifolii,* and *R. phaseoli* have identical glycoside linkages (Zevenhuizen 1973). Methylation analysis also revealed that *R. meliloti* ECPS glycoside linkages are virtually identical with those of *Agrobacterium* and *Alcaligenes faecalis* ECPS (Misaki, Saito, Ito, and Harada 1969;

TABLE 7.2
Percentage relative sugar compositions of fast-growing Rhizobium *ECPS*

Rhizobium strains	Glucose	Galactose	Mannose	Acetyl groups	Pyruvic acid	Glucuronic acid	Galacturonic acid	References
R. trifolii								
0226	56	11	0	11	9.4	12	0	Sømme 1974
Coryn	53	12	0	8.5	11	14	0	Sømme 1974; Zevenhuizen 1971
Bart A	56	12	0	10	10	13	0	Sømme 1974; Hepper 1972
Ar-3	48	10	0	14	15	13	0	Zevenhuizen 1971
In-2	50	10	0	14	15	18	0	Zevenhuizen 1971
0403	43	11	0	7	11	25	0	Hepper 1972
R. leguminosarum								
HVIII	48	11	0	8	14	19	0	Zevenhuizen 1971
P8	48	11	0	9	13	19	0	Zevenhuizen 1971
PRE	48	11	0	9	13	19	0	Zevenhuizen 1971
310a	51	9	0	9	15	16	0	Zevenhuizen 1971
313	46	14	0	11	15	14	0	Zevenhuizen 1971
402	49	10	0	10	15	16	0	Zevenhuizen 1971
404	49	10	0	17	14	17	0	Zevenhuizen 1971
U311	45	7	0	13	16	20	0	Sǿomme 1974
R. phaseoli								
Blink	48	11	0	12	13	16	0	Zevenhuizen 1971
460	50	10	0	12	11	17	0	Zevenhuizen 1971
Bokum	50	9	0	9	14	18	0	Zevenhuizen 1971
U453	51	12	0	10	12	16	0	Sømme 1974
U458	51	10	0	12	12	15	0	Sømme 1974
R. meliloti								
U27	67	14	0	7	11	0	0	Sømme 1974
K24	70	14	0	9	6	1	0	Zevenhuizen 1971
A148	75	13	Tr	6	6	0	0	Zevenhuizen 1971
BPe	77	10	0	8	5	0	0	Zevenhuizen 1971
A145	83	8	0	6	3	0	0	Zevenhuizen 1971

Tr = Trace amounts are present.

Saito, Misaki, and Harada 1970; Harada, Amemura, Jansson, and Lindberg 1979; Zevenhuizen 1973). More recently the structures of ECPS from strains of *R. trifolii*, *R. phaseoli*, *R. leguminosarum*, and *R. meliloti* were examined using several different endoglucanases (Anderson and Stone 1978). These different endoglucanases were active only if the uronic acids of the ECPS were reduced and ECPS were depyruvylated and/or deacetylated. Enzyme treatment of the modified ECPS from *R. meliloti* released only one oligosaccharide, β-D-Gal-($1\rightarrow$ 3)-D-Glc. Similar treatment of the ECPS from *R. trifolii, R. leguminosarum,* and *R. phaseoli* suggest that the ECPS from these three *Rhizobium* species are identical to each other. They all contain a D-GlcA-($1\rightarrow$4)-D-GlcA-($1\rightarrow$4)-D-Glc oligosaccharide and a D-Gal-($1\rightarrow$3)-D-Glc disaccharide which occupies a terminal position on a side chain. Furthermore, partial acid hydrolysis of the uronic acid containing ECPS of *Rhizobium* produced the following oligosaccharides: β-D-GlcA-($1\rightarrow$4)-D-Glc, β-D-GlcA-($1\rightarrow$4)-GlcA and β-D-GlcA-($1\rightarrow$4)-β-D-GlcA-($1\rightarrow$4)-Glc (Sømme 1975). These results suggest that the uronic acid containing ECPS from *Rhizobium trifolii, R. leguminosarum,* and *R. phaseoli* all contain a polysaccharide chain containing two β-($1\rightarrow$4)-linked D-glucuronic acid residues linked to D-glucose at position 4.

The structures of only a few *Rhizobium* ECPS have been determined. The structures are shown in Table 7.3. The structures of the ECPS from two strains of *R. trifolii* are identical to each other and identical to the

TABLE 7.3
Rhizobium *ECPS structures*

Rhizobium strain	Structure
R. trifolii U226[a]	$\rightarrow$4)-β-D-GlcA-($1\rightarrow$4)-α-D-Glc-($1\rightarrow$4)-β-D-Glc-($1\rightarrow$
R. trifolii NA30 and 0403[b] *R. legumino-sarum* 128C53 and 128C63[b]	

TABLE 7.3 *continued*

Rhizobium strain	Structure

R. meliloti →4)-β-D-Glc-(1→4)-β-D-Glc-(1→3)-β-D-Gal-(1→4)-β-D-Glc-(1→
U27[c]
                                      6
                                      ↑
                                      1

β-D-Glc-(1→3)-β-D-Glc-(1→3)-β-D-Glc-(1→6)-β-D-Glc
    4   6
      C
  H₃C   CO₂H

H_3C CO_2H

R. japonicum →3)-Man-(1→3)-Glc-(1→3)-GalA-(1→
3I1b138[e]
                  6
                  ↑
                  1
4-*O*-Me-Gal-(1→3)-Glc

R. japonicum →4)-α-L-Rha-(1→3)-β-L-Rha-(1→4)-β-L-Rha-(1→
71A, CC708
and
CB1795[d]
           3
           ↑
           1
4-*O*-Me-β-D-GlcA

[a] This structure is reported by Jansson *et al.* (1979). Acetyl groups are reported to be linked to the α-D-Glc residue in the main chain on positions 2 and/or 3. These workers also report that one-third of the side chains lack the terminal galactose.

[b] This structure is reported by Albersheim (personal communication). This structure differs from the above structure in that there are two adjacent GlcA residues in the main chain and that the 4,6-linked glucose is α-linked rather than β-linked. Albersheim does not find any side chains which lack the terminal galactose. The presence of two adjacent GlcA residues in the main chain is also supported by Anderson and Stone (1978) and by Sømme (1975) (see text).

[c] Jansson *et al.* (1977).

[d] Dudman (1978).

[e] Mort *et al.* (1979).

ECPS structures of two *R. leguminosarum* strains (Albersheim, personal communication). There are some differences between the ECPS structure reported by Jansson *et al.* (1979) for *R. trifolii* and that reported by Albersheim (personal communication). Albersheim reports an additional glucuronic acid in the back bone of the repeating unit and also that the 4,6-linked glucose residue is α-linked, while Jansson *et al.* (1979) suggest that it is β-linked. The work of both Anderson *et al.* (1979) and Sømme (1975) support the presence of two glucuronic acid residues in the back bone of the repeating unit. Jansson *et al.* (1979) suggest that one-third of the side chains lack the terminal galactose residue, while Albersheim suggests that all the side chains contain the terminal galactose. Jansson *et al.* (1979) report that the acetyl groups are located in the 2 and/or 3 positions of the 4-linked glucose residue in the main chain; Albersheim has not reported the location of the acetyl groups. Structural analysis of the ECPS from more strains of *R. trifolii* should determine the significance of these reported differences. These structural analyses suggest that the ECPS from *R. trifolii* and *R. leguminosarum* are very similar, but differences between the two could still be present in the location of the acetyl groups. The structure of *R. phaseoli* ECPS has not yet been determined.

The ECPS structure of *R. meliloti* strain U-27 is quite different from the *R. trifolii* and *R. leguminosarum* ECPS structures. Although the *R. meliloti* ECPS has a more simple composition than the *R. trifolii* or *R. leguminosarum* ECPS, its structure is just as complex. *Rhizobium meliloti* ECPS, like that of *R. trifolii* and *R. leguminosarum*, ECPS has an octa-saccharide repeating unit (Jansson *et al.* 1977). As mentioned previously, methylation patterns of ECPS from several strains of *R. meliloti*, *Agrobacterium*, and *Alcaligenes faecalis* are identical. While the methylation analyses suggest that these ECPS have identical repeating units, the data suggest that there may be some differences in the amount of acetyl, pyruvate, or succinyl groups attached to the repeating oligo-saccharide. For example, the ratio or succinyl to pyruvate groups for *Alcaligenes faecalis* and *Agrobacterium tumefaciens* ECPS is 1.0 and 0.45 respectively (Harada *et al.* 1979). *Rhizobium meliloti* ECPS have a ratio of acetyl to pyruvate groups of 1.05 and a ratio of acetyl to succinyl groups of 1:1.4 (Harada *et al*, 1979). Prior to the work by Harada *et al.* (1979), succinyl groups had not been reported for *R. meliloti* ECPS. At present, Zevenhuizen and co-workers have not been able to identify succinic acid in their *R. meliloti* ECPS samples (Zevenhuizen, personal communication). These results suggest that, apart from the mode of acyl-ation, the ECPS from these species of bacteria may all have a structure similar to that shown in Table 7.3 for *R. meliloti* ECPS.

The structures of ECPS from several *R. japonicum* strains are given in Table 7.3. Overall, the oligosaccharide repeating units of these two structures are very different from one another. However, the repeating units are similar in that the terminal sugar on the side chain of both ECPS contains a 4-*O*-methyl group (Dudman 1978; Mort, Slodki, Plattner, and Bauer 1979: Mort and Bauer 1980). In one case a galactose residue is methylated and in the other a glucuronic-acid residue is methylated. Methylation analysis of the ECPS from *R. japonicum* strains WH1I, WH5R, and WH6R reveal that these ECPS may be similar to the structure shown for the ECPS from strain 311b138. The WH1I, WH5R, and WH6R ECPS contain 15 per cent terminal-galactose residues, part of which are monomethylated, a

branching 1,3,6-linked glucose residue (19 per cent), and chain fragments of 1,3-linked glucose (26 per cent), mannose (24 per cent), and
galacturonic acid (17 per cent) (Zevenhuizen, personal communication).
These results suggest that it may be possible to place strains of
R. japonicum into groups based on the similarity of their ECPS structures. For example, if we consider the ECPS structures shown in Table
7.3 and the compositions shown in Table 7.1, *R. japonicum* strains
CI1b110, CI1b138, WH1I, WH5R, WH6R, and perhaps CB1809, 123,
and 127 would be in one group, while strains 71a, CC708, and CB1795
would be in another group. Although more ECPS structures from
slow-growing *Rhizobium* are not available at this time, it is obvious from
the composition studies (see Table 7.1) that these ECPS structures can
vary greatly even within a species of slow-growing *Rhizobium*.

The structural studies verify the earlier composition results and
suggest that: (1) *R. leguminosarum*, *R. trifolii*, and *R. phaseoli* ECPS have
closely related and perhaps identical structures; (2) *R. meliloti* ECPS
structures are the same among the various strains of *R. meliloti* but
different from those of other fast-growing *Rhizobium*. The *R. meliloti*
ECPS structure may also be very similar to the structure for *Agrobacterium*
and *Alcaligenes faecalis* ECPS; (3) the ECPS structures of the slow-
growing *Rhizobium*, e.g. *R. japonicum*, vary greatly even among strains of
a single species.

The *Rhizobium* homopolysaccharide most recently studied is a β-2-
linked glucan (Zevenhuizen and Scholten-Koerselman 1979; York *et al.*
1980). This glucan was isolated from the culture filtrates of
R. leguminosarum, *R. phaseoli*, *R. trifolii*, and *Agrobacterium tumefaciens*
(York *et al.* 1980). It was also found to be associated with the cell wall,
which appears to be its primary location (Zevenhuizen and Scholten-
Koerselman 1979). The β-2-linked glucan appears to have very few, if
any, branching points and has a molecular weight of 3000–3500 which
corresponds to about 20 glucose units (Zevenhuizen and Scholten-
Koerselman 1979; York *et al.* 1980). A further aspect of this β-2-linked
glucan is that a reducing terminal was not detected in the intact
molecule, but after partial acid hydrolysis the expected amounts of
reducing terminal glucose residues can be detected (York *et al.* 1980).
The β-2-linked glucans were originally found in the culture fluid of
Agrobacterium tumefaciens (McIntire, Peterson, and Riker 1942) and later
established as a β-2-linked glucan by methylation analysis (Putman,
Potter, Hodgson, and Hassid 1950; Gorin, Spencer, and Westlake
1961). In addition, a β-2-linked glucan has been reported to be
synthesized by extracts of *R. japonicum* (Dedonder and Hassid 1964).
These data suggest that the presence of β-2-linked glucan is ubiquitous
in the Rhizobiaceae family. In addition to the β-2-linked glucans,
β-4-linked glucans have been found in strains of *R. leguminosarum*,

R. trifolii, R. phaseoli, and *R. meliloti* (Zevenhuizen, Scholten-Koerselman, and Posthumus 1980; Deinema and Zevenhuizen 1971; Napoli, Dazzo, and Hubbell 1975).

Recently, Dudman and Jones (1980), have analysed a glucan fraction found during the purification of *R. japonicum* ECPS. This fraction consists of a mixture of two different polysaccharides, one glucan having a molecular weight of about 4500 and the other of about 12 000. Each of the glucans consists of 1,3-, 1,6-, and 1,3,6-linked and non-reducing terminal-glucose groups. However, the larger glucan contains equal proportions of 1,3- and 1,6-linked groups while the smaller glucan contains a 2:1 ratio of 1,3-linked to 1,6-linked glucose groups. The physiological significance of these glucans is not known.

7.3 *Rhizobium* **lipopolysaccharides**

General considerations

Rhizobium LPS are similar to the LPS of other Gram-negative bacteria. In the LPS from Gram-negative bacteria the polysaccharide usually consists of two structural regions; a repeating oligosaccharide known as the O-antigen and a core oligosaccharide (see Wilkinson 1977 for review). The O-antigen is attached to the core oligosaccharide and the core is attached to the lipid-A through an acid-labile 2-keto-3-deoxyoctanoic (KDO) acid linkage. The polysaccharide of *Rhizobium* polysaccharide is also attached to the lipid-A through an acid-labile KDO linkage (Carlson, Sanders, Napoli, and Albersheim 1978; Zevenhuizen *et al.* 1980). Thus the polysaccharide region of *Rhizobium* LPS, as with the LPS from other Gram-negative bacteria LPS, can be separated from the lipid-A by mild acid hydrolysis with acetic acid and extraction in chloroform/water (Carlson *et al.* 1978; Zevenhuizen *et al.* 1980). *Rhizobium* LPS often contains many unusual sugars such as di-*O*-methyl-6-deoxyhexoses and amino-dideoxyhexoses (Carlson *et al.* 1978; Zevenhuizen *et al.* 1980). The LPS of Gram-negative bacteria generally contain unusual dideoxyhexoses and unusual amino sugars (Lüderitz *et al.* 1966; Lüderitz *et al.* 1968; Ørskov *et al.* 1977; Wilkinson 1977). *Rhizobium* LPS are also similar to the LPS of other Gram-negative bacteria in that they are the specific antigenic determinants of *Rhizobium* (Carlson *et al.* 1978; Zevenhuizen *et al.* 1980; Humphrey and Vincent 1969). *Rhizobium* LPS are unlike those from *Salmonella* or *E. coli* in that *Rhizobium* LPS contain uronic acids and frequently do not contain heptose (Carlson *et al.* 1978; Zevenhuizen *et al.* 1980).

Purification of Rhizobium *LPS*

Rhizobium LPS are purified by the general extraction procedure of

Westphal and Jann (1965) using 45 per cent hot phenol–water. On cooling the LPS, ribonucleic acid, acidic ECPS, and the neutral β-2-linked glucan are found in the water layer (Carlson *et al.* 1978; Zevenhuizen and Scholten-Koerselman 1979; Zevenhuizen *et al.* 1980). One method of removing any nucleic acids which may be present at this stage is to incubate the water phase with ribonuclease and deoxyribonuclease enzymes (Carlson *et al.* 1978), followed by dialysis and lyophilization. The LPS can be partially separated from the ECPS by ultracentrifugation, which pellets the LPS due to its large aggregate size (Carlson *et al.* 1978). However, this method usually results in LPS which is still contaminated by some acidic ECPS (Carlson *et al.* 1978). It was found that the LPS could be converted to a lower molecular-weight form of uniform size using an EDTA–triethylamine solution. In this form it could be separated from any remaining ECPS, which stays in a high molecular-weight form, by gel-filtration chromatography (Carlson *et al.* 1978). The LPS is partially included on a Sepharose 4B column while the acid ECPS elutes with the void volume (Carlson *et al.* 1978). The symmetry of the LPS peak and the constant KDO/hexose ratio across the peak suggest that the LPS is pure (Carlson *et al.* 1978). Presumably the EDTA–triethylamine solution complexes any calcium or magnesium ions which are responsible for producing highly aggregated states of LPS (Galanos and Lüderitz 1975). The removal of calcium and magnesium ions from *Rhizobium* LPS facilitates the lowering of the molecular weight and, thus, the separation of the LPS

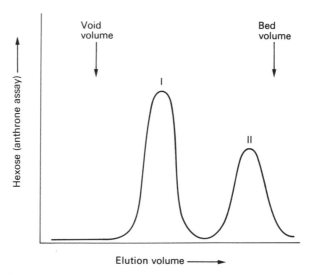

FIG. 7.1. A schematic representation of a typical fractionation on Sephadex G-50 of the polysaccharides released from *Rhizobium* LPS by mild acid hydrolysis.

from any remaining high molecular-weight acidic ECPS. The *Rhizobium* LPS have also been further purified from the water phase of the phenol–water extraction by selectively precipitating the highly acidic ribonucleic acid and acidic ECPS using cetyltrimethylammonium bromide in 0.2 M sodium chloride (Zevenhuizen and Scholten-Koerselman 1979; Zevenhuizen *et al.* 1980). Once the acidic polymers have been removed, the LPS is separated from the low molecular-weight neutral glucan by gel-filtration chromatography or by precipitating the LPS with ethanol (Zevenhuizen and Scholten-Koerselman 1979; Zevenhuizen *et al.* 1980). The LPS purified in this manner still elutes with the void volume of a Bio-Gel Agarose A-5 m column. The KDO/hexose ratio across the peak was not determined. Thus it is possible that some of the LPS purified in this manner are still contaminated by residual acidic ECPS.

The purification of *Rhizobium* LPS can be followed by assaying for KDO, a sugar which has been reported to be present in all of the *Rhizobium* LPS examined to date. The assay for KDO consists of measuring the red product formed by reacting the periodate oxidation product of KDO with thiobarbituric acid (Weissbach and Hurwitz 1958). Lipopolysaccharides can also be assayed by using carbocyanine dye (Janda and Work 1971) or by using the *Limulus* lysate assay (Yin, Galanos, Kinsky, Bradshaw, Wessler, Lüderitz, and Sarmieto 1972). *Rhizobium* LPS have been purified from over 25 different strains using the above purification procedures (Carlson *et al.* 1978; Zevenhuizen *et al.* 1980)

Analytical procedures for Rhizobium *LPS*

The procedures used for determining the composition of *Rhizobium* LPS are the same as those described above for *Rhizobium* ECPS analysis. At the time of writing, no structures of *Rhizobium* LPS have been reported. However, the procedures used for the structural determination of ECPS will undoubtedly be used for determining the structure of *Rhizobium* LPS. In fact, some of the procedures described have been used in determining the structures of LPS from other Gram-negative bacteria (Lindberg and Lönngren 1978).

A procedure which has proved useful in the analysis of LPS is the mild acid hydrolysis of the LPS followed by isolation of the released polysaccharides (Osborn 1963; Ryan and Conrad 1974). Mild acid, usually 1 per cent acetic acid at 100 °C for one hour, will hydrolyse the ketosidic bond between the KDO of the polysaccharide and the lipid-A (Osborn 1963; Ryan and Conrad 1974). The lipid-A usually precipitates during this procedure and can be separated from the polysaccharide by centrifugation or extraction with chloroform.

TABLE 7.4

Compositions[a] of LPS from R. leguminosarum, R. trifolii and R. phaseoli

Rhizobium strains	KDO	Uronic acid	Glucosamine	Glucose	Mannose	Galactose	Fucose	Rhamnose	2-O-Me-6-deoxyhexose	3-O-Me-6-deoxyhexose	3-O-Mehexose	2,3-Di-O-Me-6-deoxyhexose	2,4-Di-O-Me-6-deoxyhexose	3,4-Di-O-Me-6-deoxyhexose	N-Me-3-amino-3,6-dideoxyhexose	2-Amino-2,6-dideoxyhexose	Tetrose	Heptose
R. leguminosarum																		
PF-2[b]	9.3	7.6	1.9	9.9	18	7.5	25	18	0	0	0	0	1.7	0	0	0	0.7	0
PRE[b]	8.6	7.4	3.4	7.7	19	6.0	25	21	0	0	0	0	1.2	0	0	0	0.9	0
S-402[b]	10	7.8	1.9	4.3	17	4.0	30	22	0	0	0	0	2.6	0	0	0	0.3	0
128C53[c]	10	12	3.2	1.8	18	4.5	30	20	0	0	0	0	0	0	0	0	0	0
3HOQI[c]	6.0	8.8	4.1	25	21	2.1	19	14	0	0	0	0	0	0	0	0	0	0
P-8[b]	12	40	5.6	4.1	3.6	6.2	0.9	0	2.4	0	0	0	0	0	9.0	0	0	25
128C63[c]	9.3	23	8.3	7.4	9.0	8.7	13	0	0	0	0	4.2	0	0	0	0	0	7.7
R. trifolii																		
Ar-3[b]	1.4	2.3	0.9	28	1.0	0	22	43	1.5	0	0	0	0	0	0	0	0	0
Coryn KL[b]	4.7	19	3.4	46	13	7.9	1.3	0.3	3.4	0	0	0	0	0	0	0	Tr	1.6
K-8[b]	10	13	3.0	18	8.2	7.1	18	4.8	16	0	0	0	0	0	0	0	0	0
0403[c]	7.0	26	7.9	9.1	7.0	7.3	7.3	0	8.5	0	0	0	0	0	11	0	0	9.4
162S7	2.9	8.9	6.1	3.6	7.7	4.5	16	2.4	18	0	0	0	0	2.8	0	0	0	0
2S[c]	6.4	20	5.1	1.9	5.3	6.4	21	22	0	9.9	2.7	0	0	0	0	0	0	0
R. phaseoli																		
Blink[b]	6.6	13	3.1	6.8	4.8	8.7	0	36	12	8.5	0	0	0	0	0	0	Tr	0
Bokum[b]	8.5	15	3.6	3.9	3.6	3.3	0	43	10	8.0	0	0	0	0	0	0	0	0
S-460[b]	11	38	5.1	5.6	6.7	8.8	3.2	0	5.6	0	0	0	0	0	0	0	0	16
127K17[c]	17	33	8.1	6.6	6.9	10	7.7	0	0	0	0	0	0	0	6.6	3.9	0	0

[a] Compositions are given as relative sugar percentages.
[b] Compiled from data presented by Zevenhuizen et al. (1980).
[c] Compiled from data presented by Carlson et al. (1978).
Tr = Trace amounts present.

TABLE 7.5

Compositions[a] of LPS from R. meliloti

R. meliloti strains	KDO	Uronic acid	Glucose	Galactose	Mannose	Glucosamine	Rhamnose
BPe[b]	51	8.8	34	Tr	0	6.9	0
A-148[b]	59	12	16	3.9	0	9.8	0
102F51[c]	53	nd	29	18	Tr	nd	Tr
Rm1000[c]	50	nd	36	4.4	5.6	nd	5.0
Rm2[c]	75	nd	18	2.3	3.1	nd	1.6

[a] Compositions are relative sugar percentages.
[b] Compiled from data reported by Zevenhuizen *et al.* (1980).
[c] Data not previously published but compiled by Carlson, Sanders and Albersheim. Rm2 is a streptomycin-resistant mutant of Rm1000. The abbreviations nd and Tr represent 'not determined' and 'trace amounts present' respectively.

The chemistry of Rhizobium *lipopolysaccharides*

While there have been many chemical studies on *Rhizobium* ECPS, studies of *Rhizobium* LPS have been scarce (Humphrey and Vincent 1969; Lorkiewicz and Russa 1971; Russa and Lorkiewicz 1974; Planqué and Kijne 1977). However, recently the chemical compositions of LPS from several strains of *R. trifolii, R. leguminosarum, R. phaseoli,* and *R. meliloti* have been determined (Carlson *et al.* 1978; Zevenhuizen *et al.* 1980). The relative sugar compositions of these *Rhizobium* LPS are given in Tables 7.4 and 7.5.

The data reveal that *Rhizobium* LPS vary greatly in their composition. Gram-negative bacteria which differ in their LPS compositions are referred to as being of different chemotypes; the data given in Table 7.4 show that *Rhizobium* are of many different chemotypes. Table 7.4 shows that out of the seventeen strains of *Rhizobium* examined, there are eleven different chemotypes; four chemotypes among the seven *R. leguminosarum* strains, three among the four *R. phaseoli* strains, and five among the six *R. trifolii* strains. There is only one case in which *Rhizobium* strains from different species were of the same chemotype; *R. phaseoli* S-460 has the same LPS composition as *R. leguminosarum* P-8. However, within a species of *Rhizobium* there are a greater number of strains of the same chemotype. *Rhizobium leguminosarum* strains PF-2, PRE, and S-402 are of the same chemotype, while strains 128C53 and 3HOQ1 are of a different one. *Rhizobium phaseoli* Blink and Bokum have the same chemotype, as do *R. trifolii* strains Ar-3 and K-8. It has been stated that the LPS compositions vary as much among strains of a single *Rhizobium* species as they do among the different species of Rhizobium (Carlson *et al.* 1978). However, the data shown here suggest that this is not entirely true since there are more strains of *Rhizobium* with a common chemotype among a single *Rhizobium* species than there are common chemotypes among the different species of *Rhizobium*.

Lipoplysaccharides from different strains of a *Rhizobium* species but with the same composition may also be closely related structurally. In the case of *R. leguminosarum*, chemical studies show that the LPS from strains which are of the same chemotype also have identical or closely related immunodominant structures. Antiserum to *R. leguminosarum* PRE appears to cross-react as well with LPS preparations from strains PF-2 and S-402 as with LPS from strain PRE (Zevenhuizen *et al.* 1980). Also, antiserum to the purified LPS of *R. leguminosarum* strain 128C53 cross-reacts as well with the LPS from strain 3HOQ1 as it does with the LPS from strain 128C53 (Carlson *et al.* 1978). Other than common chemotypes, the antisera to purified LPS is very strain specific (Carlson *et al.* 1978). Antiserum to the LPS from a strain of *R. phaseoli* will cross-react only with that *R. phaseoli* strain and not with other *Rhizobium* or with other *R. phaseoli* strains. Similar results were obtained with antisera to LPS from *R. trifolii* and *R. leguminosarum*.

The data in Table 7.4 show that the LPS compositions vary greatly among strains of *R. leguminosarum*, *R. trifolii*, and *R. phaseoli*. This was not true with the ECPS from these species of bacteria, which have very similar or identical structures (see pp. 205–211). However, there are some common features in the composition of LPS from these different *Rhizobium* species. All of the strains of *R. trifolii*, *R. leguminosarum*, and *R. phaseoli* have LPS which contain KDO, uronic acid, glucosamine (hexosamine), glucose, mannose, and galactose. One exception is *R. trifolii* Ar-3, which was reported not to contain any detectable level of galactose. Another common feature of the LPS from these *Rhizobium* species is that they all contain rhamnose or fucose. In fact, all the *R. leguminosarum* and *R. trifolii* strains contain fucose. The LPS from two of the four strains of *R. phaseoli* have fucose, while the other two strains have only rhamnose. The other sugar components vary greatly from strain to strain. These highly variable components are usually *O*-methyl or di-*O*-methylhexoses, unusual amino sugars and deoxysugars.

The compositions of *R. meliloti* LPS are given in Table 7.5. *Rhizobium meliloti* LPS are very different in composition from those of other *Rhizobium*. The sugar present in the largest amount is KDO, which presumably links the polysaccharide to the lipid-A of the LPS. The compositions of LPS from the various *R. meliloti* strains are quite similar to each other; in addition to KDO the common constituents are uronic acid, glucosamine, and glucose. Some of the *R. meliloti* LPS contain mannose and/or rhamnose. There have been no reports of *O*-methylated hexoses or heptose in any of the *R. meliloti* LPS. 2-Keto-3-deoxyoctanoic acid comprises an unusually large percentage of the LPS dry weight. 2-Keto-3-deoxyoctanoic acid comprises an average of 20 per cent of the dry weight of *R. meliloti* LPS, and in some cases comprising over 30 per cent. This is significantly higher than the average of 4.3 per cent KDO for the LPS from the strains of *R. trifolii*, *R. leguminosarum*, and *R. phaseoli*. Only the rough and deep-rough mutants of *Salmonella* and *E. coli* have LPS with comparable amounts of KDO (Ørskov *et al.* 1977; Lüderitz *et al.* 1966, 1968). These mutants have LPS which lack the O-antigen polysaccharide and contain only the core polysaccharide or part of the core polysaccharide, attached to the lipid-A (for reviews see Wilkinson 1977 and Sharon 1975). These data suggest that *R. meliloti* may have LPS which lack a repeating O-antigen oligosaccharide. Immuno-chemical data also suggest this possibility. Antisera to the LPS of *R. meliloti* strain Rm2 cross-reacts equally well with the LPS from *R. meliloti* strains Rm2, Rm1, 102F51, and CU-1 (a strain not shown in Table 7.4 but which contains 36 per cent KDO by dry weight) (Carlson and Albersheim, unpublished data). In addition the antiserum to Rm2 LPS cross-reacts to a lesser degree with every strain of *R. trifolii*, *R. phaseoli*,and *R. leguminosarum* with which it has been tested; it also cross-reacts slightly with the LPS from two strains of *E. coli* (Carlson and Albersheim, unpublished data). This would be expected only if the immunodominant site of the *R. meliloti* LPS is a structure which is present in the LPS of all these bacterial strains. The most likely possibility for this structure is KDO. Antiserum to deep rough mutants of *Salmonella* also cross-reacts extensively with other bacteria. This extensive cross-reactivity is thought to be due to the immunodominance of KDO in the LPS of the deep-rough mutants (Lüderitz *et al.* 1968).

Although no structures of *Rhizobium* LPS have been reported as yet, the data do reveal some of their structural aspects. As stated previously the O-antigen of LPS has been found to vary greatly from strain to strain in *Salmonella* and *E. coli* as well as in other Gram-negative bacteria (for reviews see Wilkinson 1977; Sharon 1975; Jann and Jann 1977), however the core-polysaccharide structure does not vary among the different *Salmonella* strains and varies only slightly in comparison to the O-antigen region among the *E. coli* strains (Ørskov *et al.* 1977; Wilkinson 1977). It seems likely, therefore, that the core polysaccharides of *Rhizobium* LPS would vary less in their structure than the O-antigen polysaccharide regions. The data in Tables 7.4 and 7.5 show that KDO, uronic acid, glucosamine, glucose, galactose, and possibly mannose are common to all of the *Rhizobium* LPS examined thus far. It seems likely, therefore, that these sugars are present in the core-polysaccharide region.

The core-polysaccharide sugars and O-antigen polysaccharide sugars can be identified by separating these two LPS structural regions using Sephadex G-50 gel-filtration chromatography. It has been shown that

Sephadex G-50 column chromatography of the mild acid hydrolysed polysaccharide from *E. coli* LPS results in two major polysaccharide peaks (Müller-Seitz, Jann, and Jann 1968; Jann, Jann, and Schmidt 1970; Lüderitz *et al.* 1966; Lüderitz *et al.* 1968; Dmitriev, Knirel, Kochetkov, Jann, and Jann 1977; Wilkinson 1977; Ørskov *et al.* 1977). Polysaccharide peak I elutes with the void volume or just after the void volume, and is composed of the O-antigen polysaccharide-region attached to the core oligosaccharide. Polysaccharide peak II elutes just before the included volume and comprised only the core oligosaccharide without the O-antigen attached (see above references). These results suggested that bacterial LPS molecules are of two types: those in which both the O-antigen polysaccharide and the core oligosaccharide are attached to the lipid-A and those in which only the core oligosaccharide is attached to it. This has since been verified by separating these two LPS types by gel-electrophoresis (Jann, Reske, and Jann 1975). Analysis of the mild acid hydrolysed polysaccharide from purified *Rhizobium* LPS by Sephadex G-50 chromatography gives results similar to the *E. coli* data. A schematic representation of a typical Sephadex G-50 elution profile for the polysaccharide from *Rhizobium* LPS is given in Fig. 7.1. Table 7.6 gives the relative sugar compositions of the two polysaccharide peaks, together with the relative sugar composition of the intact LPS from *R. leguminosarum* 128C53. Polysaccharide peak I consists almost solely of mannose, rhamnose, and fucose with only a small amount of glucose. These data suggest that the O-antigen of *R. leguminosarum* 128C53 LPS is composed of mannose, rhamnose, and fucose. The composition of peak II suggests that the core oligosaccharide is composed of galactose, glucose, mannose, glucosamine, and uronic acid. 2-Keto-3-deoxyoctanoic acid was not quantified, but occurs only in peak II and also elutes with the included volume. It is interesting that the O-antigen polysaccharide (I) elutes in the partially-included volume and not at or near the void volume which is where the *E. coli* O-antigen polysaccharide (I) elutes. This suggests that the *R. leguminosarum* 128C53 O-antigen polysaccharide is of a lower molecular weight than the usual *E. coli* O-antigen polysaccharide. It is also interesting that the *R. leguminosarum* polysaccharide (I) does not contain any of the core sugars, except for a small amount of glucose. This is unlike the *E. coli* polysaccharide (I) which does contain the core sugars since the polysaccharide consists of the O-antigen attached to the core oligosaccharide (see above references). Perhaps the *R. leguminosarum* O-antigen polysaccharide contains other acid-labile linkages which cause this polysaccharide to be separated from the core oligosaccharide during the mild-acid hydrolysis of the LPS. This has been previously suggested by Wilkinson (1977) with regard to the LPS from other Gram-negative bacteria. The O-antigen polysaccharide released from the mild acid hydrolysis of *Serratia marcescens* Bizio and *S. marcesens*

08 LPS are also free of any core sugars (Tarcsay, Wang, Li, and Alanpovic 1973; Wang and Alanpovic 1973). Sephadex G-50 chromatography of the polysaccharides released by mild acid hydrolysis from *R. leguminosarum* 3HOQ1 LPS and a strain of *R. phaseoli* LPS gives results similar to the *R. leguminosarum* 128C53 data described above (Carlson and Albersheim, unpublished data).

TABLE 7.6

Percentage relative sugar composition of the Sephadex G-50 polysaccharides from R. leguminosarum *128C53 LPS*

R. leguminosarum 128C53 polysaccharides	Uronic acid	Glucosamine	Glucose	Galactose	Mannose	Fucose	Rhamnose
Intact LPS	8.0	4.4	4.6	5.2	20	34	23
PS I	0	0	3.4	0	21	46	30
PS II	55	2.3	10	15	17	0	0

These data suggest that the LPS from strains of *R. trifolii*, *R. leguminosarum*, and *R. phaseoli* consist of lipid-A linked to the core oligosaccharide through an acid-labile KDO bond. The core is linked to an O-antigen polysaccharide which, in certain cases, may also contain acid-labile bonds. The data suggest that *R. meliloti* may have an LPS similar to that of the rough mutants of *Salmonella* or *E. coli* which lack an O-antigen polysaccharide region. Thus *R. meliloti* LPS would consist of the lipid-A linked to the core oligosaccharide through the acid-labile KDO bond. The remainder of the oligosaccharide consists of galactose, glucose, glucosamine, uronic acid, and in some cases mannose or rhamnose. Of course, the large amount of KDO present in *R. meliloti* could be due to some other component which gives a positive reaction to the thiobarbituric acid test, e.g. *N*-acetylneuraminic acid. Lipopolysaccharides have been found which contain *N*-acetylneuraminic acid, but they are rare (Ørskov *et al.* 1977). Paper chromatography has shown, however, that acid hydrolysis of *R. meliloti* Rm2 LPS yields only thiobarbituric-acid-positive material which co-chromatographs with authentic KDO (Carlson and Albersheim, unpublished data). More structural work is necessary in order to determine the significance of these suggestions and to determine the interspecies and inter-strain relationships of *Rhizobium* LPS.

7.4 Biological role of *Rhizobium* surface carbohydrates

Both the ECPS and LPS of Gram-negative bacteria are known to play a role in determining the specific interactions of these bacteria with other organisms. Thus the ECPS and LPS of *Rhizobium* have been postulated to play a role in determining the specificity of *Rhizobium*–legume symbioses (for reviews see Broughton 1978; Schmidt 1979; Sequeira 1978). However, at present there are no specific aspects of the infection process, such as attachment of the bacteria to the root hair, marked root hair curling, or differentiation of the bacteria to the bacteroid, which can be unambiguously assigned to a certain *Rhizobium* cell-surface molecule(s).

Although no specific role in the symbiotic process can be assigned to any certain *Rhizobium* cell-surface molecule, there are reports in the literature which suggest some of the functions of these molecules.

The role of Rhizobium *ECPS in symbiosis*

Since *Rhizobium* produce large amounts of ECPS, these molecules have been widely studied in order to determine what role, if any, they play in the symbiotic process. Fåhraeus and Ljunggren (1959) first suggested that *Rhizobium* ECPS may determine the specificity of symbiosis. Owing to their suggestion many reports have compared the ECPS compositions among the different *Rhizobium* species. For a surface molecule to be involved in determining specificity, it was expected that its structure would be very similar among strains of single *Rhizobium* species, but would vary significantly between the different species of *Rhizobium*. The data in Table 7.2 show that the ECPS compositions of *R. trifolii, R. leguminosarum,* and *R. phaseoli* are practically identical. Based on composition and methylation data, it was concluded initially that ECPS are not involved in determining the specificity of symbiosis. In addition, the data in Table 7.3 show that the structure of ECPS from strains of *R. trifolii* is very similar, and perhaps identical, to that of ECPS from strains of *R. leguminosarum*. The similarity of these two ECPS structures could account for the fact that there is occasional cross-nodulation between the pea and clover groups (see reviews Vincent 1974 and Dart 1974; Hepper and Lee 1979). However, Table 7.3 also shows that the structure of *Rhizobium* ECPS can be very complex, consisting of octasaccharide repeating units. Since the ECPS of only a few strains have been examined, it is possible that ECPS with identical compositions or sugar linkages could vary significantly in structure. Even the identical ECPS structures shown for *R. trifolii* and *R. leguminosarum* could vary in the location of the acetyl groups. If ECPS determine specificity, only subtle differences in structure may be necessary, such as the location of acetyl or methyl groups. Future structural studies need to give careful attention to the location of these two groups.

Other reports suggest that *Rhizobium* ECPS may have a role in determining the specificity of symbiosis. The lectin hypothesis has been recently reviewed (Broughton 1978; Schmidt 1979) and is also discussed by Dazzo and Hubbel in Chapter 10. This hypothesis is based on data which suggest that the host lectin specifically binds the symbiont bacteria (see above reviews and Chapter 10). With regard to the following discussion, it must be noted that there are also reports which demonstrate a lack of correlation between lectin binding and nodulation groups (Chen and Phillips 1976; Law and Strijdom 1977).

Several reports present data indicating that the lectin receptor on the symbiont bacteria is the ECPS. Electron microscopy studies have shown that attachment of the symbiont *Rhizobium* to the host root-hair occurs in a polar fashion, i.e. an end-on attachment of the bacteria to the root hair (Sahlman and Fåhraeus 1963). Bohlool and Schmidt (1976) have shown that fluorescein labelled antibody binds largely to one end of the homologous *Rhizobium* and that fluorescein labelled lectin (soyabean) binds in a similar manner to strains of *R. japonicum*. Later Tsien and Schmidt (1977) showed that the polar site of attachment for the homologous antisera and the soyabean lectin on *R. japonicum* correlated with the polar presence of ECPS as detected by ruthenium-red staining and visualized by electron microscopy. Furthermore, this polarity was manifested in the exponential-phase of *R. japonicum* growth (Tsien and Schmidt 1977). Several reports show that the binding of soyabean lectin to *R. japonicum* depends on the presence of a capsule (Bal, Shantharam, and Ratnam 1978; Bhuvaneswari, Pueppke, and Bauer 1977; Shantharam, Gow, and Bal 1980). Recent work also shows that *Arachis* agglutinin binds to two strains of the symbiont *Rhizobium* and does not bind to non-symbiont *Rhizobium* (Bhagwat and Thomas 1980). A decrease in binding was correlated with the removal of ECPS from the bacteria using a saline wash. The peanut lectin also bound to the acetone-precipitated ECPS. These workers also showed that the peanut lectin still binds to the bacterium after the saline wash. This binding activity was recovered in the water phase of a hot phenol/water extraction of the bacteria. Based on these results, Bhagwat and Thomas (1980) suggest that both the LPS and ECPS are the lectin receptors. Similar studies by Dazzo and Brill (1977) showed that fluorescein labelled ECPS from *R. trifolii* specifically bound to the root hairs of *Trifolium repens*. Dazzo, Yanke, and Brill (1978) also present data which suggest that the binding of the ECPS to the root hair is mediated by the clover lectin, trifoliin. More recently Dazzo and Brill (1979) separated a crude ECPS preparation from *R. trifolii* into three polysaccharides, two of which bound the clover lectin trifoliin. The exact composition of these polysaccharides was not determined and therefore their relationship to the ECPS described in Table 7.3 or the LPS compositions for *R. trifolii* in Table 7.4 cannot be determined.

While all of these reports suggest that ECPS are the receptors on *Rhizobium* for the host lectins, they do not rule out the possibility of other surface molecules, such as the LPS, being the receptors. The isolation procedures for the ECPS consist of either washing the cells with saline or 1 per cent phenol, followed by precipitation with cetyltrimethyl-ammonium bromide, or with acetone, or ethanol. These procedures do not separate the ECPS satisfactorily from the LPS, or in some cases (i.e. ethanol precipitation) from the neutral glucan. In fact, Dazzo and Brill (1979) showed that one of the polysaccharides purified by cetyltri-methylammonium bromide precipitation followed by DEAE column chromatography contains KDO, a sugar which is present specifically in the LPS. This polysaccharide also binds the clover lectin (Dazzo and Brill 1979). Furthermore, in the above-mentioned work by Dazzo, ECPS from non-infective *R. trifolii*, *R. meliloti*, and *R. japonicum* were used as controls. It would seem that a better control would have been the ECPS from strains of *R. leguminosarum* or *R. phaseoli*, considering their similarity in structure to *R. trifolii* ECPS. In the work by Bhagwat and Thomas (1980) discussed above, it was suggested that the LPS is also a lectin receptor, since lectin still binds to saline washed bacteria and also to a crude hot phenol/water-extracted LPS preparation. However, it has been demonstrated that the hot phenol/water extraction of saline-washed *Rhizobium* results in crude LPS preparations which contain significant amounts of ECPS (Carlson *et al*. 1978). Thus without further purification and characterization of the ECPS and LPS it cannot be stated from the work of Bhagwat and Thomas (1980) that both the ECPS and the LPS are lectin receptors. Furthermore the fact that crude LPS preparations contain ECPS and that crude ECPS preparations contain LPS suggests that the capsule which surrounds the *Rhizobium* and which apparently binds the host lectin (see above) contains both polysaccharides. Therefore the microscopic visualization of the host lectin binding to the capsule of the symbiont bacterium does not rule out the possibility that a non-ECPS surface molecule, e.g. LPS, is the lectin receptor. Also, visualization of capsular ECPS by ruthenium red staining may not distinguish between the acidic ECPS and the acidic LPS of *Rhizobium*, since this cationic dye may stain both polysaccharides.

Obviously more careful consideration needs to be given to the purity and the characterization of the ECPS and LPS preparations used in the above studies. Mort and Bauer (1980) have recently characterized *R. japonicum* ECPS as a function of culture growth and lectin binding. Previously Bhuvaneswari *et al*. (1977) had shown that several *R. japonicum* strains showed maximum lectin binding with log growth-phase cells. Analysis of the ECPS showed that the ratio of free galactose to methylated galactose (see structure in Table 7.3) is higher in strains which bind soyabean lectin. The suggestion is that methylation of the ECPS prevents lectin binding. In another recent report *Pisum* lectin was isolated from root slime and found to agglutinate *R. leguminosarum* and to precipitate LPS (Kijne, Van der Schaal, and DeVries 1980). This LPS preparation was not pure and is known to have contained a glucan which binds pea lectin (Planqué and Kijne 1977). The binding activity of the pea lectin is inhibited by D-glucose, D-mannose, D-fructose, L-sorbose, 2-deoxy-D-glucose, N-acetyl-D-glucosamine, and 3-O-methyl-D-glucoside (Kijne *et al.* 1980). It is also known that pea lectin haemagglutinating activity is powerfully inhibited by methylated hexoses (Van Wauwe, Loontiens, and de Bruyne 1975; Allen, Desai, and Newberger 1976). Based on this Kijne *et al.* (1980) suggest that 4-O-methylation of a terminal lectin-binding sugar in the capsule may prevent pea lectin binding in an analogous manner to the *R. japonicum* soyabean system suggested by Mort and Bauer (1980). It should be stated that the ECPS from *R. leguminosarum* does not contain methylated hexoses (see Tables 7.2 and 7.3) while the *R. japonicum* ECPS contains them (Tables 7.1 and 7.3). It is possible that ECPS may serve as the lectin receptors in some species of *Rhizobium* (e.g. *R. japonicum*) but not in others (e.g. *R. leguminosarum*). Lectin binding studies using double diffusion tests also suggest this possibility. The ECPS from *R. meliloti, R. leguminosarum*, and *R. phaseoli* all bind the lectins from *Canavalia ensiformis, Pisum sativum*, and *Lens culinaris* but not the lectins from *Glycine max, Phaseolus vulgaris*, or wheat germ (Kamberger 1979). However, the ECPS from *R. japonicum* specifically bound soyabean lectin and not lectins from other legumes. The possibility that the LPS of some *Rhizobium* species may serve as the lectin receptor is discussed below (see pp. 225–228).

Another aspect of *Rhizobium*–legume symbiosis is that the contact of the *Rhizobium* with the root hairs causes the root hairs to curl and become deformed (Yao and Vincent 1969; Meijer, Chapter 10). However, only a specific type of curling, marked curling, is associated with the symbiont–host interaction (Yao and Vincent 1969). *Rhizobium* ECPS preparations have been shown to cause non-specific curling and deformation of the root hairs, but do not cause the marked curling (Hubbell 1970; Vincent 1974; Yao and Vincent 1976). Very crude ECPS preparations were used in these studies. Apparently the specific marked curling of the root hair can occur only when viable rhizobia are present (Yao and Vincent 1976). It has been suggested that this root-hair curling is caused by pectic enzymes which are induced by *Rhizobium* ECPS (Ljunggren and Fåhraeus 1961). A more recent report presents data which show that the ECPS of *R. meliloti* induces the pectic enzyme polygalacturonase (Palomares, Montoya, and Olivares, 1979). They suggest that the production of this ECPS polygalacturonase inducer is derived from extrachromosomal DNA, since wild-type ECPS is significantly more active than the ECPS from an acridine-orange-cured mutant from which the extrachromosomal DNA was absent (Palomares *et al.* 1979). However, these studies also used relatively crude ECPS preparations and therefore it cannot be stated with certainty that this inducing activity is caused by ECPS alone. Also these workers did not determine whether the induction of polygalacturonase was specific for the host–symbiont interaction by using non-symbiont ECPS. In a preliminary report, plant-root extracts were reported to contain glucosidases which specifically hydrolyse ECPS and LPS from the symbiont bacteria (Solheim, personal communication). These workers reported that the breakdown of capsular polysaccharides in a compatible association was 6.5 per cent and in an incompatible association about 0.8 per cent. Other aspects of polysaccharide degrading enzymes being involved in symbiosis are discussed by Dazzo and Hubbell in Chapter 9 of this volume.

Genetic studies have also implicated a role for *Rhizobium* ECPS in symbiosis. Sanders, Carlson, and Albersheim (1978) have shown that a spontaneous mutant of an infective *R. leguminosarum* strain lacks ECPS and is non-infective. All revertants of this mutant regain both the ability to infect and to produce ECPS. This study suggests that *Rhizobium* ECPS or some genetic factor linked to ECPS production plays a role in symbiosis. The composition and immunochemical analysis of the mutant LPS show it to have the same structure as the LPS from the parent.

Very little is known about the biological role of the neutral glucans which are produced by *Rhizobium*. A glucan present in the LPS preparation from *R. leguminosarum* has been reported to bind to pea lectin (Planqué and Kijne 1977). Recently it has been suggested that this glucan is the β-1,2-linked glucan which is ubiquitous to all *Rhizobium* examined thus far (Kijne *et al.* 1980; Zevenhuizen *et al.* 1980). Another recent report also suggests that a neutral polysaccharide from *R. japonicum* binds to the root hair of soyabean (Hughes, Lecce, and Elkan 1979). The chemical nature of this neutral polysaccharide was not determined.

The role of Rhizobium *LPS in symbiosis*

Unlike the ECPS, *Rhizobium* LPS vary greatly in structure among the strains of *R. trifolii*, *R. leguminosarum*, and *R. phaseoli* and appear to be the strain-specific antigenic determinants of these bacteria (see pp. 213–220). Because of the structural diversity of *Rhizobium* LPS it has been suggested that they are not involved in determining the specificity of symbiosis. This conclusion is based on the idea that strains of a single *Rhizobium* species should have similar or identical LPS structures if the LPS molecules are responsible for specificity (Dudman 1977). The structural diversity of LPS has been determined by chemical composition and immunochemical studies. Since it is not required that the immunodominant sites of *Rhizobium* LPS be the same as the sites which determine specificity, the above conclusion is not warranted. It is the author's opinion that the inter-strain and inter-species specificities of the *Rhizobium* LPS structures make these molecules likely candidates for a role in determining symbiotic specificity.

In addition to *Rhizobium* ECPS, LPS have been proposed to be the receptors for the lectins from the host plant (Wolpert and Albersheim 1976). Using lectin–Sepharose affinity columns and LPS purified by phenol–water extraction, Wolpert and Albersheim (1976) obtained data which suggested that the LPS from the symbiont bacteria bound only to the host lectin–Sepharose affinity column. More recently it was reported that the LPS from a strain of *R. leguminosarum* bound specifically to pea lectin, while the LPS from a strain of *R. japonicum* bound specifically to soyabean lectin (Kato, Maruyama, and Nakamura 1979). Binding of the LPS to the lectin was assayed by the ability of the LPS to inhibit the haemagglutinating activity of the lectin. The specificity of binding was not absolute, in that *R. leguminosarum* LPS also inhibited haemagglutination by soyabean lectin but at a minimum concentration which is greater than the minimum concentration required to inhibit pea-lectin haemagglutination. Kamberger (1979) has measured the interaction of lectins with *Rhizobium* LPS by using double diffusion studies. He used LPS from strains of *R. meliloti, R. leguminosarum, R. phaseoli, Rhizobium* sp, *R. lupini*, and *R. japonicum* and obtained results which differ somewhat from those obtained by Kato *et al.* (1979). *Rhizobium leguminosarum* LPS specifically forms a precipitin band with lectin isolated from the seeds of *Lens culinaris* and *Pisum sativum*, and *R. meliloti* LPS specifically forms a precipitin line with lectin isolated from *Medicago sativa*. Thus in these two cases the lectins from the hosts do specifically bind the LPS from the symbiont. However, the lectin isolated from *Phaseolus vulgaris* did not form a precipitin line with the LPS from any of the *Rhizobium* species, including *R. phaseoli*. This was also true for the lectin from *Glycine max*, i.e. it did not form a precipitin line with the LPS from any species of *Rhizobium* including *R. japonicum*. This differs from Kato *et al.*'s (1979) results which show that *R. japonicum* LPS specifically inhibits soyabean lectin haemagglutination. From these studies it appears that the specific receptors on the symbiont *Rhizobium* for the host lectin may be the ECPS in the case of *R. japonicum* (see p. 222), the LPS in the case of *R. leguminosarum* and *R. meliloti* and, as yet, an undetermined molecule(s) in the case of *R. phaseoli, R. lupini*, and *Rhizobium* sp. However, in both of these studies chemical analysis of the ECPS and LPS preparations were not carried out and the purification procedures used may not have resulted in ECPS preparations free of LPS or vice versa.

Regarding the above lectin-binding studies and those described on pp. 220–223, it should be noted that (a) many have been done with seed lectin and not with lectin isolated from the host root, (b) the ECPS and LPS have been isolated from bacteria grown in yeast extract or synthetic media and not from bacteria grown in root exudates or grown in the presence of the host root, (c) often the LPS and ECPS preparations have not been adequately characterized so that purity can be judged, and (d) there are data which suggest a lack of correlation between lectin binding and nodulation specificity (Law and Strijdom 1977; Chen and Phillips 1976). These considerations are important in that the lectin in the root may have different or additional receptor sites than the seed lectin. Dazzo (see Chapter 9) has used lectin, i.e. trifoliin, isolated from the clover root in his studies; however, the identity of the *R. trifolii* lectin receptor is still not known (Dazzo and Brill 1979). Bhuvaneswari and Bauer (1978) have shown that effective strains of *R. japonicum* which do not have receptors for the soyabean lectin when grown in culture, obtain these receptors when grown in the presence of soyabean roots or in soyabean-root exudate. Lectins in the roots of pea, clover, and soyabean have been reported and the binding of these lectins to *Rhizobium* surface molecules is under investigation (Bhuvaneswari and Bauer 1978; Dazzo, Chapter 9; Kijne *et al.* 1980).

Rhizobium LPS have been implicated in the differentiation of the bacteria to the bacteroid form. Van Brussel, Planqué, and Quispel (1977) have shown that the cell wall of *R. leguminosarum* in the bacteroid form is much more porus, sensitive to osmotic shock, and has less polysaccharide than the cell wall of the bacteria. These data suggest that there has been a change in the LPS as the bacteria differentiate to bacteroids. In a subsequent study, Planqué, van Nierop, Burgers, and Wilkinson (1979) analysed the LPS from *R. leguminosarum* bacteria and bacteroids as extracted into the water phase using the phenol–water extraction procedure. While in many cases this crude LPS preparation would contain substantial amounts of acidic ECPS (Carlson *et al.* 1978), the strain of *R. leguminosarum* used excreted it only small amounts, and thus these LPS preparations are probably free of contaminating ECPS. Comparison of the LPS preparation from the bacteria with that of the bacteroid shows that they are qualitatively the same. However, they are quite different quantitatively. The bacteroid LPS preparations contain three to ten times less 2-*O*-methylfucose, fucose, mannose, and heptose per milligram of LPS than the bacteria LPS. There were no differences in the gel electrophoresis patterns of bacteroid and bacteria LPS. While these data suggest that there are chemical differences between the bacteroid and bacteria LPS, there was substantial loss of the bacteroid LPS during purification and it is suggested by these workers that the material which was analysed may not be representative of bacteroid LPS (Planqué *et al.* 1979). Thus until further studies are carried out, these data must be interpreted with caution.

The interaction of *Rhizobium* with bacteriophages implicate the LPS as being involved in the symbiotic process. The LPS of Gram-negative bacteria are often the specific receptors for bacteriophage (for reviews see Sharon 1975; Lindberg 1977). The LPS of *Rhizobium* have also been shown to be receptors for bacteriophage (Zajac, Russa, and Lorkiewicz 1975; Pfister and Lodderstaedt 1977). Recently a series of papers has presented data which show that 50 per cent of *R. trifolii* mutants selected for their resistance to a bacteriophage are altered in several aspects of symbiosis (Evans, Barnet, and Vincent 1979*a*, *b*; Barnet 1979). One type of mutant infected the host but did not differentiate to the bacteroid form, another was non-infective, and another was non-infective and produced tumors. Some of the mutants which were infective had a much lower ability to compete for infection in the presence of the parental strain (Evans *et al.* 1979*b*). Most of the non-infective strains were altered in their somatic antigen response (Barnet 1979). Since the strain-specific somatic antigens of *Rhizobium* are the LPS (Humphrey and Vincent 1969; Carlson *et al.* 1978; Zevenhuizen *et al.* 1980) these data suggest that alterations in the LPS affect the ability of the *Rhizobium* to infect the host plant. In addition to the LPS, the ECPS of Gram-negative bacteria are also known to serve as bacteriophage receptors (Lindberg 1977). Up to 50 per cent of the bacteriophage-resistant clones of a mucoid *R. trifolii* strain were non-mucoid (Barnet 1979), suggesting an alteration in the ECPS. In addition, *Rhizobium* bacteriophage produce specific enzymes which hydrolyse the ECPS (Barnet and Humphrey 1975; Dandekar and Modi 1978). Extracellular polysaccharide preparations from *R. japonicum* have also been shown to inhibit the infection of the *R. japonicum* by the bacteriophage M-1 (Dandekar and Modi 1978). It is likely, therefore, that *Rhizobium* ECPS also serve as receptors for some bacteriophage. Thus these bacteriophage studies do not rule out a role for *Rhizobium* ECPS in determining symbiosis. However, bacteriophages will be useful in the selection of mutants which are altered in specific aspects of the nodulation process.

Role of Rhizobium *flagella in symbiosis*

Very little has been published regarding the role of flagella in the nodulation process. However, two recent reports suggest that flagella do not play a role in nodulation. It has been found that three types of mutants of *R. meliloti*, all affecting motility, still effectively nodulate *Medicago sativa* (Ames, Schluederberg, and Bergman 1980). These mutants are (a) nonflagellated, (b) nonmobile but flagellated and (c) nonchemotactic mutants. The other report by Napoli and Albersheim (1980) shows that nonmotile mutants of *R. trifolii* are as effective at nodulating the host clover as the parental *R. trifolii*. These reports essentially rule out the possibility that *Rhizobium* flagella play a role in determining symbioses.

7.5 Summary

The data regarding the role of *Rhizobium* ECPS and LPS in determining the specificity of symbiosis are somewhat confusing. Literature can be found which support a role for the ECPS while other literature support a role for the LPS. The solution to this confusion lies in the further characterization of the *Rhizobium* surface molecules.

The data available suggest that the molecule which has the largest degree of structural diversity on the surface of the fast-growing *Rhizobium* is the LPS, while the molecule on the surface of the slow-growing *Rhizobium* with the largest structural diversity may be the ECPS. The LPS from the slow-growing organisms have not been characterized in any detail to make conclusions with regard to their structural diversity. Given the structural diversity of ECPS for slow-growing strains and the structural diversity of LPS for fast-growing ones, perhaps it is not surprising then that when a specific receptor for the host lectin has been found, it is the ECPS in the case of the slow-growing *R. japonicum* and the LPS in the case of the fast-growing *R. leguminosarum* and *R. meliloti* species (Kato *et al.* 1979; Kamberger 1979).

The idea that one type of *Rhizobium* could have the ECPS as the strain-specific determinant while another type has the LPS as the strain-specific determinant would not be without precedence. Among the strains of *E. coli* are those in which the capsular ECPS determine the antigenic specificity and the specificity of infection, while in other strains it is the LPS (Ørskov *et al.* 1977; Jann and Jann 1977). In fact surface chemical and immunochemical studies have revealed that several general types of *E. coli* exist (Jann and Jann 1977). Type 'a' *E. coli* contain neutral O-antigenic LPS which vary greatly from strain to strain and contain a low molecular-weight acidic capsular ECPS, which often consists of neuraminic acid. Type 'b1' *E. coli* strains contain O-antigenic LPS which have homopolysaccharides (e.g. mannans) attached to the lipid-A and high molecular-weight acidic capsular ECPS which are heteropolysaccharides, vary greatly from strain to strain and are the strain-specific antigenic determinants. In fact, some strains of this type have the acidic ECPS polysaccharide bound to the core-lipid-A. Type 'b2' *E. coli* contain two types of LPS, those with a homopolysaccharide as the O-antigen (e.g. a mannan) and those with an acidic heteropolysaccharide as the O-antigen. Both of these LPS molecules are present on the same bacteria. Type 'c' *E. coli* have LPS with neutral heteropolysaccharides as the O-antigen and do not have ECPS. These strains are similar to *Salmonella* strains (Lüderitz *et al.* 1966), and contain many different O-groups. Type 'd' *E. coli* have only acidic heteropolysaccharides as the O-antigen. The same acidic hetero-polysaccharide can be present as high molecular-weight ECPS in strains of this type.

Based on the data described in the previous sections, it seems likely that several types of *Rhizobium* exist. These types are shown in Table 7.6. Type I *Rhizobium* have acidic ECPS which contain uronic acid, are of high molecular-weight, vary greatly in structure from strain to strain, and have unusual methylated hexoses. It is this ECPS from *R. japonicum* which appears to bind specifically to the host, soyabean lectin. The LPS of these *Rhizobium* have not been examined in any detail. Thus it is not known whether the *R. japonicum* LPS has a polysaccharide attached to the lipid-A which is the same as the acidic ECPS, a homopolysaccharide such as a mannan or a heteropolysaccharide which is different from the ECPS. Type II *Rhizobium* contain LPS in which the polysaccharide varies greatly in structure from strain to strain and also contains many unusual methylated hexoses. The ECPS of Type II is an acidic heteropolysaccharide containing uronic acid, of high molecular-weight and does not vary in structure from strain to strain. Examples of Type II *Rhizobium* are strains of *R. leguminosarum*, *R. trifolii*, and *R. phaseoli*. Type III *Rhizobium* contain LPS which have an acidic heteropolysaccharide that does not vary greatly from strain to strain and contains KDO as its principal sugar constituent. The LPS of this type differ greatly in comparison to the LPS of Type II in that there are no deoxy sugars or unusual methylated sugars present. The ECPS of Type III *Rhizobium* is an acidic heteropolysaccharide which does not contain uronic acid but does contain pyruvate groups and perhaps succinate groups. These ECPS do not vary in structure from strain to strain. *Rhizobium* in this group consist of the *R. meliloti* strains and perhaps the pathogenic *Agrobacterium* species. It is possible that *Agrobacterium* may be of this type based on the similarity of its ECPS structure to that of the *R. meliloti* strains. All of the three *Rhizobium* types contain β-1,2-glucans and β-1,4-glucans as homopolysaccharides.

The existence of these three different *Rhizobium* types is also supported by serological and bacteriophage studies. Using antiserum to whole bacteria and determining the relationships of different *Rhizobium* with agglutination tests, three broad serological groups have been defined. Group 1 consists of *R. trifolii*, *R. leguminosarum*, and *R. phaseoli*; group 2 consists of *R. meliloti*, and group 3 consists of *R. japonicum* and *R. lupini* (Graham 1969). Furthermore, internal antigens define *R. trifolii*, *R. leguminosarum*, and *R. phaseoli* as one group, *R. meliloti* as another, and *R. japonicum* and *R. lupini* as the third group (Humphrey and Vincent 1965; Vincent and Humphrey 1970; Vincent, this volume). Bacteriophage studies have also defined these three *Rhizobium* groups. Phage isolated from any individual strains of *R. trifolii*, *R. leguminosarum*, or *R. phaseoli* lyse some strains of all three of these species. These phages do not lyse strains of *R. meliloti* or *R. japonicum*. Phage isolated from strains of *R. meliloti* lyse only *R. meliloti*, while phage isolated from strains of *R. japonicum* lyse only *R. japonicum* (Napoli, Sanders, Carlson, and Albersheim 1980). Thus the three broad groups of *Rhizobium* defined by serological and bacteriophage studies are the same as the three types of *Rhizobium* defined by the chemistry of the ECPS and LPS shown in

TABLE 7.7
Types of Rhizobium

Surface carbohydrates	I	II	III
LPS	? (Highly purified molecules have not been examined chemically or immuno-chemically)	Consist of uronic acid containing heteropolysaccharides which vary greatly in structure. O-antigen polysaccharides contain many unusual methylated hexoses	Consist of uronic acid containing heteropolysaccharides which are very similar in structure from strain to strain. These LPS typically contain KDO as the major sugar component
ECPS	Uronic acid containing acidic heteropolysaccharides of high molecular weight. These ECPS are highly variable in structure and contain unusual methylated hexoses	Uronic acid containing acidic heteropolysaccharides of high molecular weight which do not vary in structure from strain to strain	Consist of acidic heteropolysaccharides which do not contain uronic acid groups but do contain pyruvic acid groups. These ECPS structures do not vary from strain to strain and appear to be very similar in structure to the ECPS of *Agrobacterium tumefaciens*
	Low molecular weight neutral β-2-linked and -4-linked glucans are present as homopolymers	Low molecular weight neutral β-2-linked and β-4-linked glucans are present as homopolymers	Low molecular weight neutral -2-linked and -4-linked glucans are present as homopolymers
Examples of *Rhizobium* strains	*R. japonicum* and *R. lupini*	*R. leguminosarum R. trifolii* and *R. phaseoli*	*R. meliloti,* and perhaps *Agrobacterium*

Table 7.6.

The surface molecule(s) which determines symbiotic specificity could vary depending on the *Rhizobium* type. For example, it seems possible that in Type I *Rhizobium* the ECPS determine symbiotic specificity, while in Type II the LPS determine specificity. Detailed chemical and immunochemical investigations of the purified surface polysaccharides are needed for many more strains of *Rhizobium* in order to understand the inter-strain and inter-species relationships of these molecules and also to understand the relationship between the LPS and the ECPS of these three *Rhizobium* types. Analysis of the surface chemistry of *Rhizobium* mutants which have been altered in certain aspects of the nodulation process (e.g. Evans *et al.* 1979*a*, *b*; Barnet 1979; Sanders *et al.* 1978) or which have altered nodulation specificities (see Chapter 5 by Beringer *et al.*) should reveal much about which of these molecules are involved in determining symbiotic specificity.

Acknowledgements

The author wishes to acknowledge the following, who offered critical reviews of the manuscript: Dr P. Albersheim, Dr A. Darvill, Dr T. L. Graham, and Dr L. P. T. M. Zevenhuizen. In addition, the author acknowledges personal communication and manuscript preprints from Dr P. Albersheim, Dr L. P. T. M. Zevenhuizen, Dr W. F. Dudman, and Dr. B. Solheim.

References

Albersheim, P., Nevens, D. J., English, P. D., and Karr, A. (1967). *Carbohydr. Res.* **5,** 340–5.

Allen, A. K., Desai, N. N., and Newberger, A. (1976). *Biochem. J.* **155,** 127–35.

Ames, P., Schluederberg, S. A., and Bergman, K. (1980). *J. Bacteriol.* **141,** 722–7.

Anderson, M. A. and Stone, B. A. (1978). *Carbohydr. Res.* **61,** 479–92.

Ayers, A. R., Ayers, S. B., and Goodman, R. N. (1979). *Appl. Environ. Microbiol.* **38,** 659–66.

Bailey, R. W., Greenwood, R. M., and Craig, A. (1971). *J. gen. Microbiol.* **65,** 315–24.

Bal, A. K., Shantharam, S., and Ratnam, S. (1978). *J. Bacteriol.* **133,** 1393–400.

Barnet, Y. M. (1979). *Can. J. Microbiol.* **25,** 979–86.

—— and Humphrey, B. (1975). *Can. J. Microbiol.* **21,** 1647–50.

Bhagwat, A. A. and Thomas, J. (1980). *J. gen. Microbiol.* **117,** 119–25.

Bhuvaneswari, T. V. and Bauer, W. D. (1978). *Plant. Physiol.* **62,** 71–4.

—— Pueppke, S. G., and Bauer, W. D. (1977). *Plant Physiol.* **60,** 486–91.

Bitter, T. and Muir, H. M. (1962). *Analyt. Biochem.* **4,** 332–4.

Björndall, H., Hellerquist, C. B., Lindberg, B., and Svensson, S. (1970). *Angew. Chem. Int. Ed.* **9,** 610.

Blumenkrantz, N. and Asboe-Hansen, G. (1973). *Analyt. Biochem.* **54,** 484–9.

Bohlool, B. B. and Schmidt, E. L. (1976). *J. Bacteriol.* **125,** 1188–94.

Broughton, W. J. (1978). *J. appl. Bacteriol.* **45,** 165–94.

Carlson, R. W., Sanders, R. E., Napoli, C., and Albersheim, P. (1978). *Plant. Physiol.* **62,** 912–17.

Chen, A. T. and Phillips, D. A. (1976). *Physiol. Plant.* **38,** 83–8.
Dandekar, A. M. and Modi, V. V. (1978). *Can. J. Microbiol.* **24,** 685–8.
Dart, P. J. (1974). In *The biology of nitrogen fixation* (ed. A. Quispel). pp. 381–429. North Holland, Amsterdam.
Dazzo, F. B. and Brill, W. J. (1977). *Appl. Environ. Microbiol.* **33,** 132–6.
—— —— (1979). *J. Bacteriol.* **137,** 1362–73.
—— Yanke, W. E., and Brill, W. J. (1978). *Biochim. biophys. Acta* **539,** 276–86.
Dedonder, R. and Hassid, W. Z. (1964). *Biochim. biophys. Acta* **90,** 239–48.
Deinema, M. H. and Zevenhuizen, L. P. T. M. (1971). *Arch Mikrobiol.* **78,** 42–57.
Dmitriev, B. A., Knirel, Y. A., Kochetkov, N. K., Jann, B., and Jann, K. (1977). *Eur. J. Biochem.* **79,** 111–15.
Dudman, W. F. (1976). *Carbohydr. Res.* **46,** 97–110.
—— (1977). In *Surface carbohydrates of the procaryotic cell* (ed. I. Sutherland), p. 357. Academic Press, London.
—— (1978). *Carbohydr. Res.* **66,** 9–23.
—— and Jones, A. J. (1980). *Carbohydr. Res.,* in press.
Evans, J., Barnet, Y. M., and Vincent, J. M. (1979*a*). *Can. J. Microbiol.* **25,** 968–73.
—— —— —— (1979*b*). *Can. J. Microbiol.* **25,** 974–8.
Fåhraeus, G. and Ljunggren, H. (1959). *Physiol. Planta.* **12,** 145–54.
Galanos, C. and Lüderitz, O. (1975). *Eur. J. Biochem.* **54,** 603–10.
Gorin, P. A. J., Spencer, J. F. T., and Westlake, D. W. S. (1961). *Can. J. Chem.* **39,** 1067–73.
Graham, P. H. (1969). In *Analytical serology of microorganisms,* Vol. 2. (ed. F. Kwapinski). pp. 353–78. Interscience, New York.
Graham, T. L., Sequeira, L., and Huang, T. R. (1977). *Appl. Environ. Microbiol.* **34,** 424–32.
Harada, T., Amemura, A., Jansson, P. E., and Lindberg, B. (1979). *Carbohydr. Res.* **77,** 285–8.
Hepper, C. M. and Lee, L. (1979). *Pl. Soil* **51,** 441–5.
Hestrin, S. (1949). *J. biol. Chem.* **180,** 249–61.
Hubbell, D. H. (1970). *Bot. Gaz.* **131,** 337–42.
Hughes, T. A., Lecce, J. G., and Elkan, G. H. (1979). *Appl. Environ. Microbiol.* **37,** 1243–4.
Humphrey, B. A. and Vincent, J. M. (1959). *J. gen. Microbiol.* **21,** 477–84.
—— —— (1965). *J. gen. Microbiol.* **41,** 109–18.
—— —— (1969). *J. gen. Microbiol.* **59,** 411–25.
Janda, J. and Work, E. (1971). *FEBS Lett.* **16,** 343.
Jann, B., Jann, K., and Schmidt, G. (1970). *Eur. J. Biochem.* **15,** 29–39.
—— Reske, K., and Jann, K. (1975). *Eur. J. Biochem.* **60,** 239–46.
Jann, K. and Jann, B. (1977). In *Surface carbohydrates of the procaryotic cell* (ed. I. Sutherland). pp. 247. Academic Press, London.
Jansson, E., Lindberg, B., and Ljunggren, H. (1979). *Carbohydr. Res.* **75,** 207–20.
Jansson, P., Kenne, L., Lindberg, B., Ljunggren, H., Lönngren, J., Ruden, U., and Svensson, S. (1977). *J. Am. Chem. Soc.* **99,** 3812–5.
Kamberger, W. (1979). *Archs Microbiol.* **121,** 83–90.
Kato, G., Maruyama, Y., and Nakamura, M. (1979). *Agric. Biol. Chem.* **43,** 1085–92.
Kennedy, L. D. (1978). *Carbohydr. Res.* **61,** 217–21.
—— and Bailey, R. M. (1976). *Carbohydr. Res.* **49,** 451–4.
Kijne, J. W., Van der Schaal, I. A. M., and DeVries, G. E. (1980). *Plant Sci. Lett.* **18,** 65–74.
Law, I. J. and Strijdom, B. W. (1977). *Soil. Biol. Biochem.* **9,** 79–84.
Lindberg, A. A. (1977). In *Surface carbohydrates of the prokaryotic cell* (ed. I. Sutherland). pp. 289–356. Academic Press, London.

Lindberg, B. and Lönngren, J. (1978). In *Methods in enzymology* Vol. L (ed. V. Ginsberg), pp. 3–33. Academic Press, New York.

Ljunggren, H. and Fåhraeus, G. (1961). *J. gen. Microbiol.* **26**, 521–28.

Lorkiewicz, Z. and Russa, R. (1971). *Pl. Soil*, Special Volume 105–9.

Lüderitz, O., Jann, K., and Wheat, R. (1968). *Comp. Biochem.* **26A**, 105–228.

—— Staub, M. A., and Westphal, O. (1966). *Bacteriol. Rev.* **30**, 192–255.

McIntire, F. C., Peterson, W. H., and Riker, A. J. (1942). *J. biol. Chem.* **143**, 491–6.

Misaki, A., Saito, H., Ito, T., and Harada, T. (1969). *Biochemistry* **8**, 4645–50.

Mort, A. J. and Bauer, W. D. (1980). *Plant Physiol.*, *Lancaster*, **66**, 158–63.

—— Slodki, M. E., Plattner, R. D., and Bauer, W. D. (1979). *Plant Physiol.*, *Lancaster* **63**, 135, Abs. 746.

Müller-Seitz, E., Jann, B., and Jann, K. (1968). *FEBS Lett.* **1**, 311–14.

Napoli, C. and Albersheim, P. (1980). *J. Bacteriol.* **141**, 979–80.

—— Dazzo, F., and Hubbell, D. (1975). *Appl. Microbiol.* **30**, 123–32.

—— Sanders, R., Carlson, R., and Albersheim, P. (1980). In *Nitrogen fixation*, Vol. 2 (ed. W. H. Orme-Johnson and W. E. Newton). University Park Press, Baltimore.

Osborn, M. J. (1963). *Proc. natn. Acad. Sci. U.S.A.* **50**, 499–506.

Ørskov, I., Ørskov, F., Jann, B., and Jann, K. (1977). *Bacteriol. Rev.* **41**, 667–710.

Palomares, A., Montoya, E., and Olivares, J. (1979). *Microbios.* **22**, 7–13.

Pfister, H. and Lodderstaedt, G. (1977). *J. gen. Virol.* **37**, 337–47.

Planqué, K. and Kijne, J. W. (1977). *FEBS Lett.* **73**, 64–6.

—— van Nierop, J. J., Burgers, A., and Wilkinson, S. G. (1979). *J. gen. Microbiol.* **110**, 151–9.

Putman, E. W., Potter, A. L., Hodgson, R., and Hassid, W. Z. (1950). *J. Am. Chem. Soc.* **72**, 5024–6.

Russa, R. and Lorkiewicz, Z. (1974). *J. Bacteriol.* **119**, 771–5.

Ryan, J. M. and Conrad, H. E. (1974). *Arch. Biochem. Biophys.* **162**, 530–5.

Sahlman, K. and Fåhraeus, G. (1963). *J. gen. Microbiol.* **33**, 425–7.

Saito, H., Misaki, A., and Harada, T. (1970). *Agr. Biol. Chem.* **34**, 1683–9.

Sanders, R., Carlson, R. W., and Albersheim, P. (1978). *Nature, Lond.* **271**, 240–2.

Schmidt, E. (1979). *A. Rev. Microbiol.* **33**, 355–76.

Sequeira, L. (1978). *A. Rev. Phytopath.* **16**, 453–81.

—— and Graham, T. L. (1977). *Physiol. Plant Pathol.* **11**, 43–54.

Shands, J. W. and Chun, P. W. (1980). *J. biol. Chem.* **255**, 1221–6.

Shantharam, S., Gow, J. A., and Bal, A. K. (1980). *Can. J. Microbiol.* **26**, 107–14.

Sharon, N. (1975). In *Complex carbohydrates. Their chemistry, biosynthesis and functions*, pp. 318–90. Addison-Wesley, Reading, MA.

Sómme, R. (1974). *Carbohydr. Res.* **33**, 89–96.

—— (1975). *Carbohydr. Res.* **43**, 145–9.

Sutherland, I. W. (1969). *Biochem. J.* **115**, 935–45.

—— (1977). In *Surface carbohydrates of the prokaryotic cell* (ed. I. Sutherland), p. 27. Academic Press, London.

Svensson, S. (1978). In *Methods in enzymology* Vol. L (ed. V. Ginsberg), pp. 33–8. Academic Press, New York.

Tarcsay, L., Wang, C. S., Li, C. S., and Alanpovic, P. (1973). *Biochemistry* **12**, 1948–55.

Taylor, R. L. and Conrad, H. E. (1972). *Biochemistry* **11**, 1383–8.

Tsien, H. C. and Schmidt, E. L. (1977). *Can. J. Microbiol.* **23**, 1274–84.

Valent, B. S., Darvill, A. G., McNeil, M., Robertsen, B. K., and Albersheim, P. (1980). *Carbohydr. Res.* **79**, 165–92.

Van Brussel, A. A. N., Planqué, K., and Quispel, A. (1977). *J. gen. Microbiol.* **101**, 51–6.

Van Wauwe, J. P., Loontiens, F. G., and de Bruyne, C. K. (1975). *Biochim. biophys. Acta* **379**, 456–61.

Vincent, J. M. (1974). In *The biology of nitrogen fixation* (ed. A. Quispel), pp. 265–341. North Holland, Amsterdam.

—— and Humphrey, B. (1970). *J. gen. Microbiol.* **63,** 379–82.

Wang, C. S. and Alanpovic, P. (1973). *Biochemistry* **12,** 309–15.

Weissbach, A. and Hurwitz, J. (1958). *J. biol. Chem.* **234,** 705–9.

Westphal, O. and Jann, K. (1965). In *Methods in carbohydrate chemistry*, Vol. 5 (ed. R. L. Whistler) pp. 83–91. Academic Press, London.

Whatley, M. H. and Spiess, L. D. (1977). *Plant Physiol., Lancaster* **60,** 765–6.

—— Bodwin, J. S., Lippincott, B. B., and Lippincott, J. A. (1976). *Infect. Immunol.* **13,** 1080–3.

—— Hunter, N., Cantrell, M. A., Hendrick, C., Keegstra, K., and Sequeira, L. (1980). *Plant. Physiol., Lancaster* **65,** 557–9.

Wilkinson, S. G. (1977). In *Surface carbohydrates of the prokaryotic cell* (ed. I. Sutherland), pp. 97–175. Academic Press, London.

Wolpert, J. S. and Albersheim, P. (1976). *Biochem. Biophys. Res. Commun.* **70,** 729–37.

Yao, P. Y. and Vincent, J. M. (1969). *Aust. J. Biol. Sci.* **22,** 413–23.

—— —— (1976). *Pl. Soil* **45,** 1–16.

Yin, E. T., Galanos, C., Kinsky, S., Bradshaw, R. A., Wessler, S., Lüderitz, O., and Sarmiento, M. T. (1972). *Biochim. biophys. Acta* **261,** 284–9.

York, W. S., McNeil, M., Darvill, A. G., and Albersheim, P. (1980). *J. Bacteriol.* **142,** 243–8.

Zajac, E., Russa, R., and Lorkiewicz, Z. (1975). *J. gen. Microbiol.* **90,** 365–7.

Zevenhuizen, L. P. T. M. (1971). *J. gen. Microbiol.* **68,** 239–43.

—— (1973). *Carbohydr. Res.* **26,** 409–19.

—— and Scholten-Koerselman, H. J. (1979). *Anton. van Leeuwen.* **45,** 165–75.

—— —— and Posthumus, M. A. (1980). *Archs Microbiol.* **125,** 1–8.

8 Serology

J. M. VINCENT

8.1. Introduction

Serological reactions, as commonly utilized with the rhizobia, involve two major reactants: *antigen*, a large molecule, such as protein or complex polysaccharide, and *antibody*, produced by an animal in response to an immunodominant portion of a foreign antigen. The marked degree of specific conformity between the antibody and its antigenic determinant, gives the method particular value for recognition of identity or relatedness in a test system. When the antigen is attached to the surface of a cell, reaction with antibody can be detected by clumping (agglutination) of the cells. Antigens associated with flagella cause loosely flocculent agglutination, whereas those directly on the body of the cell (somatic) result in compact finely granular aggregation. Antibody occurs in several molecular forms of globulin which share the same specificity for antigen. The most common of these, and those most utilized *in vitro*, are the larger (IgM) and smaller (IgG) immunoglobulins. The balance between the two forms in an antiserum varies according to such factors as the nature of the antigen, its attachment to or separation from the cell, and the length and nature of the immunization regimen. Although the two forms of antibody share the same antigen specificity, they differ in ease of demonstration.

Removal (absorption) of antibody from antiserum can be used as a technique for demonstrating specificity between the absorbing compound or complex antigen, as a way of determining antigenic identity, and to provide monospecific antisera by removing cross-reacting antibody.

Some serological, or serological-like reactions can occur without the one reactant having resulted from prior exposure to the other. An example of this phenomenon is the ability of antisera to some pneumococcus polysaccharides to precipitate other polysaccharides of diverse origin, but having some sufficiently similar chemical groupings for combination to occur. Another is the capacity of some plant proteins (lectins) to act as 'pseudo-antibodies' and combine with various cell-surface components, including those of some rhizobia.

The application of several serological methods to the study of the root-nodule bacteria goes back to the first decade of this century, but neither the methods used nor the number of strains examined was adequate for the purposes of these early investigations. Fred, Baldwin, and McCoy (1932) provide a brief review to that date. Agglutination was most widely used, and within these limitations it was found that cross agglutination outside several of the established inoculation groups was rare (and of low titre) but that there were two or several non cross-agglutinating subgroups within each inoculation group (Stevens 1923, 1925; Wright 1925). Bushnell and Sarles (1939) recognized the need for improved techniques and used three variously prepared suspensions: unheated whole cells, heated cells, and saline extracted. The results they gave were, however, almost all restricted to those with unheated whole cells and they attempted no distinction between flagellar and somatic reactions. In fact their statement that unheated cells gave, in general, the same results as those that were heated makes it seem unlikely that their unheated whole-cell reaction included the flagellar type. Keys to some of the earlier tabulations (such as Wright 1925) suggest that floccules characteristic of flagellar agglutination had at times been observed without their significance being appreciated.

Distinction between agglutination due to flagellar and somatic antigens and the application of antibody absorption tests have improved understanding of the antigenic constitution of rhizobia and the ability of the test to distinguish serotypes and improve strain recognition (Vincent 1941, 1942; Kleczkowski and Thornton 1944). More recently application of the Ouchterlony gel immune-diffusion technique for the detection of soluble, or solubilized, antigens (Dudman, 1964) has assisted in more direct strain recognition and the demonstration of antigens not detectable by agglutination. These possibilities are being extended further by various modifications of more sensitive immune electrophoretic methods. Fluorescent labelled antibody has been successfully applied to the direct demonstration of the nature of the strain, or strains, occupying a nodule (Trinick, 1969), and as a basis for quantitative determination of specific rhizobia in the soil (Bohlool and Schmidt 1972; Schmidt 1974). Both agglutination and gel immune-diffusion methods can be applied directly to larger nodules (Means, Johnson, and Date 1964; Skrdleta 1969a; Dudman 1971) and, with some modification to agglutination procedure, to small nodules as well (Parker and Grove 1970). Alternatively smaller nodules may be handled directly by a passive haemagglutination technique (Cloonan and Humphrey 1976).

Objectives in studying rhizobial antigens are variously to (i) obtain insights into the nature of the bacterium and location of its serologically detectable components, (ii) secure evidence as to likely taxonomic relationships, (iii) define strains in terms of their antigenic composition and to recognise them when used experimentally or routinely, (iv) provide markers for genetic studies and for authentication of mutants, (v) probe possible relationship between antigenic composition and symbiotic capacity.

Several recent reviews have dealt fairly comprehensively, if briefly, with aspects of this topic (Graham 1969; Vincent 1974, 1977; Dudman 1977); for details of older, but still widely used, techniques, see Vincent (1970).

8.2. Techniques

Serological techniques used for the rhizobia are similar to those used with other Gram-negative bacteria, with some modifications associated with the cultivation and properties of these particular organisms. Techniques associated with the simplest and most direct procedures (agglutination and gel diffusion) are described in some detail in the handbook prepared for the International Biological Programme (Vincent 1970). Others which have been subsequently developed will be described in this section. Dudman (1977) has provided a concise account of serological methodology together with a useful background bibliography.

Development of antisera

The principles of developing antisera in an experimental animal are substantially the same whatever form their reaction with antigen takes. Detailed methodology varies from worker to worker. The variations are generally of little practical significance, but, some particular effects and requirements need to be noted.

Choice of animal

Mature laboratory rabbits provide a suitable balance between demand on space and yield of antiserum and of these New Zealand white rabbits produce more blood. Between-animal variation in immunological response can be encountered however, both in antibody quantity and quality. This is particularly likely in the degree of cross-reactivity involving 'minor' antigens and makes it desirable to use two or several animals for each rhizobial strain being utilized and to record the identity of each antiserum. Alternatively, antisera to the same rhizobial strain may be pooled. However, although such pooling of antisera may be convenient, it may lead to some confusion by obscuring variation between animals which ought to be known to the operator.

Preparation of immunizing suspension

Cells can be harvested from the surface of agar slopes or flats based on a complex medium (such as yeast-extract mannitol agar) when one is concerned only with surface-located antigens (agglutination, reaction with FITC-conjugated antibody, or haemagglutination based on somatic lipopolysaccharide). On the other hand tests involving soluble antigens may be confused if antibody is developed to complex constituents of the medium. For that purpose it will be safer to use cells raised on a simpler defined liquid medium or, at least, with yeast extract separated within a dialysis sac.

Development of an adequate titre against flagellar antigens seems to require sufficient flagella-bearing cells in the inoculating suspension. This will be facilitated by suspending the growth of young cells grown on the moist surface of an agar medium in 0.85 per cent sodium chloride. Suspensions for the development of antibodies to somatic antigens may be similarly prepared, or by using the growth from adequately aerated (shaken) liquid culture. In the latter case, the suspended cells may be deposited by centrifugation and resuspended in the saline to provide an appropriate cell density. In either case the inoculating suspension should be freed of large clumps and should contain about $1 \times 10^9 - 10 \times 10^9$ cells ml^{-1}. Heated cells of rhizobia seem generally satisfactory for the production of antibody to somatic antigens and may avoid any complication due to interference by a 'Vi-like' antigen (p. 254). On the other hand, it is possible that with some strains antibody production against heated surface antigen may be reduced, as can happen with other Gram-negative bacteria. (Neter, Hi, and Mayer 1973). Mechanically broken (Mickled) cells are likely to increase antibody response to internal antigens without loss of surface immunogenicity.

Inoculation route and procedure

Of the many detailed procedures which have been used, the intramuscular route with adjuvant, followed by a booster dose (intramuscular as before, or intravenously without adjuvant) is one which can be recommended to provide high titre antisera, with most antibody in the IgG form and therefore suitable for gel immune-diffusion as well as for other techniques.

When a satisfactory level of antibody has been attained (agglutination titre commonly 3200 or better) the blood can be collected by bleeding the animal out via the jugular vein, successive bleeding from the marginal ear vein, or by heart puncture. The latter method requires more skill and experience on the part of the operator but permits collection, under aseptic conditions, of 10–20 ml at 2–3 day intervals from rabbits. The animal can then be kept, with appropriate boosting injections over a longer period. Under these circumstances antibodies to 'minor' antigen have the chance to develop; this may be disadvantageous, if cross-reactivity is thereby extended, or advantageous, if it permits the detection of such an antigen having some particular significance in the organism's behaviour. Antigens reported as being related to specific invasiveness in *R. trifolii* (Dazzo and Hubbell 1975*a*) may have been demonstrable as a result of an extraordinarily extended inoculation procedure, whereas normal regimens have failed to reveal any such relationship, even in one of the same pairs studied by Dazzo and Hubbell (Humphrey and Vincent, unpublished).

After bleeding, the antiserum is separated from the blood clot and either preserved (with 0.5 per cent phenol or with merthiolate) or held as deep-frozen or lyophilized (0.5–1 ml) aliquots. Setting aside small volumes is advantageous in that it reduces the risk of loss of antibody activity as a result of repeated freezing and thawing, or the need to hold an unused portion of reconstituted antiserum for too long a period.

Agglutination

Cells of the testing antigen are developed along the same lines as those for the inoculating antigen, but are harvested with 0.85 per cent sodium chloride from slope or liquid culture based on a complex or simple medium, without aseptic precautions. An unheated suspension, containing relatively young motile cells, will provide for both flagellar and somatic reactions; heating at 100 °C for 30 min destroys flagellar antigens. The test suspension should be moderately turbid (about 5×10^8 cells ml^{-1}), free of clumps, and stable in the saline controls. Most suspensions give stable controls, at least for the 4 h at 52 °C required for most positive somatic reactions. Final confirmation can generally be read after standing overnight at room temperature. A few cultures are, however, relatively unstable and require careful comparison with the saline control even after 4 h incubation. If this is still unsatisfactory, substitution of 0.5 per cent sodium chloride as the suspending solution may help or an alternative method (such as gel immune-diffusion) may have to be used.

Antiserum is dispensed in a series of dilutions in saline, commonly successively two-fold, to cover a range suitable to the purpose of the test (determination of titre or simple recognition of reaction). Small volumes of the serially diluted antiserum are set out in suitable small tubes (such as Dreyer tubes, about 1 ml capacity) arranged in a prerecorded pattern in suitable racks. The test suspension is then added (generally in equal volume) and allowed to mix by convection, induced by having the tubes submerged in the water bath to about a third of their height. A result for flagellar agglutination is produced after 1–2 hours at 52 °C,

ally in 4 h. Thirty-seven degrees Celsius may be used,
ion considerably.

e test suspension contains enough flagella-bearing
action results in the formation of large flocculent,
gates which, when not accompanied by a somatic
leave a distinctly turbid supernatant. A relatively young
culture (2–3 days for a fast-growing *Rhizobium*, 5–6 days for a slow-growing one) grown on a moist agar surface is likely to be most satisfactory for the flagellar reaction. The suspension should be directly prepared in saline from the agar culture; if deposited and separated from the supernatant they are likely to lose most of the flagella (Vincent 1941). Formalin treatment of the cells does not reliably suppress the somatic reaction, as it seems to do with other bacteria carrying abundant peritrichate flagella. Mild heat treatment of antiserum has been used to selectively inactivate somatic-reacting antibodies (Kleczkowski and Thornton 1944). Its general applicability has not been established however.

Somatic reactions, though slower, should be well advanced after 4 h at 52 °C, having passed through a finely granulated stage to having a compact deposit and completely clear supernatant. Towards the limiting dilution, these may remain scintillatingly granular. The final titre can be determined from tubes held overnight at room temperature.

Shortened tests may be used for qualitative purposes such as strain recognition (Vincent 1970). In such quick typing procedures, using somatic agglutination only, relatively crude suspensions obtained directly as the growth in liquid medium and heated at 100 °C for 30 min, can serve as the test antigen—always compared with the corresponding saline control. Slide or tray agglutinations may be used, but with rather special precautions. The juice of crushed nodules can directly provide the test antigen (Means *et al.* 1964 with *Glycine max*; Parker and Grove 1970 with *Trifolium* spp and other pasture species).

Precipitation tests

Whereas antigens responsible for agglutination still function when attached to the bacterial cell, and need to be exposed at that surface to react with antibody, those involved in immunological precipitation need to be solubilized, either before or during the course of the test, and need not be located at the cell surface. The test can be performed by simply giving antigens the opportunity of diffusing from the well into the saline-containing gel. Diffusion methods are likely to reveal more antigens than are available for agglutination; the position of the precipitation bands, as well as the phenomenon of spurring, will give more immediate information as to the identity or partial relatedness of antigens. On the other hand precipitation reactions on their own will

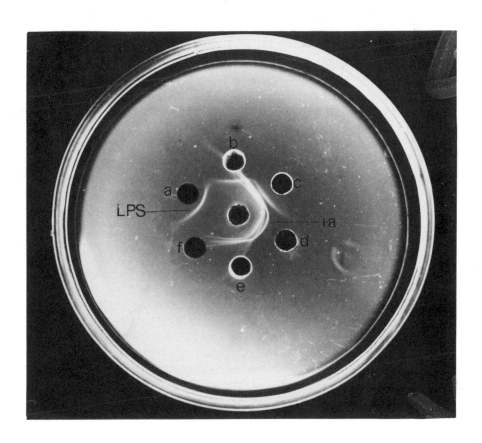

Fig. 8.1. Ouchterlony gel immune-diffusion precipitation lines with rhizobial antigens.

Centre well: Antiserum to *R. trifolii*, TA1

Outer well	Species	Strain	Treatment	Result
a	*R. trifolii*	TA1	LPS	Somatic line
b	*R. trifolii*	TA1	Whole cells	Somatic line
c	*R. trifolii*	TA1	Broken cells	Somatic + internal lines
d	*R. trifolii*	SU297/31	Broken cells	All internal lines shared TA1
e	*R. meliloti*	U45	Broken cells	Some internal lines shared TA1
f	*R.* sp (cowpea)	CB756	Broken cells	No internal lines shared TA1

(Material supplied by B. A. Humphrey.)

give little information about the location of antigens on, or in, the bacterial cell. Precipitation bands due to flagella have been detected with *R. meliloti* (Humphrey and Vincent 1975) but have not often been seen.

Gel immune-diffusion

The Ouchterlony double-diffusion method, developed for rhizobia first by Dudman (1964), has been widely utilized for the detailed study of rhizobial antigens (Humphrey and Vincent 1965, 1969*a*, 1973, 1975; Skrdleta 1969*b*; Vincent and Humphrey 1970; Dudman 1971; Vincent, Humphrey, and Skrdleta 1973), and as a means of strain recognition by direct diffusion of crushed nodules (Skrdleta 1969*a*; Padmanabhan and Broughton 1979; Ikram and Broughton 1980*b*). Procedural details are described in the IBP Manual (Vincent 1970) and in references already quoted; they will not be repeated here. Some points and implications of the method will need to be mentioned however.

When the antigen source is a cell suspension, this needs to be sufficiently concentrated to provide clear lines. However some internal antigen lines have been better resolved with a less concentrated suspension than that best suited to the prominent surface (lipopoly-saccharide) line. Care needs to be taken to standardize the concentration of antigen suspension both in surveys of many cultures and in more detailed studies. In our experience a cell dry weight of 2 mg per well is satisfactory for the demonstration of surface antigens; 0.5 mg is better for resolution of internal lines. Extracted lipopolysaccharide can be detected with as little as 0.1–1.6 μg per well, according to strain (Humphrey and Vincent 1969*a*). More critical work should not be limited to single arbitrary levels of antigen. The demonstration of some heat-resistant antigens will be facilitated by using a heated suspension (Skrdleta 1969*b*; Dudman 1971; Humphrey and Vincent 1975), pro-vided the suspension is tested in the same supernatant as that in which it has been heated. Internal antigens are best demonstrated with broken cells (Vincent and Humphrey 1970; Vincent *et al.* 1973) but some may appear in cells which are leaky because of calcium deficiency (Humphrey and Vincent 1965), age, storage or as a result of freezing and thawing or lyophilization. Ultrasonic disintegration also liberated some faster diffusing, internal, antigens but gave anomalous results with a strain of *R. meliloti* with respect to the homologous surface antigen (Gibbins 1967). Such a sonicated suspension also interfered with the surface antigen reaction of whole cells; this effect was removed by boiling.

To the less experienced, extra precipitation lines due to internal antigens may confuse strain identification. On the other hand, the wider cross reactivity of such antigens accords with taxonomically valid the somatic generally in 4 h.Thirty-seven degrees Celsius may be used, but slows the reaction considerably.

Provided that the test suspension contains enough flagella-bearing cells, the flagellar reaction results in the formation of large flocculent, slow-settling aggregates which, when not accompanied by a somatic reaction, leave a distinctly turbid supernatant. A relatively young culture (2–3 days for a fast-growing *Rhizobium*, 5–6 days for a slow-growing one) grown on a moist agar surface is likely to be most satisfactory for the flagellar reaction. The suspension should be directly prepared in saline from the agar culture; if deposited and separated from the supernatant they are likely to lose most of the flagella (Vincent 1941). Formalin treatment of the cells does not reliably suppress the somatic reaction, as it seems to do with other bacteria carrying abundant peritrichate flagella. Mild heat treatment of antiserum has been used to selectively inactivate somatic-reacting antibodies (Kleczkowski and Thornton 1944). Its general applicability has not been established however.

Somatic reactions, though slower, should be well advanced after 4 h at 52 °C, having passed through a finely granulated stage to having a compact deposit and completely clear supernatant. Towards the limiting dilution, these may remain scintillatingly granular. The final titre can be determined from tubes held overnight at room temperature.

Shortened tests may be used for qualitative purposes such as strain recognition (Vincent 1970). In such quick typing procedures, using somatic agglutination only, relatively crude suspensions obtained directly as the growth in liquid medium and heated at 100 °C for 30 min, can serve as the test antigen—always compared with the corresponding saline control. Slide or tray agglutinations may be used, but with rather special precautions. The juice of crushed nodules can directly provide the test antigen (Means *et al.* 1964 with *Glycine max*; Parker and Grove 1970 with *Trifolium* spp and other pasture species).

Precipitation tests

Whereas antigens responsible for agglutination still function when attached to the bacterial cell, and need to be exposed at that surface to react with antibody, those involved in immunological precipitation need to be solubilized, either before or during the course of the test, and need not be located at the cell surface. The test can be performed by simply giving antigens the opportunity of diffusing from the well into the saline-containing gel. Diffusion methods are likely to reveal more antigens than are available for agglutination; the position of the precipitation bands, as well as the phenomenon of spurring, will give more immediate information as to the identity or partial relatedness of antigens. On the other hand precipitation reactions on their own will longer period. Under these circumstances antibodies to 'minor' antigens have the chance to develop; this may be disadvantageous, if cross-reactivity is thereby extended, or advantageous, if it permits the detection of such an antigen having some particular significance in the organism's behaviour. Antigens reported as being related to specific invasiveness in *R. trifolii* (Dazzo and Hubbell 1975a) may have been demonstrable as a result of an extraordinarily extended inoculation procedure, whereas normal regimens have failed to reveal any such relationship, even in one of the same pairs studied by Dazzo and Hubbell (Humphrey and Vincent, unpublished).

After bleeding, the antiserum is separated from the blood clot and either preserved (with 0.5 per cent phenol or with merthiolate) or held as deep-frozen or lyophilized (0.5–1 ml) aliquots. Setting aside small volumes is advantageous in that it reduces the risk of loss of antibody activity as a result of repeated freezing and thawing, or the need to hold an unused portion of reconstituted antiserum for too long a period.

Agglutination

Cells of the testing antigen are developed along the same lines as those for the inoculating antigen, but are harvested with 0.85 per cent sodium chloride from slope or liquid culture based on a complex or simple medium, without aseptic precautions. An unheated suspension, containing relatively young motile cells, will provide for both flagellar and somatic reactions; heating at 100 °C for 30 min destroys flagellar antigens. The test suspension should be moderately turbid (about 5×10^8 cells ml^{-1}), free of clumps, and stable in the saline controls. Most suspensions give stable controls, at least for the 4 h at 52 °C required for most positive somatic reactions. Final confirmation can generally be read after standing overnight at room temperature. A few cultures are, however, relatively unstable and require careful comparison with the saline control even after 4 h incubation. If this is still unsatisfactory, substitution of 0.5 per cent sodium chloride as the suspending solution may help or an alternative method (such as gel immune-diffusion) may have to be used.

Antiserum is dispensed in a series of dilutions in saline, commonly successively two-fold, to cover a range suitable to the purpose of the test (determination of titre or simple recognition of reaction). Small volumes of the serially diluted antiserum are set out in suitable small tubes (such as Dreyer tubes, about 1 ml capacity) arranged in a prerecorded pattern in suitable racks. The test suspension is then added (generally in equal volume) and allowed to mix by convection, induced by having the tubes submerged in the water bath to about a third of their height. A result for flagellar agglutination is produced after 1–2 hours at 52 °C, groupings amongst the rhizobia (see p. 261), and provides insight into other likely relationships. These include the bacterium responsible for nodulation of *Parasponia* (Trinick 1973), which shows the precipitation pattern of a typical slow-growing *Rhizobium*, and the non-invasible cohabitant found in soyabean nodules (van Rensburg and Strijdom 1972) which gives a pattern typical of fast-growing rhizobia (Humphrey and Vincent, unpublished).

Well patterns may be provided in agar within Petri dishes and, for large scale work, this may be facilitated by a special well-cutting device (Gault, Koci, and Brockwell 1974). Small-well arrangements with agar spread on glass slides may also be used. The recently reported use of the gel immune-diffusion method for detection of rhizobial antigen in soil (Kremer and Wagner 1978) seems suited to highly artificial conditions rather than any ecologically realistic situation.

Immunoelectrophoresis

The older method of immunoelectrophoresis has had only occasional application to the rhizobia as, for example, in comparing the charge characteristics of lipopolysaccharide antigens of strains of *R. trifolii* (Humphrey and Vincent 1969*b*). However new techniques, such as those in which the antigen is moved into an area of even antibody concentration (Axelson, Kroll, and Weeke 1973), have considerable potential for the resolution and identification of antigens. These are now being adapted to the rhizobia. 'Rocket-line immunoelectrophoresis' (RLI) is particularly promising; some typical results are shown in Fig. 8.2.

Fluorochrome-conjugated antibody

The availability of good quality fluorescein isothiocyanate,(FITC), the relative ease of conjugation with antibody, and improved fluorescence microscopy equipment makes this a convenient and reliable method of strain recognition based on surface antigens. The method is particularly useful for the direct study of nodule contents (Trinick 1969; van der Merwe, Strijdom, and van Rensburg 1972; van der Merwe and Strijdom 1973; Lindeman, Schmidt, and Ham 1974). On the other hand it should not be attempted with inferior reactants or optical equipment, and the operator needs to have regular known negative and positive checks so arranged as to ensure a completely objective assessment.

There are variations in the preparation of the conjugated antibody, all of which seem able to give a satisfactory product (as, for example, the use of celite-adsorbed fluorescein isothiocyanate (Rinderknecht 1962), to speed up the operation). Essentially the following steps are involved: (i) globulin fractionation and dialysis, all at low temperature; (ii) determination of protein concentration, by the Biuret reaction stan-

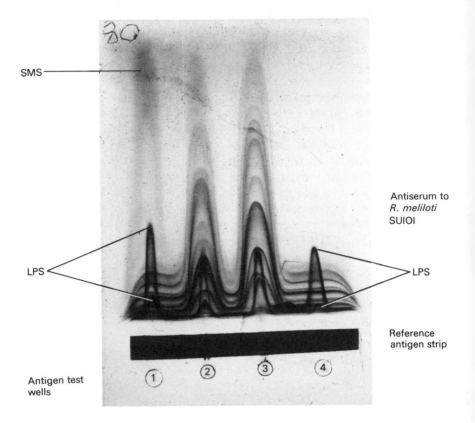

SMS

Antiserum to
R. meliloti
SUIOI

LPS

LPS

Reference
antigen strip

Antigen test
wells

① ② ③ ④

Fig. 8.2. Antigen resolution with rocket-line immunoelectrophoresis (RLI). Reference antigen strip: broken cells of *R. meliloti* SU101/W2. Antigen test wells (left to right):

1. Whole cells of SU101/W2
2. Broken cells of SU101/W2
3. Broken cells of SU101/W1
4. Whole cells of SU101/W1

Peaks due to surface antigens breaking through multiple lines due to internal antigens of broken cells in test strip (1 and 4).

Elevation of the many antigen lines above wells containing broken cells as test antigen (2 and 3).

The high peak due to 'specific masking substance' (SMS) on the strain-specific form (SU101/W2 in well 1) and its absence in the group reactive form (SU101/W1 in well 4).

(Material supplied by M. Taylor.)

dardized against beef globulin; (iii) gradual conjugation with fluorescein isothiocyanate; (iv) separation of faster-moving conjugate from unreacted FITC on a Sephadex column; (v) clarification of conjugate and storage, deep-frozen or lyophilized.

Rhodamine is another fluorochrome suitable for conjugation with antibody; it is mostly used to provide a contrast stain when more than one rhizobial strain is involved. In principle its use is similar to FITC; details differ and are given in references quoted below.

Material to be examined may involve a bacterial suspension, juice from a nodule, or a specially prepared soil suspension. In the first two cases, the material is spread thinly on a good quality microscope slide and fixed by mild heat or, preferably, with acetone. The smears are flooded with conjugated antibody (commonly at 1/8 or 1/16 concentration), thoroughly washed with buffered saline and covered for examination. Soil suspension requires stages of fairly intricate preparation and concentration on a bacterium-retaining membrane (Bohlool and Schmidt 1972; Schmidt 1974).

Examination of the stained preparations requires the use of a microscope equipped for fluorescence microscopy, either with transmitted light (using a dark field condenser) or reflected light. In both cases the light source has to have sufficient output at the short wave-lengths needed to excite the fluorochrome and to be used with an exciter (primary)filter which blocks longer wave-length light whilst permitting sufficient transmission of exciting wavelength. A mercury vapour or halogen lamp is commonly used with a BG12 or KP500 exciter filter. The secondary filter complements the exciter and blocks excitation wavelengths, but passes the longer wave-length light resulting from fluorescence. The reflected light system utilizes a lateral source of light above the objective and has its filter system built into the tube of the microscope. Light passing down through the objective to the object is restricted to the necessary exciter range; that fluoresced by the object is selectively allowed to pass to the ocular. This latter system must be used with the membrane-deposited cells of Bohlool and Schmidt's technique and, in any case, has the advantage that it can be used alternately with substage phase illumination. This has consequent advantages in locating and identifying fields, and conveniently providing for recognition of non-stained as well as fluorescing bacteria.

The method undoubtedly lends itself to direct serological typing of nodule contents and is probably the best one available for an approximate quantitative assessment of specific rhizobia in soil, (lower working limit of about 10^4-10^5 g^{-1}). Fluorescein-conjugated lectin has also been useful in providing evidence of attachment to specific rhizobia (Bohlool and Schmidt 1974), and the attachment of rhizobial capsular material to root-hairs (Dazzo and Brill 1977). Experience in the author's laboratory (Edgley, unpublished) has given disappointing results in the apparently simple task of securing differential quantitation of strains in a mixed suspension. The sum of the proportion of each, determined in parallel counts, generally, and often considerably, exceeded unity. This indicated either a significant degree of cross-staining of one strain in the presence of the other, or differential loss of stained and unstained cells in washing after staining. It was also found that some cells of some strains of *R. trifolii* failed to take up the antibody—an effect which could by obviated by prior mild heat treatment of the suspension (52 °C, 5 min). This last condition is reminiscent of the 'Vi-like' phenomenon responsible for anomalous agglutination failures (Vincent 1953) and could be due to blocking by (possibly capsular) material.

The 'sandwich' method is also available for use with the rhizobia and avoids the need to label a range of antisera. The latter still provide the specific reactions but the attachment of antibody (rabbit) globulin is demonstrated with antibody to this globulin which has been developed in another animal (such as sheep or goat) and labelled with fluorochrome. Such labelled anti-rabbit-globulin is readily available commercially.

For more detail as to procedures, see Nairn (1976) for a comprehensive account and such references as Trinick (1969), and Bohlool and Schmidt (1972) for application to the rhizobia.

Passive haemagglutination (Cloonan and Humphrey 1976).

Passive haemagglutination is a more specialized technique in that it functions in respect of lipopolysaccharide (LPS) antigens and depends on the production of LPS-conjugated red cells. It is capable of typing relatively small individual nodules and depends on LPS on the surface of the nodule bacteria absorbing the specific antibody and so blocking combination with, and agglutination of, specific LPS-conjugated red cells.

Enzyme-linked immunosorbent assay (ELISA)

In this case antibody is coupled to an enzyme and combination between antigen and conjugated antibody is demonstrated by a colour-producing reaction with appropriate substrate. It seems a practical way of readily identifying the occupants of a large number of nodules by means of a macroscopic observation without having to resort to specialized microscopy. As few as 10^4-10^5 cells per ml and as little as 0.4 mg nodule crushed in 1 ml phosphate buffered saline could be typed in this way (Kishinevsky and Bar-Joseph 1978).

Antibody absorption

Methods for establishing qualitative identity, or incomplete sharing of antigens responsible for cross-reactivity, have been successfully used for agglutination reactions (Vincent 1941, 1942; Kleczkowski and Thornton 1944). Quantitative studies, though tedious to perform, can be useful in particular cases, as in the close examination of the effect on the rhizobial cell surface of calcium deficiency (Vincent and Humphrey 1968) and lysogenic conversion (Barnet and Vincent 1970).

Antibody absorption can also be used with gel immune-diffusion, either by pretreating antiserum with the antigen preparation or, more conveniently, by 'in-well' absorption *viz.* the addition of the absorbing antigen to the well containing the antiserum at the time the test is set up (Humphrey and Vincent 1965). In the latter case juxtaposition of an unabsorbed antiserum well will demonstrate the desired condition of antigen excess in the absorbed.

Cross absorption tests also help to determine whether a 'substrain' is truly related to a parent and not a replacement or a contaminant in the original immunizing suspension (Vincent 1944; Kleczkowski and Thornton 1944).

Assessment of methods

Table 8.1 provides a brief comparison of the available methods.

Agglutination remains a basic technique for detecting surface-located antigens. Simplified prodedures permit the occupants of larger nodules to be typed directly; smaller nodules can be handled the same way with some modification, but the smallest require isolation and cultivation before typing by this method. Instability of some cultures in saline and non-agglutinability of a proportion of subcultures of some strains are undesirable features.

The Ouchterlony gel immune-diffusion technique permits other antigens as well as surface antigens, to be detected and, in conjunction with agglutination data, provides information as to location on or in the cell. The relative insensitivity of the method, particularly towards the IgM form of antibody, helps strain recognition by avoiding some of the cross-reactions detectable with agglutination. Positive identification is assisted by the separation of precipitation lines for different antigens and spur-formation in cases of qualitative non-identity. However, in the case of *R. meliloti*, somatic agglutination may provide easier specific recognition of a strain than does gel immune-diffusion. With immune-diffusion the prominent precipitation line is apparently due to a widely shared lipopolysaccharide which, in a recently isolated culture, is likely to be masked by a fast-diffusing, less readily detected, agglutinogen. Care also needs to be exercised to avoid confusion with shared internal antigens. Direct nodule typing has the same possibility, and size limitations as with agglutination.

TABLE 8.1
Comparison of principal serological techniques

Method	Antigens involved	Equipment	Manipulations	Reaction time (h)	Application to Nodule typing	Soil count	Other points
Agglutination	Surface (flagellar and/or somatic)	Simple	Simple repetitive	2–4	Large and medium nodules, direct	n.a.[a]	Detects major and minor antigens; reacts well with IgM and IgG; determines titre
Gel immunodiffusion	Soluble or solubilised	Simple	Simple repetitive	24–72	Large nodules, direct	n.a.	Less sensitive with IgM and minor antigens; permits antigen separation and distinction
Rocket-line immuno electrophoresis	Soluble or solubilised	Electrophoresis unit	Moderately skilled	24	n.t.[b]	n.a.	Permits more detailed analysis
Passive haemagglutination	Surface LPS	Simple	Moderately skilled	24	Large to small nodules	n.a.	
Fluorescent antibody	Surface	Fluorescence microscope	Moderately skilled	c. 1	All nodules	Moderately quantitative	Some distinction between major and minor cross-reactive antigens; distinction in mixed populations

[a] n.a.: not applicable   [b] n.t.: not tested.

The improved immunoelectrophoresis methods have only recently been applied to the rhizobia. They seem to be a useful refinement of simple diffusion methods for detailed antigenic analysis. Whether the degree of technical sophistication involved in their use is justified for routine nodule typing remains to be determined.

Passive haemagglutination is an interesting, if somewhat specialized, means of securing greater sensitivity and hence reducing the size at which direct nodule typing is possible.

Labelled antibody techniques are especially useful in the direct typing of small, as well as large, nodules. The now well-established FITC-conjugated antibody method requires some technical sophistication; this could be reduced if the ELISA method proves as useful for rhizobia as it has for other bacteria. However the fluorescence method, with some considerable supplementation, seems at present the only one which comes anywhere near a direct quantitative assessment of rhizobia in the soil.

8.3 The nature and range of rhizobial antigens

The antigens detected by agglutination are those attached to the cell, and sufficiently close to the surface to be accessible to antibody. They either comprise the flagella or are part of the surface of the cell body (somatic antigens). The old, non-informative, abbreviations 'H' and 'O' are still in use for the flagellar- and somatic-antigens respectively; 'O' is often restricted to lipopolysaccharide (LPS) antigens. Somatic antigens may be 'major' or 'minor', according to their immunogenic (antibody-stimulating) capacity, they may be LPS, some other form of complex polysaccharide, or protein in nature. So long as they are diffusible, or can be prepared in diffusible form, they are demonstrable also by precipitation (such as the Ouchterlony agar method or by immunoelectrophoresis).

The surface somatic antigens of the rhizobia are the most 'strain' specific so that any collection of even the one species is likely to provide many distinctive patterns of agglutination. Flagellar antigens are generally more widely shared amongst strains, generally within a species or amongst closely related species, but some specificities do occur and the combination of somatic and flagellar agglutinability can be even more definitive of a serotype. Internal, non-agglutinating, antigens are widely shared within, but not outside, slow-growing rhizobia on the one hand, and the fast-growing strains, as well as *Agrobacterium*, on the other.

Flagellar antigens
These can be recognized by the flocculent nature of the quick-forming

agglutinated aggregates and confirmed by susceptibility to heat (antigen suspension heated at 100 °C, 30 min). Whilst the antigens of rhizobial flagella can be expected to be broadly similar to those of other Gram-negative bacteria, there is in fact little information available for them, and what there is largely restricted to the fast-growing forms (Vincent 1941, 1942; Kleczkowski and Thornton 1944; Purchase, Vincent, and Ward 1951*a;* Drozańska 1959; Scheffler and Louw 1967). Two distinctive flagellar groups covered 12 strains of *R. trifolii* studied in detail; a minimum of three distinctive and one shared minor antigen was needed for 16 strains of *R. meliloti* (Purchase *et al.* 1951*a*). Larger collections showed that 104 out of 126 strains of *R. trifolii* (Purchase and Vincent 1949) and 127 out of 156 *R. meliloti* (Vincent 1941; Hughes and Vincent 1942; Purchase, Vincent, and Ward 1951*b*) fitted into the few flagellar groupings revealed in the more detailed study. There is no doubt however, that more flagellar types will be revealed as the source of cultures is widened. Indeed more flagellar antigens have to be postulated to explain some data (Graham 1963; Scheffler and Louw 1967). In Drozańska's experience *R. lupini, R. japonicum,* and other slow-growing rhizobia commonly failed to yield even a homologous reaction— presumably due to some technical difficulty. On the other hand Graham reported high homologous flagellar titres with all his slow-growing strains.

For the most part flagellar antigens require the agglutination reaction for their demonstration but it has been possible to recognize a flagellar precipitation line with strains of *R. meliloti* (Humphrey and Vincent 1975). Heating (100 °C, 30 min), although it prevented flagellar agglutination, improved detection of this gel-diffusion line, evidently by fragmentation of the flagella into smaller (more diffusible) units of intrinsically heat stable flagellin.

Somatic antigens

Strains of a given rhizobial species, fast- or slow-growing, fall into many serologically distinct types based on agglutination reactions of surface antigens. Antibody absorption tests and evidence of non-identity in some cross-reacting antigens often extend any minimal antigenic formula. This has been demonstrated with *R. trifolii, R. leguminosarum, R. meliloti, R. japonicum,* and other, diverse, slow-growing forms (Vincent 1941, 1942; Hughes and Vincent 1942; Kleczkowski and Thornton 1944; Purchase and Vincent 1949; Purchase *et al.* 1951*a, b;* Koontz and Faber 1961; Drozańska 1963, 1964,1965; Graham 1963; Loos and Louw 1964; Date and Decker 1965; Skrdleta 1965; Holland 1966; Scheffler and Louw 1967; Means and Johnson 1968; Dadarwal, Singh, and Subba Rao 1974; Padmanabhan and Broughton 1979; Ikram and Broughton 1980*a*).

Combinations of flagellar and somatic formulations further distinguish antigenic types. The relationship between antigens revealed by agglutination and gel immune-diffusion need some further consideration. In the case of homologous reactions with some species (e.g. *R. trifolii* and *R. japonicum*) precipitation lines formed near and concave to the antigen well correspond to major antigens revealed by agglutination, and are evidently due to LPS (Humphrey and Vincent 1969*a*; Skrdleta 1969*b*). The position with *R. meliloti* is more complex. Freshly isolated strains of this species are likely to depend on a faster-diffusing 'specific masking substance' for specific agglutination, whereas the typical LPS-line is likely to be widely shared between strains and cause considerable cross-agglutinability in variants which develop in many cultures after long maintenance in the laboratory (Humphrey and Vincent 1975; Wilson, Humphrey, and Vincent 1975).

Minor, cross-reactive, antigens are more likely to be detected in agglutination than in gel-diffusion. In some cases the negative result with gel-diffusion can be attributed to its relative insensitivity to low antibody concentration; in others it reflects the inferior precipitating property of IgM which can dominate IgG in the antibody response to a minor (less immunogenic) more widely shared antigen (Humphrey and Vincent 1973).

A collection of 152 strains and substrains tested by agglutination and gel immune-diffusion showed that four of the nine antisera were markedly specific for their homologous strains by both methods (3–9 per cent heterologous agglutinations which generally correlated with gel immune-diffusion, 3–5 per cent). The remaining antisera were however more cross-reactive in agglutination (23–45 per cent) than they were in precipitation (12–24 per cent) (Humphrey and Vincent, unpublished). As failure of antisera to precipitate with cross-agglutinating can be due to low titre or form of antibody, gel immune-diffusion may be more discriminating in strain recognition but less informative as to total antigenic constitution.

Rhizobial lipopolysaccharides

Lipopolysaccharides which constitute the specific somatic-antigens of most rhizobia that have been investigated have not been as exhaustively studied as they deserve. The constituent sugars of lipopolysaccharides of a fairly wide range of rhizobia have been reported on briefly (Graham and O'Brien 1968). Two strains of *R. trifolii* (Humphrey and Vincent 1969*a*) had many features in common with other Gram-negative bacteria (hexoses, methylpentose, 2 keto-3-deoxyoctonate (KDO)).

Some (hexosamine and uronic acid) were shared with *Xanthomonas* but not with the Enterobacteriaceae. Phosphorus content was low and heptose was abundant in one, though not detected in the other. They were fully antigenic with rabbits when used with an intramuscular injection schedule with Freund's complete adjuvant followed by an intravenous booster dose. Antibody was chiefly in the IgM form. The isolated LPS was highly active in gel diffusion giving, in a dose ranging from $0.5-16 \mu g \ ml^{-1}$, the typical concave line near the antigen well which corresponded to the line associated with the specific agglutinating antigen from whole cells. A minor second diffusion line appeared to be due to a smaller fragment from the main molecule and having the same specificity, as judged by antibody-absorbing capacity.

The fact that cells having multiple antigenic determinants, as revealed by agglutination data, commonly produce a single LPS line indicates that the different determinants are on the one macromolecule, or that some of the agglutinating antigens are not released in diffusible form when tested by the Ouchterlony method. The restriction of some cross-reacting minor antigens to the production of IgM antibody could also be indicative of such determinants being immunologically obscured. Difficulty in achieving complete absorption of the total antibody complement (Vincent and Humphrey 1968) could result from the same condition. Less readily immunogenic configurations having a role in nodulation specificity could also explain obvious lack of relationships between this property and the major LPS determinants, the need for an exhaustive immunization schedule to produce a correlated antibody response (Dazzo and Hubbell 1975*a*) and the low concentration of antibody attributable to an 'invasiveness' antigen (Maier and Brill 1978). Modifications observed as a result of lysogenic conversion (Barnet and Vincent 1970) could be similarly explained.

The very low phosphorus content of rhizobial LPS, and the fact that it can be even further reduced in a different method of preparation and purification without interfering with activity in gel diffusion, makes it seem unlikely that this element plays a role in rhizobial antigens, the major structural use that has been assigned to it in the Entero-bacteriaceae. Uronic and glucuronic acids are common in rhizobial strains, but they are rarely reported in the Enterobacteriaceae. In this family they are regarded as being of capsular polysaccharide origin, with the suggestion that sometimes the somatic (O) and capsular (K) antigens are bound together with unusual firmness. Although the antigens of both SU 297/31 and TAI contained the same concentration (6 per cent) of glucuronic acid, its linkage in the two strains coud well be different, as evidenced by the additional negative charge of the TAI antigen. The differences in antigen charge reflected the respective electrophoretic mobility patterns of the two strains (Humphrey and Vincent 1969*b*, Marshall 1967).

The low phosphorus content, presence of acidic sugars and, in strain TAI, a high percentage of neutral sugars relates these rhizobial antigens chemically to the capsular (K) antigens of *Escherichia coli*, but they differ in that they contain KDO and a higher percentage of firmly bound lipid: both properties characteristic of enterobacterial LPS. That the extracted antigen was not simply a mixture of the two forms of antigen was indicated by inability to separate two forms by procedures which had achieved such a separation of somatic antigen and capsular polysaccharide in *E. coli* and the homogeneity of the precipitation reaction in gel immune-diffusion both by the Ouchterlony method and by immunoelectrophoresis. Although the LPS of strain TAI used in this study was prepared from cells which showed no sign of the 'capsules' seen in older preparations (Dudman 1968), the extracted antigen was very gelatinous and would seem to represent a surface capsular-like form of LPS which accumulates as the cell ages, and takes on the appearance of a capsule.

Lipopolysaccharide of *R. leguminosarum* has been subject to the more rigorous examination now possible with these preparations (Planqué, van Nierop, Burgers, and Wilkinson 1979). As far as comparisons are possible, the analysis of the aqueous phase (nucleic acid free) material and the two polysaccharide fractions (PS II and III) of the soluble material left after acetic acid treatment agrees with the earlier results with less exhaustively treated LPS of *R. trifolii* (Humphrey and Vincent 1969*a*). Particularly noteworthy in both cases are the low phosphorus contents and persistence of hexuronic acid in the more drastically treated fractions. Lipid A included three β-hydroxy fatty acids and glucosamine. Planqué *et al.* (1979) concluded that the high molecular-weight polysaccharide (PS I) is neither the extracellular polysaccharide nor part of the LPS. A surface β-1,2 glucan, closely associated with the LPS-like fraction, was separable by fractionation with cetavlon (Zevenhuisen and Scholten-Koerselman 1979). It is unfortunate that these detailed chemical studies were not accompanied by serological data.

Comparison of the wall of cultured cells and bacteroid forms of *R. leguminosarum* (van Brussel *et al.* 1977) showed that the latter contained less LPS. In cultivated cells the concentration of LPS was affected by composition of the media, apparently by shortage of calcium, due particularly to the inclusion of the chelating agent— EDTA. The morphological changes and extra lysozyme sensitivity associated with calcium deficiency are reminiscent of those already recorded for *R. trifolii* (Vincent and Colburn 1961; Vincent and Humphrey 1963).

Non-lipopolysaccharide surface antigens

Although surface lipopolysaccharides are so often major surface antigenic determinations, there are occasions when other non-LPS substances can modify or prevent reactions involving LPS. Experience with *R. meliloti* (Humphrey and Vincent 1975; Wilson *et al*. 1975) clearly demonstrates such an effect. The common condition in this species is for the LPS to be widely cross-reactive amongst strains and for specific agglutination to be due to a strain-specific masking substance (SMS), which seems to be located so as to prevent reaction between group reactive LPS and its antibody. SMS could be regarded as micro-capsular; it is not the water soluble exopolysaccharide.

Distinctly capsulated cells can be seen amongst a non-capsulated majority in very young cultures of *R. trifolii* (Napoli, personal com-munication). Such cells might relate to the reported Vi-like condition in rhizobia (Vincent 1953) and the K (capsular) antigen of Entero-bacteriaceae. In some cases the inhibitory effect of this condition on LPS-mediated agglutination can be removed by heating (52 °C, 30 min) but in others even heat fails to activate the cells of a culture evidently in a capsule-dominated phase. Such negative cultures regularly produce a majority of reactive sub-cultures and a similar minority of non-reacting ones (Vincent and Waters 1953 and unpublished data). The most direct explanation of these observations is that some cultures yield cells with sufficient masking (capsular?) antigen to block the specific LPS reaction but that most have the somatic LPS sufficiently exposed for agglutination. Structural poly-saccharides of the β-1,2 glucan type (Zevenhuisen and Scholten-Koerselman 1979) could fill such a blocking role. However the suggested identification of such a 'capsular' polysaccharide with the 'specific masking substance (SMS)' of *R. meliloti* (Humphrey and Vincent 1975) leaves unexplained the marked antigenic specificity characteristic of the 'masked' strain, which is evidence of distinctive molecular composition or configuration. Involvement of SMS in any lectin-binding encounters the same difficulty as does the equal nodulating capacity of SMS and non-SMS substrains (Wilson, unpublished data).

Cases of lack of correlation between phage adsorption and LPS antigens (Kleczkowski and Thornton 1944; Barnet 1972) and the inability of LPS-coupled red cells to remove all agglutinations from homologous antiserum (Humphrey, unpublished) indicates the presence of other kinds of specific receptor or antigenic sites.

The antigenic status of extracellular polysaccharides

Most strains of rhizobia, particularly the fast-growing species, produce a large amount of gum which is excreted into the growth medium and readily separated from the cell proper. Some progress has been made in determining qualitative composition and unit sub-structure (Hopkins, Peterson, and Fred 1930; Schluchterer and Stacey 1945; Haworth and

Stacey 1948; Humphrey and Vincent 1959; Humphrey 1959; Graham 1965; Skrdleta 1966; Amarger, Obaton, and Blachère 1967; Hepper 1972; Zevenhuisen 1971, 1973; Linevich, Lobzhanidze, and Stepanenko 1976; Jansson, Kenne, Lindberg, Ljunggren, Lönngren, Ruden, and Svensson 1977; Dudman 1976, 1978; Kennedy 1976, 1978; Kennedy and Bailey 1976). The exopolysaccharides of the fast-growing rhizobia are readily separated from the cells and have a relatively simple composition, which they share with *Agrobacterium*, *viz* glucose, galactose, uronic acid (traces only in *R. meliloti* and *Agrobacterium*) with pyruvyl and acetyl side chains (Dudman and Heidelberger 1969). Acetyl and pyruvyl groups are thought to explain the ability to precipitate with pneumococcus type 27 antiserum. However there is no evidence to show that the rhizobial extracellular polysaccharide is fully antigenic in the sense of being able itself to induce antibody formation. Nor do they block reaction between antibody and precipitating antigens. Precipitation lines formed by partially purified preparations are attributable to contaminating amounts of highly active somatic or internal antigens (Humphrey and Vincent 1965).

Definition of surface antigenic constitution

It is difficult to give a defintitive description of the surface antigenic determinants of a *Rhizobium*. This is so even in the one laboratory with a well studied collection. Its wider application between centres, in the one or several countries, requires decision as to reference strains and antisera, and nomenclature, and has hardly been attempted. The frequent use of the United States set of serogroups of *R. japonicum* is a start in this direction but is likely to mask fundamental weaknesses as a basis for strain definition (Gibson, Dudman, Weaver, Horton, and Anderson 1971).

It is possible to arrive at a 'minimal antigenic constitution' for a relatively small collection of strains when these are studied in a full reciprocal test pattern, and antibody absorption tests are included. Results of this approach have been reported (Purchase *et al.* 1951a) for 12 isolates of *R. trifolii* and 16 of *R. meliloti*. Two flagellar and nine somatic antigens had to be postulated for *R. trifolii*, four and 15 respectively for *R. meliloti*. Two pairs of the 12 *R. trifolii* appeared to be qualitatively identical and one group of three isolates of *R. meliloti*. Using the same approach with a collection of 28 isolates of *R. japonicum*, it was necessary to postulate 24 somatic antigens. Various combinations of these as determined by antibody absorption tests provided the basis for 21 serotypes, of which only five contained more than one isolate.

More recently in the author's laboratory (Humphrey and Vincent, unpublished) we have combined the information given by agglutination and gel immune-diffusion to develop a different approach involving 'major' and 'minor' antigenic determinants of *R. trifolii*. We have taken the major antigenic determinants to be characterized by high titre somatic agglutination and a lipopolysaccharide precipitation line, consistent in the reciprocal test and from animal to animal. Minor antigens, on the other hand, are associated with relatively low titre, weak or negative lipopolysaccharide line, and results likely to be inconsistent between animals and reciprocal tests. The nine cultures used included six strains used also in the 1951 report (including 'large' and 'small' colony variants of one of them) and two more recently isolated strains. Strains SU 36, SU 91, SU 157, and SU 204 (Rothamsted Cl. F) retained their relative somatic specificity; SU 61 and SU 94, together with the new cultures TAI and SU 297/31, showed wide cross reactivity. A minor antigenic determinant common to all strains, but which seemed particularly sporadic in its expression, could be detected with some antisera. From these results we would like to propose a new system of minimal antigenic determinants to describe the constitution of the strains of *R. trifolii* we have studied (A, B, etc., major antigens; k, l etc., minor):

Specific serotypes		Cross-reactive serotypes	
SU 36	A s	SU 61	E H l n s
SU 91	B s	TAI	E J n s
SU 157	C s	SU 94S	F l n p s
SU 204	D k s	SU 94L	F k l n p s
		SU 297/31	G l m p s

The second approach, which is less troublesome but less informative, is to group a collection of strains according to patterns of reactivity with antisera to a small number of selected strains. Table 8.2 gives examples of groupings obtained in this way. Such groups are quite empirical and liable to modification when additional antisera are brought into the test programme. They also group together strains antigenically unrelated to each other, but each having an antigen possessed by the strain responsible for developing the antiserum.

Internal antigens

During the course of studying the calcium nutrition of rhizobia it was observed that a deficiency of this element, which reduced growth and caused an abnormal morphological condition (Vincent and Colburn 1961; Bergersen 1961; Vincent 1962a—probably due to a weakening of the envelope structure) permitted the outward diffusion of more deeply located non-agglutinating antigens whose reactivity crossed the normal

Table 8.2
Diversity of rhizobial antigens and serogroups

Rhizobium	Number of Strains	Antisera	Flagellar Antigens	Groups	Somatic Antigens	Groups	Combined groups	Techniques	Reference
R. meliloti	6	6	3	2	7	5	5	Agglutination	Vincent 1941
	42	5		2(1)[a]		6(1)[a]	8(3)[a]	Agglutination	Vincent 1941
	20	5	4	2(1)	15	4(1)	6(3)	Agglutination	Hughes and Vincent 1942
	16	16		3		13	14	Agglutination	Purchase et al. 1951a
	89	13		4(1)		9(1)	19(3)	Agglutination	Purchase et al. 1951b
R. trifolii	6	6	2	2	3	3	4	Agglutination	Vincent 1942
	32	3		2(1)		4(1)	7(2)	Agglutination	Vincent 1942
	36	3	2	3(1)		4(1)	3(4)	Agglutination	Hughes and Vincent 1942
	12	12		2	9	8	10	Agglutination	Purchase et al. 1951a
	126	6		4(1)		7(1)	16(3)	Agglutination	Purchase and Vincent 1949
	9	9	n.s.[b]		9 major 6 minor	7	n.s.[b]	Agglutination and gel diffusion	Humphrey and Vincent unpublished
	152	9				13(1)		Agglutination	
R. trifolii and R. leguminosarum	190	6		2(1)		9(1)	12(9)	Agglutination	Kleczkowski and Thornton 1944
R. japonicum	13	14	n.s.			11	n.s.	Agglutination	Koontz and Faber 1961
	28	28	n.s.		24	17	n.s.	Agglutination	Date and Decker 1965
R. sp.	62	6	n.s.			24(1)	n.s.	Agglutination	Ikram and Broughton 1980a
						23(1)		Gel diffusion	Ikram and Broughton 1980a

[a] Negative-reacting throughout: at least one additional group.   [b] n.s.: not studied.

boundaries of species differentiation. The same additional precipitinogens were released from calcium-adequate rhizobia by mechanical disintegration, freezing, drying, and after treatment with lysozyme or chloroform (Humphrey and Vincent 1965), as well as by sonication (Gibbins 1967). Widely cross-reactive antigens found with *R. japonicum* lyophilized cells (Skrdleta 1969*b*) would also have been due to internal antigens.

Reactions involving the internal antigens provide valuable insights into the taxonomic distinction between fast- and slow-growing species of rhizobia and, within the fast-growing forms support the grouping of some of the present species (see pp. 261–262).

Stability of antigenic constitution

Not much work has been carried out to determine the influence of growth media and conditions on the serological behaviour of rhizobia. Some effects have been noted, in that a shaken liquid culture is inferior to one grown on a moist agar surface for the demonstration of flagellar agglutination, and calcium deficiency causes some modification of surface antigens (Vincent and Humphrey 1968) as well as permitting outward diffusion of internal components. Media which favour excessive production of gum can give a suspension in which, if the gum is not removed by centrifugation and washing, high viscosity may interfere with agglutination. Otherwise the growth medium does not seem critical; it has been possible to switch from a medium containing yeast extract to a relatively defined one having nitrogen in a simple combined form, without any apparent effect on antigenic constitution. Moreover, the extremely different conditions during the development of the cultured cell and the bacteroid form generally leave at least the major surface antigens qualitatively unchanged (Means *et al.* 1964) and independent of the species from which the nodules are obtained (Pankhurst 1979).

Mutations, which occur in some strains quite frequently and which affect many features of rhizobia, generally have little effect on their antigenic characteristics. These include cases of lost symbiotic capacity (Almon and Baldwin 1933; Vincent 1944 1954; Kleczkowska 1950).

It has been possible, in the author's experience, to compare the serological behaviour of strains of *R. trifolii* and *R. meliloti* over periods of 10 years in the first case (Purchase *et al.* 1951*a*) and subsequently for a further 30 years. With *R. trifolii*, two out of 31 comparisons revealed new flagellar- and one out of 51 new somatic-cross reactions. These new reactions were consistent in reciprocal tests. Later (a further 30 years) experience with *R. trifolii* has been consistent with this high degree of antigenic stability. Flagellar reactions with *R. meliloti* were similarly consistent but experience with somatic antigens of this species has been quite different (Wilson *et al.* 1975). They showed that a trend, (already evident to some extent in 1951) for wider somatic cross reactivity became very obvious for a substantial portion of strains which were investigated in some detail. Although increased cross reactivity of one strain seemed to depend on whether the test antigen was grown on liquid or solid medium, this did not explain results with most of the strains which were exhaustively tested with several technical variations.

Freshly isolated, completely unrelated, cultures showed a degree of specificity in their agglutination reactions similar to that first recorded in 1941. However, the fact that many of these specific cultures shared a cross-reacting lipopolysaccharide-type somatic antigen and depended on another faster diffusing antigen ('specific masking substance') for its specificity provided a clue as to the reason for loss of agglutinating specificity in strains which had been held in culture for a long time. Cross-agglutinating and specific substrains of the one parent strain confirmed this view. Attempts, over the short term, to narrow cross-reactivity by plant passage or widen it by repeated sub-cultivation, have failed. Nor has the specific substran shown the competitive advantage in nodulation over its cross-reactive substran that might explain the greater tendency for it to be found in freshly isolated strains. However the time span for such experiments would be far too short to explain a small degree of bias that could be operative over many years of cultivation or selection in the field.

Genetic changes in some other strains associated with colony form and symbiotic efficiency have been responsible for detectable changes in antigenicity. Agglutination and absorption data (Bloomfield 1959 *cit* Vincent 1962*b*) show that in each of the cases of two strains of *R. trifolii* (SU 297 and SU 298) the small colony (non-gummy) variant lacked an antigen possessed by the parent (large gummy colony) type. Other gummy or mucoid large colony forms of strain SU 298 were cross-reactive with, but failed to absorb, all antibodies from the parent-type antiserum; in the case of one variant the antigenic change was more drastic. The somatic antigenic behaviour of the two serologically related but distinguishable substrains has been tabulated and discussed in the reference quoted. Conclusions at the time also took account of bacteriophage sensitivity. In the light of other evidence one would not now attribute the antigenic difference between large and small colony forms to lack of exopolysaccharide. A small colony form of *R. trifolii* SU 94 seems to have lost one of the several minor antigenic determinants found in the parent, large colony strain (Humphrey and Vincent, unpublished). None of the variants of this set showed any difference in their flagellar antigens. Antigenic differences have been reported between 'smooth' and 'rough' forms of *R. trifolii* (Lorkiewicz and Dusinski 1963).

Another set of genetic variants involves *R. trifolii* NA 34 which has regularly produced two colonial mutants. One, a small, non-gummy form though fully invasive, is non-effective in nitrogen fixation. The second variant, indistinguishable in size or gum production from the invasive, effective parent, has become non-invasive. In our hands (data due to B. A. Humphrey) exhaustive quantitative absorption has failed to show any difference in somatic antigens amongst the parent and variant types. It should be noted however that other experience, comparing the small (invasive though ineffective) and the large-colony non-invasive variants of the same strain, led to the conclusion that an antigen correlating with invasiveness could be demonstrated in the invasive form but not in the non-invasive substrain (Dazzo and Hubbell 1975*a*). The different result might be attributed to a heavy and sustained immunization treatment, which could permit the expression of a minor antigenic determinant which had a role in invasion. Kaushik, Dadarwal, and Venkatraman (1973) compared the surface and internal antigens of a strain each of *R. trifolii* and *R. leguminosarum* for the respective wild types and after genetic modification. Some ultraviolet induced mutants were indistinguishable in their antigenic contribution; others were changed to partial homology in the case of surface antigens or varied in internal antigens. Two non-invasive mutants showed more drastic loss of homology at the surface and in more of the internal antigens. A recombined strain had the surface antigen modified; a transformed mutant has had the homology of its surface antigen restored. The most consistent change detected in preparations of rough mutants of *R. trifolii* was loss of rhamnose (Lorkiewicz and Russa 1971). Lysogenization of a strain of the same species resulted in modification of its surface antigens (Barnet and Vincent 1970).

It has to be concluded that the detailed nature of the antigens of rhizobia can be subjected to serious genotype modification. This raises doubts as to the suitability of this marker for long-term ecological studies. Reassuring evidence relates to the stability under field conditions of the antigenic label on a strain nodulating *Lotononis bainesii* over a 12 year period (Diatloff 1977). It should be noted, however, that this organism is taxonomically atypical of the rhizobia.

Nature of antibody response

It is a well known and accepted hazard that the same antigen suspension can evoke a different immunological response from rabbit to rabbit. These differences can be reflected in titre, and in balance between forms of immunoglobulin (IgG and IgM). Route and regimen of injection, presence of adjuvant and form of antigen also have effects (Humphrey and Vincent 1973).

Lectins

'Pseudoantibodies' found in plants and able to react with large molecules having appropriate groups, were first recognized by their ability to cause agglutination of red cells. A partial correlation between seed lectins of soyabean and ability to attach to cells of *R. japonicum* (Bohlool and Schmidt 1974) led to the proposal that a specific inter-action between host lectin and specific rhizobia might be a step in securing root invasion leading to nodulation. This proposition has been actively promoted and investigated as will be dealt with briefly in section 4 (p. 264) and in other chapters. The nature and possible role of lectins have been comprehensively reviewed in Broughton (1978).

8.4 Biological significance of rhizobial antigens

The earliest serological work with rhizobia was largely concerned with establishing its validity as a means of recognizing species. Because speciation and other groupings of rhizobia have been almost entirely based on capacity to nodulate designated host plants, a nexus between antigenic and symbiotic properties has been sought. Such a connection has, however, proved elusive and only recently has shown some promise of achievement.

Taxonomic relevance

A strain of *R. trifolii* can be expected to cross-agglutinate with some other strains of the same species, and even with some strains of *R. leguminosarum* and *R. phaseoli*, but not with strains of *R. meliloti* or with slow-growing rhizobia. The frequency of such cross-reactions is, in fact, in line with other established taxonomic affinities of these major rhizobial groups (Graham 1963; Drozańska 1966; Vincent, Nutman, and Skinner 1979), but there remain the majority of cases with which there is no cross-agglutination within such groups. Strain recognition by serological means depends on just this preponderance of negative cross-reactivity within a species. It is still possible however that by emphasizing the immunization programme and paying more attention to widely cross-reactive low-titre antibodies will reveal a species-shared minor antigen even in strains otherwise regarded as very specific. This has indeed already been indicated with a well studied group of *R. trifolii* strains (see p. 255).

It is when we come to the internal antigens, however, that clear relationships can be seen between antigenic constitution and other taxonomic evidence (Graham 1964; Vincent and Humphrey 1970; Graham 1971; Dadarwal and Sen 1973; Vincent 1974, 1977; Vincent *et al.* 1973).

Together the evidence points to the following conclusions: (i) fast-

and slow-growing rhizobia do not fit comfortably into a single genus; they differ more from each other than fast-growing species (notably *R. meliloti*) differ from *Agrobacterium*, (ii) amongst the fast-growing ones *R. trifolii* and possibly *R. phaseoli* could advantageously be included in the species *R. leguminosarum*, (iii) *R. meliloti* should maintain its own specific status, (iv) the present separate speciation of *R. lupini* and *R. japonicum* is difficult to justify and they should by merged with other slow-growing rhizobia as symbiotypes of *R. japonicum*, (v) as yet unnamed fast-growing rhizobia, such as those isolated from *Leucaena leucocephala*, *Lotus corniculatus*, *Cicer arietinum*, and *Lupinus densiflorus*, require further taxonomic investigation in which their internal antigens can be expected to provide valuable insights, (vi) the large miscellany of slow-growing strains also requires more taxonomic study. A recent investigation of fast- and slow-growing *Lotus* rhizobia (Pankhurst 1979) shows very clearly the taxonomic relevance of internal antigens.

Strain definition and recognition

It is tempting to use the distinctive nature of rhizobial surface antigens to define a strain and to think of terms like 'serotype' or 'serogroup' as something more than an abridged statement of certain imcompletely determined, open-ended, serological reactions. The risk is that the relative convenience of obtaining such a designation will lead to the expectation that fresh isolations will, when so grouped, share other properties, particularly those relating to symbiotic performance. In fact any such relationships are incomplete or no better than chance. The error is worse when grouping depends merely on a pattern of reactivity with a group of antisera, because it can be expected that the extension of the test patterns to extra antisera,will result in further subdivisions, and that the use of different test antisera could lead to entirely different groupings. Difficulties attached to the use of a serotype label to study the ecology of naturally occuring heterogeneous populations (e.g. Caldwell and Weber 1970; Caldwell and Hartwig 1970) are clearly shown in the futher detailed analysis of 'serogroup 123' of *R. japonicum* (Gibson *et al.* 1971).

It is, however, valid to use a serological label for the recognition of a strain which has been used experimentally against the appropriate control. The technique can also be used for the detection of contamination, strain substitution, and the authenticity of a non-invasive mutant. The more general reactivity associated with internal antigens can at least be used as a presumptive test for *Rhizobium* or a taxonomically related form.

Relationship to bacteriophage adsorption

Experience as to the relationship between specific surface antigens and phage susceptibility has been contradictory. In some cases there appeared to be no connection (Kleczkowski and Thornton 1944; Kowalski, Ham, Frederick, and Anderson 1974); in others the possession of certain surface antigens was necessary for phage susceptibility, although not a guarantee of infection (Marshall and Vincent 1954). Both conditions are demonstrated in the trifoliphage collection studied by Barnet (1972, Table 2).

Lysogenization of strain of *R. trifolii*, besides causing resistance to adsorption of the homologous bacteriophage, modified the surface (lipopolysaccharide) antigen significantly. This suggests that in this case the surface antigen is the receptor site (Barnet and Vincent 1970). Similarly Atkins and Hayes (1972) found changes in lipopolysaccharides and lipoprotein of *R. trifolii* mutants which adsorbed homologous bacteriophage poorly. More recently too, it has been found that cell walls and lipopolysaccharides of sensitive strains were able to combine with, and hence inactivate, the bacteriophage to which they were sensitive (Zajac, Russa, and Lorkiewicz 1975). Preparations from phage-resistant substrains did not so inactivate the bacteriophage.

Relationship to symbiotic capacity

Ability to invade a particular legume may coincide with the possession of certain internal antigens, as in the case of distinctive internal antigens shared within, but not between, such species as *R. trifolii, R. meliloti*, and *R. japonicum*. On the other hand, shared internal antigens need not mean cross-invasiveness, as, for example, those between *R. trifolii* and *R. leguminosarum, R. japonicum*, and *R. lupini*, or *R. meliloti* and *Agrobacterium*.

Surface antigens can be even less predictive. Strains of *R. trifolii* cross-agglutinating with some of those from *R. leguminosarum* need not nodulate the heterologous host and *vice-versa*. Strains fully nodulating within an inoculation group are not likely to share any major somatic antigen, and non-invasive mutants need not reveal any readily detected antigenic change (Kleczkowski and Thornton 1944; Vincent 1944).

Some aspects of the matter do, however, require further comment. Dazzo and Hubbell (1975a) reported that a collection of strains of *R. trifolii* each contained a diffusing antigen not possessed by respective non-invasive mutants. A possible explanation of this positive finding, which contrasted with previous failure to demonstrate an 'invasive-related' antigen, lies in the extraordinarily exhaustive immunizing procedures which seem to have been needed to develop antibody to a minor antigenic determinant. The widely cross-reactive minor antigen which has been demonstrable with some antisera to *R. trifolii* (see p. 265) could be a related condition. Relevant too is the comparison involving non-invasive mutants of *R. japonicum* lacking a minor antigenic determinant possessed by the parent strain (Maier and Brill 1978). Sanders, Carlson, and Albersheim (1978) interpreted experience with a mutant of *R. leguminosarum*, which produced markedly less extra-cellular polysaccharide at the same time as it lost infectivity, as evidence of causal relationship. However the claim made for lipopolysaccharide identity between parent and mutant does not convincingly dispose of the possibility of a changed minor antigenic determinant. Cases with *R. trifolii* where a similar mutational loss of gum production leaves the mutant fully invasive (e.g. Vincent 1962*b*), were not taken into account. Dazzo and Brill (1979) have reported on rhizobial polysaccharides which seem to be involved with lectin, in the binding of the bacteria to root hairs (see p. 277).

Another aspect of the suggested relationship between invasive *Rhizobium* and leguminous host concerns a postulated degree of homology between antigens shared by both partners. Some antigens were indeed found to be shared between rhizobia and leguminous, but not non-leguminous, hosts (Charudattan and Hubbell 1973). This relationship was only relatively specific however to the extent that legume antisera cross-reacted with certain other bacterial genera *Xanthomonas*, *Pseudomonas*, *Erwinia*, and *Agrobacterium* and cut across established nodulation groups. Similarity between root and rhizobial antigens has been incorporated into a lectin-mediated attachment theory, which will be dealt with briefly in the next section and more fully in another chapter (see Ch. 9).

Retained infectivity but loss of ability to complete an effective nitrogen-fixing symbiosis may or may not be associated with qualitative change in surface antigens (Vincent 1954; Pankhurst 1974). The ineffective mutants studied by Pankhurst leaked internal antigens more readily than their respective parent strains. There was no consistent common surface antigen or exopolysaccharide component in a group of slow-growing strains which shared the capacity to develop nitrogenase activity in culture (Kennedy and Pankhurst 1979).

The lectin hypothesis of specific attachment

Fluorescein-labelled seed lectin of soyabean was seen to attach itself, evidently to the polar tip, to most but not all, strains of *R. japonicum*, and not to other diverse rhizobia which were unable to nodulate that host (Bohlool and Schmidt 1974, 1976). This observation led to the sug-gestion that host lectin may play a role by facilitating attachment to the root, seemingly as a multivalent bridge between plant and bacterial antigen. (Dazzo and Hubbell 1975*b*). However the non-reacting homologous strains and several reports of other anomalies left a large area of doubt as to the validity of the proposed mechanism (Dazzo and Hubbell 1975*c*; Chen and Phillips 1976; Law and Strijdom 1977).

Further work has considerably strengthened the hypothesis and explains at least some of the anomalies. This has involved purification of multivalent clover-root lectin (trifoliin) and its partial characterization (Dazzo, Yanke, and Brill 1978), distribution, and localization of soyabean lectin (Pueppke and Bauer 1976; Pueppke, Bauer, Keegstra, and Ferguson 1978), preferential lectin-mediated adsorption to roots of homologous (clover) host (Dazzo, Napoli, and Hubbel 1976), binding of FITC-labelled rhizobial capsular material to root hairs with evidence for cross-linking a complementary polysaccharide through specific binding sites (Dazzo and Brill 1977), binding of pea lectins to a glycan type polysaccharide (Planqué and Kijne 1977), differential binding of soyabean lectin to *R. japonicum* to an extent affected by the growth conditions, including that associated with the root environment (Bhuvaneswari, Pueppke and Bauer 1976, 1977; Bhuvaneswari and Bauer 1978). Whilst variability associated with growth conditions might be invoked to explain failures with attachment of lectin to homologous rhizobia, cases of non-specific combination (Law and Strijdom 1977) are more difficult to dispose of, short of postulating a multifactor sequence, of which facilitated attachment is only one.

Particularly important are the ultrastructural studies which show marked localization of ferritin-labelled soyabean lectin bound in a biochemically specific manner to the capsular material of *R. japonicum* (Calvert, Lalonde, Bhuvaneswari, and Bauer 1978) and seem to run counter to the proposed involvement of lipopolysaccharide as site of lectin attachment (Wolpert and Albersheim 1976). At the same time, note should be taken of the capacity of outer membrane lipopolysaccharides to diffuse from their normal location. This is the basis of their demonstration by the Ouchterlony method of gel immune-diffusion. It is also possible that the 'capsular' material may restrict inward access of labelled antibody, although this could be contra-indicated by failure to obtain ferritin labelling of the outer membrane of non-capsulated cells. A possible reconciliation of the differing views (lipopolysaccharide v. capsular antigen) could be along these lines, (i) any reaction between lipopolysaccharide and lectin would have to involve specific centres different from those associated with strain-specific major antigenic response, (ii) lectin-specific centres may be obscured to some degree, so long as the lipopolysaccharide is attached to the outer cell membrane and its liberation wholly or in part may be necessary for combination with lectin, (iii) localized aggregation of such a lectin-active moiety, constituting the whole or of the observed capsule, could be expected to play the postulated role in specific attachment. The more detailed study of bacterial polysaccharide binding between *R. trifolii* and clover root hairs (Dazzo and Brill 1979) showed that one of the two lectin-combining fractions included typical lipopolysaccharide components and shared its other identified constituents with the other active fraction. The non-active third fraction was quite different. [Comparative analysis of lipopolysaccharides of *R. leguminosarum* (Planqué *et al.* 1979).]

Some genetic evidence has also been provided in *R. trifolii*-induced transformants of *Azotobacter* which were agglutinated both by the clover lectin and by antiserum to clover root antigens (Bishop, Dazzo, Applebaum, Maier, and Brill 1977).

Broughton (1978) has provided a useful, non-partisan, review of the possible role of lectins in determining nodulating specificity.

8.5 Serological techniques as a tool in the study of rhizobia

The degree of specificity that exists between antigen and its antibody makes serology a convenient means for determining whether a particular organism conforms to a certain serotype or, in the case of more specialized biochemical studies, whether a preparation comes up to expectation in its nature and homogeneity.

Serological typing mostly depends on the nature of surface antigens, generally those on the cell itself, which are more strain specific, rather than the more cross-reactive ones associated with the flagella. In this way, reactivity with antisera of known specificity can be used to determine the authenticity and purity of a culture or legume inoculant, to estimate the proportions of experimentally mixed serotypes in mixed populations, to type the rhizobia found within a nodule and, to some extent, study the ecology of natural populations. The technique is particularly useful as a means of recognizing a non-invasive form as a direct mutant of an invasive, serologically typed, parent culture.

Shared internal antigens can provide information about non-invasive forms and about organisms unusual in origin and properties. Such use is demonstrated by experience with two cultures (Humphrey and Vincent, unpublished). The internal antigens of a non-invasive fast-growing bacterium (WBK) found in soyabean nodules along with the independently invasive typical slow-growing strain (van Rensburg and Strijdom 1972) showed a close relationship with *R. trifolii* and none with *R. japonicum*. On the other hand the organism responsible for the nodulation of *Parasponia* ('Trema'), isolated by Trinick (1973), yielded a pattern of internal antigens characteristic of slow-growing rhizobia.

Advantages of the technique are that the identifying reactant (antibody) can be produced to match a particular rhizobial strain and that differentially recognizable strains need not, on that account, be biased in their experimental use. Attention needs, however, to be given to such points as, (i) specificity in relation to other bacteria or components likely to be encountered, (ii) sensitivity, (iii) reliability, (iv) stability of antigen on which recognition is based, (v) convenience, particularly if many identifications are involved.

Application to ecological studies

Antigenically similar, and in some case identical, strains can occur in widely different parts of the world. This has been found between the United States of America, the United Kingdom and cultures of *R. meliloti* and *R. trifolii*, naturalized, though not indigenous, in Australia. Antigenically similar strains have also been found in widely separated regions within Australia (Vincent 1954). Further serological character- ization of our collection of *R. trifolii* has, however, revealed a tendency for some strains received as cultures from the United Kingdom to fall into a particular group (Humphrey and Vincent, unpublished).

When the rhizobia from the specified localities have been grouped according to their serological patterns, the common finding is that there is considerable heterogeneity amongst strains in close proximity, and even from different nodules of the same plant. This has been documented for pasture species (Hughes and Vincent 1942; Purchase and Vincent 1949; Purchase *et al.* 1951*b*; Loos and Louw 1964; Holland 1966) and for soyabean (Johnson and Means 1963; Damirgi, Frederick, and Anderson 1967; Ham, Frederick, and Anderson 1967, 1971; Caldwell and Hartwig 1970; Caldwell and Weber 1970; Weber, Caldwell, Sloger, and Vest 1971).

Attempts to relate the occurrence of particular serotypes (as judged by reaction with a range of antisera) have had only limited success with pasture legumes, and relative serotype representation has also been fluc- tuated between seasons (Purchase and Vincent 1949; Purchase *et al.* 1951*b*). Much more effort has been put into this approach with soya- beans. Correlations have been reported between serogroup distribution and soil type and factors such as soil, nitrogen, pH and exchangeable magnesium (Ham *et al.* 1967, 1971; Bezdicek 1972), temperature (Weber and Miller 1972), planting time (Caldwell and Weber 1970), and host genotype (Damirgi *et al.* 1967; Caldwell and Vest 1968; Caldwell and Hartwig 1970). Interpretation of serogroup relationships remains difficult, however, (Gibson *et al.* 1971) and a more critical study of sampling variability raises other doubts (Vest, Caldwell, and Peterson 1971).

As judged by serological patterns, rhizobia within a nodule are much more likely to be homogeneous than nodules of the same or different plants (Dunham and Baldwin 1931; Hughes and Vincent 1942). Exclusion of a second strain, even under field conditions, is not absolute (Brockwell, Schwinghamer, and Gault 1977) and has been frequently reported in experimental situations likely to maximize the opportunity for double infection (Vincent 1954; Marques-Pinto, Yao, and Vincent 1974). Detection of the presence of a fast-growing rhizobial-like, but non-invasive, bacterium in nodules of soyabean, together with the invasive slow-growing *R. japonicum* is another example of a serologically-based demonstration of double infection (van der Merwe *et al.* 1972).

Application to agronomic studies

Serological typing of nodules has been extensively used to determine the success of applied inocula in field and greenhouse experiments (Thornton and Kleczkowski 1950; Pochon, Manil, Tchan, Bonnier and Chalvignac 1950; Read 1953; Vincent and Waters 1954; Jenkins, Vincent, and Waters 1954; Skrdleta 1968; Dudman and Brockwell 1968; Brockwell and Dudman 1968; Kapusta and Rouwenhorst 1973; Dorosinkii and Makarova 1977; Ikram and Broughton 1980*b*). It has been utilized in more defined laboratory competition experiments (Vincent and Waters 1953; Israilski and Rushkova 1967; Jones and Russell 1972; Marques-Pinto *et al.* 1974).

Other applications

Because of the relative specificity of serological reactions and their sensitivity, they have considerable potential in the study and purification of enzymes and in their intracellular location. The method has been used to obtain evidence for synthesis of a portion of the nitrogenase molecule by cultured *R. japonicum* (Bishop, Evans, Daniel, and Hampton 1975).

Applicability of various techniques

When strain recognition is the objective, use needs to be made of surface, generally somatic, antigens. These can be typed by all the common techniques, and the operator's choice will be determined by such factors as facilities, cost and work load, familiarity, but also by specificity and suitability for a particular purpose. Gel immune-diffusion is likely to be more strain specific than somatic agglutination with *R. trifolii*, but the reverse can be true with *R. meliloti*. Both can be used directly with squashes of larger nodules (Means *et al.* 1964; Skrdleta 1969*a*); with the smaller nodules there could be a case for passive haemagglutination (Cloonan and Humphrey 1976)—provided the antigen involved is lipopolysaccharide—but none has a lower limit of detection than fluorescein-labelled antibody. The only direct method of quantitative study of soil rhizobia is that involving the reaction of cells concentrated on a membrane filter with such labelled antibody (Schmidt 1974). It is not easy to critically evaluate the accuracy of the method in that no data have been presented comparing estimates so obtained in non-sterile soil with viable counts based on a quantitative plant infection test. Recovery of *R. japonicum* growing in sterile soil varied from 25 per cent after one day to 127 per cent after 50 days. Detection of soluble antigens in soil by immunodiffusion (Kremer and Wagner 1978) requires an unrealistically large number of specific rhizobia and does not seem to have any ecologically significant application.

A combination of techniques is called for when more information is required as to the nature and location of antigens. This is illustrated by the extra information obtainable from either agglutination or gel immune-diffusion when it is backed up by the other. In this kind of study, too, there is scope for antibody absorption, qualitative and/or quantitative. Labelled antibody can refine cellular location of antigen, particularly if it can be combined with electron microscopy making use of ferritin or peroxidase as a label.

No method is entirely free of risk of false negative results (culture with cells having the antigen in question masked by non-reactive material or the inclusion of non-rhizobial cultures in the test material) or misleading positive data (due to a background strain with similar antigenic composition or the loss of a more specific surface antigen); appropriate controls must always be provided and correctly interpreted.

Although antigenic variation can occur in some long-maintained cultures (notably some strains of *R. meliloti*) strains of *R. trifolii* and the *Rhizobium* of *Lotononis* have maintained their recognizable surface antigens for several to many years in the field (Brockwell *et al.* 1977; Diatloff 1977).

Acknowledgements

I wish to express my thanks to the several collaborators, particularly to Mrs. B. A. Humphrey, for their considerable contribution to the study of rhizobial serology, and to my wife, without whose help the preparation of this article could not have been undertaken.

References

Almon, L. and Baldwin, I. L. (1933). *J. Bacteriol.* **26,** 229–50.
Amarger, N., Obaton, M., and Blachère, H. (1967). *Can. J. Microbiol.* **13,** 99–105.
Atkins, G. J. and Hayes, A. H. (1972). *J. gen. Microbiol.* **73,** 273–8.
Axelson, N. H., Kroll, J., and Weeke, N. (1973). A manual of quantitative immuno-electrophoresis. Methods and applications. *Scand. J. Immunol.* **2,** Suppl. No. 1. Universitets forlaget, Oslo.
Barnet, Y. M. (1972). *J. gen. Virol.* **15,** 1–15.
—— and Vincent, J. M. (1970). *J. gen. Microbiol.* **61,** 319–25.
Bergersen, F. J. (1961). *Aust. J. Biol. Sci.* **14,** 349–60.

Bezdicek, D. F. (1972). *Soil Sci. Soc. Am. Proc.* **36,** 305–7.
Bhuvaneswari, T. V. and Bauer, W. D. (1978). *Pl. Physiol.* **62,** 71–4.
—— Pueppke, S. G., and Bauer, W. D. (1976). *Pl. Physiol., Lancaster* **57,** suppl. item 418, p. 80.
—— —— (1977). *Pl. Physiol.* **60,** 486–91.
Bishop, P. E., Dazzo, F. B., Applebaum, E. R., Maier, R. J., and Brill, W. J. (1977) *Science, N.Y.* **198,** 938–40.
—— Evans, H. J., Daniel, R. M., and Hampton, R. O. (1975). *Biochim. biophys. Acta* **381,** 248–56.
Bohlool, B. B. and Schmidt, E. L. (1972). *Abs. Ann. Meet. Am. Soc. Microbiol.* **72,** 13.
—— —— (1974). *Science, N.Y.* **185,** 269–71.
—— —— (1976). *J. Bacteriol.* **125,** 1188–94.
Brockwell, J. and Dudman, W. F. (1968). *Aust. J. Agric. Res.* **19,** 749–57.
—— Schwinghamer, E. A., and Gault, R. R. (1977). *Soil Biol. Biochem.* **9,** 19–24.
Broughton, W. J. (1978). *J. appl. Bacteriol.* **45,** 165–94.
Bushnell, O. A. and Sarles, W. B. (1939). *J. Bacteriol* **38,** 401–10.
Caldwell, B. E. and Hartwig, E. E. (1970). *Agron. J.* **62,** 621–2.
—— and Vest, G. (1968). *Crop Sci.* **8,** 680–2.
—— and Weber, D. F. (1970). *Agron. J.* **62,** 12–14.
Calvert, H. E., Lalonde, M., Bhuvaneswari, T. V., and Bauer, W. D. (1978). *Can. J. Microbiol.* **24,** 785–93.
Charudattan, R. and Hubbell, D. H. (1973). *Antonie van Leeuwenhoek.* **39,** 619–27.
Chen, A. P. T. and Phillips, D. A. (1976). *Physiol. Plant.* **38,** 83–8.
Cloonan, M. J. and Humphrey, B. (1976). *J. appl. Bacteriol.* **40,** 101–7.
Dadarwal, K. R. and Sen, A. N. (1973). *J. Microbiol.* **13,** 1–12.
—— Singh, C. S., and Subba Rao, N. S. (1974). *Pl. Soil.* **40,** 535–44.
Damirgi, S. M., Frederick, L. R., and Anderson, I. C. (1967). *Agron. J.* **59,** 10–12.
Date, R. A. and Decker, A. M. (1965). *Can. J. Microbiol.* **11,** 1–8.
Dazzo, F. B. and Brill, W. J. (1977). *Appl. Environ. Microbiol.* **33,** 132–6.
—— —— (1979). *J. Bacteriol.* **137,** 1362–73.
—— and Hubbell, D. H. (1975a). *Appl. Microbiol.* **30,** 172–7.
—— —— (1975b). *Appl. Microbiol.* **30,** 1017–33.
—— —— (1975c). *Pl. Soil.* **43,** 717–22.
—— Napoli, C. A., and Hubbell, D. H. (1976). *Appl. Environ. Microbiol.* **32,** 166–71.
—— Yanke, W. E., and Brill, W. J. (1978). *Biochim. biophys. Acta* **539,** 276–86.
Diatloff, A. (1977). *Soil Biol. Biochem.* **9,** 85–8.
Dorosinkii, L. M. and Makarova, N. M. (1977). *Mikrobiologiya* **46,** 143–8.
Drozańska, D. (1959). *Acta Microboil. Polon.* **8,** 249–52.
—— (1963). *Acta Microbiol. Polon.* **12,** 163–4.
—— (1964). *Acta Microbiol. Polon.* **13,** 69–79.
—— (1965). *Acta Microbiol. Polon.* **14,** 275.
—— (1966). *Acta Microbiol. Polon.* **15,** 323–34.
Dudman, W. F. (1964). *J. Bacteriol.* **88,** 782–94.
—— (1968). *J. Bacteriol.* **95,** 1200–01.
—— (1971). *Appl. Microbiol.* **21,** 973–85.
—— (1976). *Carbohydr. Res.* **46,** 97–110.
—— (1977). In *A treatise on dinitrogen fixation.* Vol. IV *Agronomy and ecology* (ed. R. W. F. Hardy and A. H. Gibson) pp. 487–508. John Wiley, New York.
—— (1978). *Carbohydr. Res.* **66,** 9–23.
—— and Brockwell, J. (1968). *Aust. J. agric. Res.* **19,** 739–47.
—— and Heidelberger, M. (1969). *Science, N.Y.* **104,** 945–55.
Dunham, D. H. and Baldwin, I. L. (1931). *Soil Sci.* **32,** 235–48.

Fred, E. B., Baldwin, I. L., and McCoy, E. (1932). *Root nodule bacteria and leguminous plants*. University of Wisconsin Press, Madison.
Gault, R. R., Koci, J., and Brockwell, J. (1974). *Laboratory Practice* **23**, 11–13.
Gibbins, L. N. (1967). *Can. J. Microbiol.* **13**, 1375–9.
Gibson, A. H., Dudman, W. F., Weaver, R. W., Horton, J. C., and Anderson, I. C. (1971). *Pl. Soil* Special Volume, 33–7.
Graham, P. H. (1963). *Antonie van Leeuwenhoek* **29**, 281–91.
—— (1964). *J. gen. Microbiol.* **35**, 511–17.
—— (1965). *Antonie van Leeuwenhoek* **31**, 349–54.
—— (1969). In *Analytical serology of microorganisms* Vol. II, (ed. J. R. G. Kwapinski). John Wiley, New York.
—— (1971). *Arch. Mikrobiol.* **78**, 70–5.
—— and O'Brien, M. (1968). *Antonie van Leeuwenhoek* **34**, 326–30.
Ham, G. E., Frederick, L. R., and Anderson, I. C. (1967). *Bact. Proc.* **A19**, 4.
—— —— —— (1971). *Agron. J.* **63**, 69–72.
Haworth, N. and Stacey, M. (1948). *A. Rev. Biochem.* **17**, 97–114.
Hepper, C. (1972). *Antonie van Leeuwenhoek* **38**, 437–45.
Holland, A. A. (1966). *Antonie van Leeuwenhoek* **32**, 410–18.
Hopkins, E. W., Peterson, W. H., and Fred, E. B. (1930). *J. Am. Chem. Soc.* **52**, 3659–68.
Hughes, D. Q. and Vincent, J. M. (1942). *Proc. Linn. Soc. (N.S.W.)* **67**, 142–152.
Humphrey, B. A. (1959). *Nature, Lond.* **184**, 1802.
—— and Vincent, J. M. (1959). *J. gen. Microbiol.* **21**, 477–84.
—— —— (1965). *J. gen. Microbiol.* **41**, 109–18.
—— —— (1969a). *J. gen. Microbiol.* **59**, 411–25.
—— —— (1969b). *J. Bacteriol.* **98**, 845–6.
—— —— (1973). *Microbios* **7**, 87–93.
—— —— (1975). *Microbios* **13**, 71–6.
Ikram, A. and Broughton, W. J. (1980a). *Soil Biol. Biochem.* **12**, 83–7.
—— —— (1980b). *Soil Biol. Biochem.* **12**, 203–9.
Israilski, V. P. and Rushkova, A. S. (1967). *Izvest. Akad. Nauk USSR*, Ser. *Biol.* **4**, 567–74.
Jansson, P.-E., Kenne, L., Lindberg, B., Ljunggren, H., Lönngren, J., Ruden, U., and Svensson, S. (1977). *J. Am. Chem. Soc.* **99**, 3812–15.
Jenkins, H. V., Vincent, J. M., and Waters, L. M. (1954). *Aust. J. Agric. Res.* **5**, 77–89.
Johnson, H. W. and Means, U. M. (1963). *Agron. J.* **55**, 269–71.
Jones, D. G. and Russell, P. E. (1972). *Soil Biol. Biochem.* **4**, 277–82.
Kapusta, G. and Rouwenhorst, D. L. (1973). *Agron. J.* **65**, 916–19.
Kaushik, B. D., Dadarwal, K. R., and Venkatraman, G. S. (1973). *Curr. Sci.* **42**, 508–9.
Kennedy, L. D. (1976). *Carbohydr. Res.* **52**, 259–61.
—— (1978). *Carbohydr. Res.* **61**, 217–21.
—— and Bailey, R. W. (1976). *Carbohydr. Res.* **49**, 451–4.
—— and Pankhurst, C. E. (1979). *Microbios* **23**, 167–73.
Kishinevsky, B. and Bar-Joseph, M. (1978). *Can. J. Microbiol.* **24**, 1537–43.
Kleczkowska, J. (1950). *J. gen. Microbiol.* **4**, 298–310.
Kleczkowski, A. and Thornton, H. G. (1944). *J. Bacteriol.* **48**, 661–72.
Koontz, F. P. and Faber, J. E. (1961). *Soil Sci.* **91**, 228–32.
Kowalski, M., Ham, G. E., Frederick, L. R., and Anderson, I. C. (1974). *Soil Sci.* **118**, 221–8.
Kremer, R. J. and Wagner, G. H. (1978). *Soil Biol. Biochem.* **10**, 247–55.
Law, I. J. and Strijdom, B. W. (1977). *Soil Biol. Biochem.* **9**, 79–84.
Lindeman, W. C., Schmidt, E. L., and Ham, G. E. (1974). *Soil Sci.* **18**, 274–9.
Linevich, L. I., Lobzhanidze, S. KH., and Stepanenko, B. N. (1976). *Dokl. Biochem.* **226**, 69–71.

Loos, M. A. and Louw, H. A. (1964). *S. Afr. J. Agric. Sci.* **7,** 135–46.

Lorkiewicz, Z. and Dusinski, M. (1963). *Acta Microbiol. Polon.* **12,** 119–24.

—— and Russa, R. (1971). *Pl. Soil* Special Volume, 105–9.

Maier, R. J. and Brill, W. J. (1978). *J. Bacteriol.* **133,** 1295–9.

Marques-Pinto, C., Yao, P. Y., and Vincent, J. M. (1974). *Aust. J. Agric. Res.* **25,** 317–29.

Marshall, K. C. (1967). *Aust. J. Biol. Sci.* **20,** 429–38.

—— and Vincent, J. M. (1954). *Aust. J. Sci.* **17,** 68–9.

Means, U. M. and Johnson, H. W. (1968). *Appl. Microbiol.* **16,** 203–6.

—— —— and Date, R. A. (1964). *J. Bacteriol.* **87,** 547–53.

Nairn, R. C. (ed. 1976). *Fluorescent protein tracing* 4th edn. Churchill Livingstone, Edinburgh.

Neter, E., Hi, Y. W., and Mayer, H. (1973). *J. Infec. Diseases* **128,** Suppl. S56–S60.

Padmanabhan, S. and Broughton, W. J. (1979). In *Soil microbiology and plant nutrition* (ed. W. J. Broughton *et al.*), pp 240–9 University of Malaysia Press.

Pankhurst, C. E. (1974). *J. gen. Microbiol.* **82,** 405–13.

—— (1979). *Microbios.* **24,** 19–28.

Parker, C. A. and Grove, P. L. (1970). *J. appl. Bacteriol.* **33,** 248–52.

Planqué, K. and Kijne, J. W. (1977). *FEBS Lett.* **73,** 64–6.

—— van Nierop, J. J., Burgers, A., and Wilkinson, S. G. (1979). *J. gen. Microbiol.* **110,** 151–9.

Pochon, J., Manil, P., Tchan, Y. T., Bonnier, C., and Chalvignac, M. A. (1950). *Ann. Inst. Pasteur* **79,** 757–62.

Pueppke, S. G. and Bauer, W. D. (1976). *Pl. Physiol., Lancaster* **57,** suppl. item 417, 80.

—— —— Keegstra, K., and Ferguson, A. L. (1978). *Pl. Physiol., Lancaster,* **61,** 779–84.

Purchase, H. F. and Vincent, J. M. (1949). *Proc. Linn. Soc. N.S.W.* **74,** 227–36.

—— —— and Ward, L. M. (1951*a*). *Proc. Linn. Soc. N.S.W.* **76,** 1–6.

—— —— —— (1951*b*). *Aust. J. agric. Res.* **2,** 261–72.

Read, M. P. (1953). *J. gen. Microbiol.* **9,** 1–14.

Rinderknecht, H. (1962). *Nature, Lond.* **193,** 167–8.

Sanders, R. E., Carlson, R. W., and Albersheim, P. (1978). *Nature, Lond.* **271,** 240–2.

Scheffler, J. G. and Louw, H. A. (1967). *S. Afr. J. Agric. Sci.* **10,** 161–74.

Schluchterer, E. and Stacey, M. (1945). *J. Chem Soc.* 776–83.

Schmidt, E. L. (1974). *Soil Sci.* **118,** 141–9.

—— Bankole, R. O., and Bohlool, B. B. (1968). *J. Bacteriol.* **95,** 1987–92.

Skrdleta, V. (1965). *Pl. Soil* **23,** 43–8.

—— (1966). *Rostl. výroba* **39,** 23–7.

—— (1968). *Rostl. výroba* **14,** 969–78.

—— (1969*a*). *Fol. Microbiol.* **14,** 32–5.

—— (1969*b*). *Antonie van Leeuwenhoek* **35,** 77–83.

Stevens, J. W. (1923). *J. Infect. Diseases* **33,** 557–66.

—— (1925). *Soil Sci.* **20,** 45–66.

Thornton, H. G. and Kleczkowski, J. (1950). *Nature, Lond.* **166,** 1118–19.

Trinick, M. J. (1969). *J. appl. Bacteriol.* **32,** 181–6.

—— (1973). *Nature, Lond.* **244,** 459–60.

Van Brussell, A. A. N., Planqué, K., and Quispel, A. (1977). *J. gen. Microbiol.* **101,** 51–6.

Van Der Merwe, S. P. and Strijdom, B. W. (1973). *Phytophylactica* **5,** 163–6.

—— —— and Van Rensburg, H. J. (1972). *Phytophylactica* **4,** 97–100.

Van Rensburg, H. J. and Strijdom, B. W. (1972). *Phytophylactica* **4,** 1–8.

Vest, G., Caldwell, B. E., and Peterson, H. D. (1971). *Crop Sci.* **11,** 780–2.

Vincent, J. M. (1941). *Proc. Linn. Soc., N.S.W.* **66,** 145–54.

—— (1942). *Proc. Linn. Soc., N.S.W.* **67,** 82–6.

—— (1944). *Nature, Lond.* **153,** 496–7.

—— (1953). *Aust. J. Sci.* **15,** 133–4.

—— (1954). *Proc. Linn. Soc., N.S.W.* **79,** iv–xxxii.

—— (1962*a*). *J. gen. Microbiol.* **28,** 653–63.

—— (1962*b*). *Proc. Linn. Soc., N.S.W.* **87,** 8–38.

—— (1970). *A manual for the practical study of the root-nodule bacteria.* IBP Handbook No. 15, Blackwell Scientific, Oxford.

—— (1974). In *The biology of nitrogen fixation*, North-Holland Research Monographs: *Frontiers of Biology*, Vol. 33 (ed. A. Quispel), pp. 265–341, North-Holland, Amsterdam.

—— (1977). In *A treatise on dinitrogen fixation:* section III *Biology* (ed. R. W. F. Hardy and W. S. Silver), pp. 277–366. John Wiley, New York.

—— and Colburn, J. R. (1961). *Aust. J. Sci.* **23,** 269–270.

—— and Humphrey, B. (1963). *Nature, Lond.* **199,** 149–51.

—— —— (1968). *J. gen. Microbiol.* **54,** 397–405.

—— —— (1970). *J. gen. Microbiol.* **63,** 379–82.

—— —— and Škrdleta, V. (1973). *Arch. Mikrobiol.* **89,** 79–82.

—— Nutman, P. S., and Skinner, F. A. (1979). In *Identification methods for microbiologists*, 2nd edn. Society for Applied Bacteriology Technical Series No. 14 (ed. F. A. Skinner and D. N. Lovelock) pp. 49–69. Academic Press, London.

—— and Waters, L. M. (1953). *J. gen. Microbiol.* **9,** 357–70.

—— —— (1954). *Aust. J. agric. Res.* **5,** 61–76.

Weber, D. F., Caldwell, B. E., Sloger, C., and Vest, H. G. (1971). *Pl. Soil* Special Volume 293–304.

—— and Miller, V. L. (1972). *Agron. J.* **64,** 796–8.

Wilson, N. H. M., Humphrey, B. A., and Vincent, J. M. (1975). *Arch. Microbiol.* **103,** 151–4.

Wolpert, J. S. and Albersheim, P. (1976). *Biochem. Biophys. Res. Commun.* **70,** 729–37.

Wright, W. H. (1925). *Soil Sci.* **20,** 95–129.

Zajac, E., Russa, R., and Lorkiewicz, Z. (1975). *J. gen. Microbiol.* **90,** 365–67.

Zevenhuizen, L. P. T. M. (1971). *J. gen. Microbiol.* **68,** 239–43.

—— (1973). *Carbohydr. Res.* **26,** 409–19.

—— and Scholten-Koerselman (1979). *Antonie van Leeuwenhoek* **45,** 165–75.

9 Control of root hair infection

F. B. DAZZO AND D. H. HUBBELL

Recognition and adsorption

9.1 Introduction

The root-hair infection process consists of several specific events which precede nodule formation in the *Rhizobium*–legume symbiosis. These include mutual host–symbiont recognition of the *Rhizobium* species on the legume rhizoplane, rhizobial adherence to differentiated epidermal cells called root hairs, root-hair curling, root-hair infection, root nodulation, and transformation of vegetative bacteria into enlarged, pleomorphic bacteroids which fix nitrogen (reviewed in Dart 1977; Broughton 1978; Schmidt 1979; Dazzo 1980c). Many steps of the infection process have been more clearly understood since the development of the slide culture technique of Fåhraeus (1957), and its use in the production of an elegant time-lapse cinema at the light microscope level (Nutman, Doncaster, and Dart 1973). The distinctive feature of an infected root-hair is the infection thread, a tube which confines the rhizobia as they penetrate into the root (Fig. 9.1).

The symbiosis is characterized by a high degree of host specificity. For instance, *R. trifolii* infects and nodulates *Trifolium repens* roots but not *Medicago* or *Glycine* roots. Host specificity is expressed at a very early step of infection of the root hair by infective rhizobia, prior to the penetration of the root-hair cell wall (Li and Hubbell 1969) and formation of the root-hair infection thread (Napoli and Hubbell 1975).

The intent of Section 1 is to draw attention to the interaction of legume lectins (carbohydrate-binding proteins or glycoproteins) with their specific saccharide receptors on the bacterium and the plant, and the possible importance of this interaction in controlling host specificity in the *Rhizobium*–legume symbiosis. Recent reviews on this rapidly developing subject have been prepared by Bauer (1977), Broughton (1978), Sequeira (1978), Dazzo (1979, 1980*a, b, c, d*), Schmidt (1979), and Solheim and Paxton (1980).

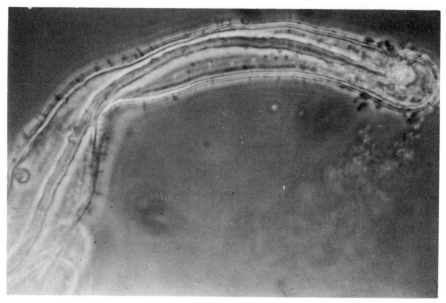

FIG. 9.1. *T. repens* root hair infected with *R. trifolii* 0403. Note the clump of rhizobia at the root-hair tip, the polarly attached rhizobia along the lateral shaft of the root hair, and the refractile infection thread (I.T.) which carries the infective rhizobia into the root. 2120 ×. Phase-contrast photomicrograph.

The lectin-recognition hypothesis

The lectin-recognition hypothesis states that recognition at infection sites involves the binding of specific legume lectins to unique carbohydrates found exclusively on the surface of the appropriate rhizobial symbiont.

A first step towards testing the lectin-recognition hypothesis is to determine if the legume has a lectin which binds to specific receptors that are expressed only on cells of the rhizobial symbiont. One approach has been to test the lectin-recognition hypothesis by examining the 'classical' lectin in the leguminous plant which is reported to agglutinate erythrocytes. In some cases, e.g. soyabean, the classic phytohaemagglutinin displays the specificity of binding to rhizobial species which correlates with their host-nodulating specificities(Bohlool and Schmidt 1974; Wolpert and Albersheim 1976; Bhuvaneswari, Pueppke, and Bauer 1977). However, in other cases where several anomalous lectin-binding interactions were found (reviewed in Dazzo 1980*a*, *b*, *c*, *d*), it remains to be determined if the legume in question contains several different lectins, some of which may be on root hairs and recognize unusual saccharides on *Rhizobium* which are not present on erythrocytes.

The lectin recognition hypothesis is being tested in the *R. trifolii–T. repens* association because of the more complete understanding of the infection process in this symbiosis (Napoli and Hubbell 1975; Dart 1977; Callaham 1979; Dazzo 1980c). Because clover is a small-seeded legume, this model system is easily amenable to quantitative microscopic analysis (Dazzo, Napoli, and Hubbell 1976; Dazzo, Yanke, and Brill 1978; Dazzo 1980a). Our approach has been to combine ultrastuctural, biochemical, immunochemical, genetic, and quantitative light and fluorescent microscopic techniques to examine clover seedlings for proteins which bind specifically (presumably in a complementary fashion) to unique surface receptors on *Rhizobium*. There is evidence to suggest that trifoliin, a clover lectin, is a cellular recognition molecule involved in determining host specificity in this symbiosis. Trifoliin binds specifically to *R. trifolii* and accumulates at clover root-hair tips where infections most frequently occur (Dazzo *et al.* 1978).

A second requirement of the lectin-recognition hypothesis is for the lectin to be present at infection sites on the legume root. Bowles, Lis, and Sharon (1979) have proposed that lectins in some developing plant roots (e.g. soyabean) may become intercalated in membranes, as suggested by the requirement of detergents for their solubilization. In contrast, lectins of other plant roots (e.g. *Arachis hypogeae*) can be solubilized by buffers containing effective sugar haptens, suggesting that they are anchored to glycosylated receptors. The removal of trifoliin from intact seedling roots with the aid of the hapten 2-deoxyglucose (Dazzo and Brill 1977; Dazzo *et al.* 1978) suggests that at least a portion of this lectin falls within the latter category. The root-associated lectin from white clover was localized on root hair tips by immunofluorescence microscopy and formed a fused precipitin reaction of identity with the purified seed lectin when these antigens were analysed by Ouchterlony immunodiffusion using anti-seed lectin antiserum (Dazzo *et al.* 1978), provided strong immunological evidence for the presence of trifoliin on clover root surfaces at sites of rhizobial attachment and infection. Similar hapten-facilitated elution of root-associated lectin has been found for peas (Kijne and Kamberger, personal communications). Lectins situated on root hairs of *Pisum sativum* have the same sugar binding specificity and are serologically very similar to lectins isolated from the seeds (Kamberger 1979b, and personal communication). Other reports of similarities between soyabean seed and root lectin have been made (Schmidt 1979). Stacey, Paau, and Brill (1980) have recently localized the root hair tips of the wild soyabean, *Glycine soja*, as a site of accumulation of the soyabean lectin.

Rhizobial attachment to legume root hairs

The focus has been on the possible involvement of trifoliin in the attachment of *R. trifolii* to clover root-hairs, although it is possible that legume lectins may be involved in other events which precede or follow cell–cell adhesion. Light- and electron-microscopic studies (Dazzo and Hubbell 1975*c*; Chen and Phillips 1976; Dazzo *et al.* 1976; Dazzo and Brill 1979; Dazzo 1980*a, b, d*; Dazzo and Hrabak 1980) have revealed at least two stages in the process of microsymbiont attachment. During the initial attachment stage, Phase I attachment, docking of the bacteria is initiated by contact of the fibrillar capsule of *R. trifolii* with electron-dense globular aggregates lying on the outer periphery of the clover root-hair wall [Fig. 9.2(a) and 2(b)]. If a large inoculum is used, the bacteria will attach first at the tips of the root hair and later along the shaft of these differentiated epidermal cells. This process occurs within a few hours. Quantitative microscopic studies of Phase I attachment, using lower inoculum densities, indicated a correlation between the ability of rhizobial cells to attach in high numbers to clover root-hairs and the ability of these bacteria to infect clover root-hairs (Dazzo *et al.* 1976; Dazzo 1980*a*). Similar results were reported for soyabeans (Stacey, Paau and Brill 1980) and peas (Kato, Maruyama and Nakamura 1980). After 12 h of incubation, rhizobial adhesion (Phase II) is less localized on the root surface and fibrillar material associated with the bacteria adhering to root hairs is more easily resolved in scanning electron micrographs (Fig. 9.3) (Dart 1971; Dazzo 1980*b, c, d*).

The first clue that the lectin, trifoliin, may be involved in Phase I attachment came from the observation that the sugar, 2-deoxyglucose, specifically inhibited the attachment of *R. trifolii* to clover root-hairs (Dazzo *et al.* 1976). By contrast, the attachment of *R. meliloti* to root hairs of its host, *M. sativa* was unaffected by 2-deoxyglucose. Subsequent studies have shown that this sugar is an effective hapten inhibitor of trifoliin binding to *R. trifolii* and its surface polysaccharides, as well as the specific binding of capsular polysaccharide preparations from *R. trifolii* to root hairs of clover seedlings (Fig. 9.4), (Dazzo and Brill 1977; Dazzo *et al.* 1978). In related studies, the binding of *R. japonicum* and its surface polysaccharides to root hairs of the host, *G. soya*, was inhibited by *N*-acetylgalactosamine, a hapten inhibitor of soyabean lectin (Stacey *et al.* 1980; Hughes and Elkan 1979; Hughes, Leece, and Elkan 1979).

In such hapten-inhibition studies it is believed that the sugar acts by combining with the site on the lectin which is normally occupied by the polysaccharide. This implies a close, but not necessarily identical structure of the inhibitor and the glycosylated receptor for the lectin. These saccharides may, therefore, only represent an analogue of the

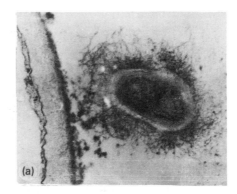

Fig. 9.2(a). Transmission electron micrograph of *R. trifolii* NA30 in the docking stage of Phase I attachment to a clover root hair. The fibrillar capsule of the bacterium is in contact with electron-dense globular aggregates on the outer periphery of the root-hair cell wall. From Dazzo and Hubbell (1975*c*), courtesy of Carolyn Napoli and the American Society for Microbiology. 34 000 ×.

native haptenic determinant on the surface polysaccharide of rhizobia which binds specifically to the corresponding lectin on the homologous root-hair tip. Since some lectins undergo conformational changes upon saccharide binding (Reeke, Becker, Cunningham, Wang, Yahana, and Edelman 1975), this possibility must also be considered in the interpretation of hapten inhibition studies.

We suggested that the ability of *R. trifolii* to adhere to clover root hairs should be influenced by conditions that affect the accumulation of trifoliin on the root surface and the saccharide receptor on the bacterium. Four lines of evidence supporting this hypothesis have been found. Firstly, antigenically altered mutant strains of *R. trifolii* (Dazzo and Hubbell 1975*a*) had significantly fewer or nondetectable levels of trifoliin receptors (Dazzo and Hubbell 1975*c*) and attached to clover root-hairs only at low background levels (Dazzo *et al.* 1976). Secondly, immunologically detectable levels of trifoliin and the attachment of *R. trifolii* to root hairs decreased in a parallel fashion as the concentrations of either NO_3^- or NH_4^+ were increased in the rooting medium (Dazzo and Brill 1978). Thirdly, under certain growth condi-

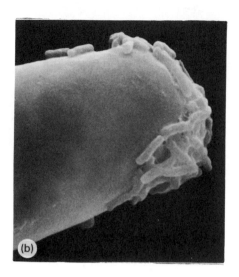

FIG. 9.2(b). Phase I selective attachment of *R. trifolii* 0403 to a clover root hair tip after 15 min of incubation, as examined by scanning electron microscopy. From Dazzo and Brill (1979), courtesy of the American Society for Microbiology. 7000 ×.

tions, the transient appearance of trifoliin receptors on *R. trifolii* coincided with their ability to attach in greatest quantity to clover roots (Dazzo, Urbano, and Brill 1979). And fourthly, intergeneric transformation of *Azotobacter vinelandii* with DNA from *R. trifolii* yielded hybrid transformants which could bind trifoliin (Bishop, Dazzo, Appelbaum, Maier, and Brill 1977). When incubated with clover seedling roots, these intergeneric hybrids had also acquired the ability to attach to clover root-hair tips by a Phase I process which could be inhibited by 2-deoxyglucose (Dazzo and Brill 1979).

One of the possible consequences of this lectin–polysaccharide interaction would be the initiation of an adhesive contact interface which would permit the bacterial cell to attach and then firmly adhere to the target host cell. By mediating the adhesion of these specific cells, the

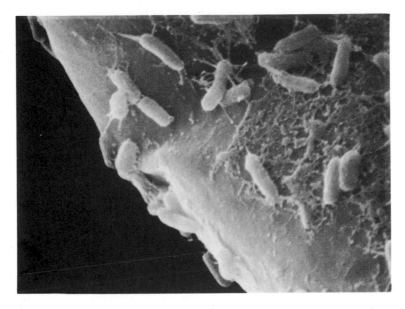

Fig. 9.3. Phase II adherence of *R. trifolii* 0403 to a clover root hair after 2 days incubation, as examined by scanning electron microscopy. Note the fibrillar material associated with the adherent bacteria. 10 000 ×.

lectin may also function as a 'cell recognition molecule' since it could feasibly influence which cells associate in sufficient proximity to the root hairs at the proper time in development to allow subsequent specific recognition steps to occur. Specific steps which follow attachment of infective rhizobia include a marked curling of the clover root-hair tip around the adherent clump of bacteria, followed by wall penetration, and initiation of the tubular infection thread which confines the microsymbionts as they enter the root.

With low densities of inoculum, *R. trifolii* attaches in a polar end-on fashion to clover root hairs after a few hours incubation with the root in hydroponic culture (Fig. 9.5) (Dazzo *et al.* 1976). This orientation is quite distinct from the typical random orientation of the rhizobia attached to root hair tips when higher inoculum densities are employed, as in short-term Phase I adherence assays (Dazzo and Brill 1977, 1979; Dazzo *et al.* 1978). Recent studies have suggested that this striking polarity in adhesion may be due to an alteration in distribution of trifoliin receptors on the bacteria in the clover root environment, prior to cell–cell adhesion (Dazzo *et al.* 1981). Trifoliin was detected by immunofluorescence in aseptically collected clover-root exudate, and its presence was confirmed by immunoaffinity chromatography followed by polyacrylamide gel electrophoresis (Dazzo and Hrabak 1981).

Trifoliin in root exudate bound uniformly to fully encapsulated cells after 1 h of incubation. However, after extended incubation ($\geq$ 4 h) of the root exudate with the rhizobia, trifoliin was found to be bound to only one of the cell's poles (Fig. 9.6). This was the case for cells heat-fixed to microscope slides, in suspension, or *in situ* in the clover-root environment of Fåhraeus slide-culture assemblies. The substance(s) in root exudate which caused this alteration were of high molecular weight and did not bind to the immunoaffinity column of immobilized anti-trifoliin immunoglobulin. These results suggest that this progressive alteration in distribution of trifoliin receptors on cells in the root environment favours their subsequent polar orientation of attachment after contact with the root surface. The detection of trifoliin in clover root exudate is consistent with the hypothesis of Solheim (1975), who proposed that the lectin is released from the roots and binds

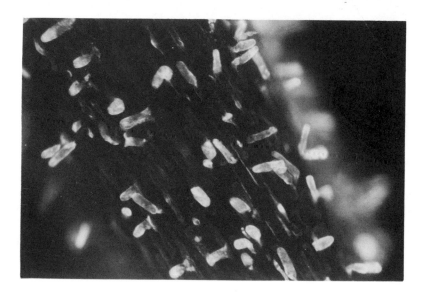

FIG. 9.4. An exclusive relation between a legume and a nitrogen-fixing bacterium is demonstrated by the binding of polysaccharides from *R. trifolii* to root hairs of its plant host *T. repens*. A fluorescent dye was first conjugated to the bacterial capsular polysaccharide and this labelled polysaccharide was then incubated with clover roots. Fluorescence of the roots showed that binding of the bacterial polysaccharide was restricted to the root hairs, the differentiated epidermal cells that serve as target cells for infection by *R. trifolii*. Similar results have been obtained with other rhizobia and their corresponding specific legume hosts. Fluorescence photomicrograph. From Dazzo and Brill (1977), courtesy of the American Society for Microbiology. 370 ×.

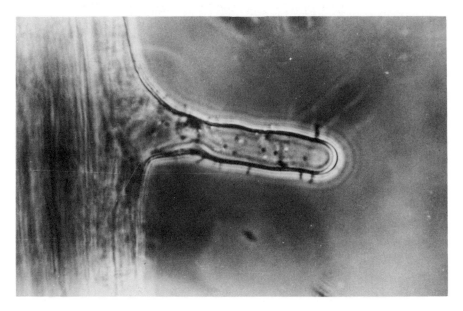

FIG. 9.5. Polar attachment of *R. trifolii* 0403 to a clover root hair after 12 h of incubation with a low inoculum density. Phase contrast photomicrograph. 1500 ×.

to the rhizobial symbiont, thereby increasing its affinity for the root surface.

Some strains of *R. japonicum* growing exponentially have soyabean lectin receptors accumulated at one cell pole (Bohlool and Schmidt 1976; Tsien and Schmidt 1977), but it is not known if this polar distribution is maintained when the cells are exposed to the root environment of the host. Interestingly, some strains of *R. japonicum* that lacked soyabean lectin receptors when grown in laboratory culture media would acquire this property when grown in root exudate (Bhuvaneswari and Bauer 1978).

Despite the extensive support in favour of the lectin-recognition hypothesis, the proof is not conclusive. In particular, we assume the cautious attitude that adhesion of the infective rhizobia is but one event in a sequence of cellular recognition steps, which comes into play and ultimately results in successful infection. Numerous examples can be cited to support this case. Heterologous rhizobia can adhere in small numbers to clover root hairs, but do not infect them (Dazzo *et al.* 1976; Dazzo 1980*a*), and very few of the root hairs to which infective rhizobia adhere subsequently become infected. The intergeneric hybrids of *A. vinelandii* transformed with DNA from *R. trifolii* (Bishop *et al.* 1977; Dazzo and Brill 1979) have acquired the ability to adhere specifically to

FIG. 9.6. Polar binding of trifoliin in root exudate to capsular material on *R. trifolii* 0403 after 4 h incubation. Trifoliin bound to one of the cell's poles is visualized by indirect immunofluorescence. Fluorescence photomicrograph. 12 500 ×.

clover root-hair tips, but do not advance the infection process to the stage of penetration. Two classes of noninfective mutant strains of *R. meliloti* have been isolated (Leps, Paau, and Brill 1980; Kamberger, personal communication). One class will not attach to root hairs of its host, *M. sativa*, and has lost the ability to be agglutinated by an *M. sativa* lectin. A second class of noninfective mutants does attach to the root hairs to *M. sativa* and is agglutinated by an *M. sativa* lectin. These results are consistent with the hypothesized role of the lectin receptor on the microsymbiont, in its adherence to the differentiated host cell. In addition, the second class of mutant strains clearly underscores the importance of pre- or post-adhesion steps essential to the infection process. These points emphasize the gaps in our understanding of the biochemistry of the lectin receptors and their interaction with the lectin in order to determine whether they play a role in cellular recognition in this nitrogen-fixing symbiosis.

The lectin receptor on Rhizobium

The biochemical nature of the saccharide receptor on *Rhizobium* which interacts specifically with the host lectin has remained a complex problem since the earliest successful attempts at detecting specific interactions of host proteins and rhizobial polysaccharides freed from the cells (Dazzo and Hubbell 1975c). The problem is complex for several reasons. First of all, several different polysaccharides are made by the rhizobia (reviewed by Carlson, Chapter 7), including acidic heteroploysaccharides, lipopolysaccharides, wall-associated glucans, β-1,2 linked glucans, and β-1, 4 linked glucans. In addition, there are reportedly other exopolysaccharides 'consisting largely of glucose with a small amount of galactose' (Wolpert and Albersheim 1976). Another complication is that the lectin may interact well with the *Rhizobium* cell at one culture age, but poorly at another age (Bhuvaneswari *et al.* 1977; Dazzo *et al.* 1979). Partial characterization of lectin-binding polysaccharides isolated from cells at various culture ages revealed compositional differences which could relate directly to their lectin-binding properties (Mort and Bauer 1980; Hrabak, Urbano and Dazzo 1981). A third factor is that, under certain growth conditions, more than one type of polysaccharide made by the same rhizobial strain can interact with the host lectin (Dazzo and Brill 1979; Kamberger 1979*a*). This phenomenon is not restricted to *Rhizobium* and legume lectins, but may occur in other plant–phytopathogen systems as well (Sequeira 1978). Last, but not least, is the synthesis and/or modification of the lectin-receptor on rhizobia which may be affected by the root environment of the host (Bhuvaneswari and Bauer 1978; Dazzo *et al.* 1981).

The numerous publications on the carbohydrate composition and structure of rhizobial polysaccharides reviewed by Carlson (Chapter 7 p. 199) do not state whether these structures can interact specifically with lectins from the corresponding host symbiont. It must not be assumed, without testing, that they have this property, especially in view of the fact that several extracellular polysaccharides are made by the cells, some of which change composition with culture age.

Based on fluorescence microscopy and subsequent cell-free studies (Dazzo and Hubbell 1975*c*; Dazzo and Brill 1977,1979; Dazzo *et al.* 1978), it was found that trifoliin could combine with saccharide receptors in the capsule of *R. trifolii* 0403. Transmission electron microscopy unequivocally documented the presence of the acidic fibrillar capsule as a dominant surface structure of *R. trifolii* (Dazzo and Brill 1979) and of the contact interface during Phase I attachment of *R. trifolii* to clover root hairs (Dazzo and Hubbell 1975*c*, and Fig. 9.2). Similar interactions were reported subsequently with soyabean lectin and *R. japonicum* (Bal, Shantharam, and Ratnam 1978; Calbert, Lalonde, Bhuvaneswari, and Bauer 1978; Tsien and Schmidt 1977).

Legume lectins have also been reported to combine specifically with lipopolysaccharides (LPS) from the outer membrane of the homologous rhizobial symbionts (Wolpert and Albersheim 1976; Kamberger 1979*a*; Kato, Maruyama, and Nekamura 1979). However subsequent studies have shown that the compositions and immunodominant structures of the LPS vary as widely among strains of a single *Rhizobium* species as among the different species of *Rhizobium* (Carlson, Saunders, Napoli, and Albersheim 1978). Therefore, in these cases, lectins may be interacting with a portion of the O-antigen or core region of the LPS which is common to different strains of the *Rhizobium* species (Albersheim, Ayers, Valent, Ebel, Wolpert, and Carlson 1977).

The pea lectin will interact with both extra-cellular polysaccharides (ECPS) and LPS from *R. leguminosarum* (Kamberger 1979*a*, Kato Maruyama and Nakamura 1980). Based on studies on mutant strains of *R. leguminosarum* defective in LPS components, Kamberger (1979*b*) has suggested that the ECPS could be responsible for adsorption of high numbers of rhizobial cells to the plant root hairs via cross-bridging lectins as a primary recognition event. This would be followed by secondary recognition events requiring the host-range specific binding of lectin on the pea root hairs to LPS, which triggers subsequent invasive steps. In this case, a loss by mutation of one of the two lectin receptors results in a loss in infectibility, yet the cell can still combine with the lectin since one of the two lectin receptors continues to be made.

It is also possible to apply this interpretation to the reverse example of non-nodulating EXO-1 mutant strain of *R. leguminosarum* (Saunders, Carlson, and Albersheim 1977) which is reported to produce the same LPS structure as the wild type nodulating strain (but see below*), but is defective in ECPS production. This mutant strain has lost the ability to attach to pea root hairs in high numbers (C. Napoli, personal communication). If Kamberger's hypothesis is correct, then the EXO-1 mutant strain may fail to accomplish the first recognition step as a prerequisite for the second recognition step, despite the presence of the second lectin receptor.

Van der Schaal and Kijne (1979) recently obtained a noninfective mutant strain of *R. leguminosarum* (RBL 101) which lost a 1.1×10^8 dalton plasmid and had a different sugar ratio in its LPS, ECPS, and wall-associated glycan (fraction PSI) previously reported to bind pea

---

* The mutant strain, EXO-1, does not nodulate peas, and Saunders *et al.* (1978) reported that 27 per cent of its total LPS mass is anthrone-reactive carbohydrate, as compared to 63 per cent of the total LPS mass from the wild-type *R. leguminosarum* strain from which it is derived. The glycosyl and antigenic compositions of the O-antigen of the mutant and wild-type strain do not seem to be different, and both strains are lysed by the same bacteriophages. The hypothesis advanced by the authors was that the mutant strain EXO-1 has reduced its production of ECPS and not of LPS (Saunders *et al.* 1978). However, an alternative hypothesis is that the EXO-1 phenotype could be due to a defective O-antigen polymerase, which would fail to polymerize in a block fashion the repeating O-antigen on the polyisoprenoid acyl carrier lipid (Osborne *et al.* 1972). When this O-antigen oligosaccharide is transferred to the R-core lipid A via the translocase reaction, an LPS with reduced O-antigenic polymerization would result. Bacteriophage with receptors for the O-antigens could still recognize those defective LPS structures, their immunodominant antigens would be immunochemically identical when isolated, and their glycosyl composition would be the same or similar, exactly matching the phenotype of these non-nodulating *R. leguminosarum* mutant strains described by Saunders *et al.* (1978). Studies which distinguish these two different possibilities remain to be performed. The reduction in O-antigen contribution to LPS in mutant strains of Gram-negative bacteria causes a concomitant pleiotropic decrease and sometimes virtual loss of outer membrane proteins (Gmiener and Schlecht 1979). This pleiotropic negative effect complicates any direct interpretation of the significance of EXO-1 mutant strain to the infection process. Other pleiotropic negative effects are also likely since EXO-1 mutant strain has a large deletion of DNA (M. Nuti, personal communication).

lectin (Planqué and Kijne 1977). The PSI fraction glycan of the mutant strain was still able to interact with the lectin, but the mutant strain itself was not, suggesting that this polysaccharide may not be exposed on the cell surface in culture.

In another study (Dazzo and Brill 1979), the capsule of *R. trifolii* 0403 was extracted from pelleted cells in early stationary phase, and then fractionated into three component polysaccharides by ion exchange and gel filtration chromatography. The major trifoliin-binding polysaccharide in the capsular extract was an acidic heteropolysaccharide comprising approximately 85 per cent of the total carbohydrate equivalents of the starting material. The second trifolin-binding polysaccharide contained approximately 12 per cent of the carbohydrate

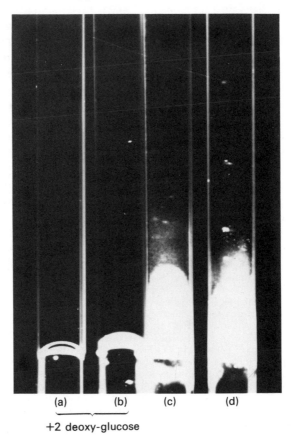

(a)   (b)   (c)   (d)

+2 deoxy-glucose

Fig. 9.7. Precipitation of trifoliin with capsular polysaccharide (C) and lipopolysaccharide (D) from *R. trifolii* 0403 in early stationary phase of growth. Hapten-inhibited controls in (A) and (B) are with 30 mM 2-deoxyglucose. Reactions are conducted in capillary tubes incubated vertically for 12 h.

equivalents. The precipitation of both polysaccharides with trifoliin was inhibited by 2-deoxyglucose (Fig. 9.7). In addition to being immunogenic and containing heptose, 2-keto 3-deoxyoctanoate and endotoxic lipid A as chemical markers for LPS (Dazzo and Brill 1979), the second polysaccharide was shown by immunoelectrophoresis to have an immunodominant antigen linked to the native structure through a weakly acid-labile linkage (unpublished observations). Its composition was suggestive of either LPS or a polysaccharide associated with LPS. (The third polysaccharide contained only 3 per cent of the total carbohydrate equivalents, contained saccharide markers which were found in neither of the other two polysaccharide fractions, and did not bind to trifoliin.)

The two trifoliin-binding polysaccharide fractions carried antigenic determinants in common, but differed from one another in electrophoretic mobility and size as shown by immunoelectrophoresis and sodium dodecylsulphate–polyacrylamide gel electrophoresis, respectively (Dazzo and Brill 1979). Descending paper chromatography of acid hydrolysates suggested that the two fractions had similar or the same sugar components. These data are consistent with the hypothesis that the major trifoliin-binding polysaccharide in the capsular material isolated from cells at this culture age is an endotoxin-free acidic heteropolysaccharide which carries the same O-antigenic determinants found in the cells' LPS at this same culture age. Because of its larger apparent size (as suggested by electrophoresis, and the absence of heptose, 2-keto 3-deoxyoctanoate and endotoxic lipid A, this major polysaccharide is not considered a smaller precursor or partial degradation product of LPS.

The antigenic difference between wild-type *R. japonicum* 61A76 and non-nodulating mutant strains SM-1 and SM-2 is a result of component alterations in their surface polysaccharides, which share this same relationship (Maier and Brill 1978). ECPS which carry the same O-antigenic determinants as the LPS have been described for several Gram-negative bacteria in the family Enterobacteriaceae (Jann, Jann, Schmidt, and Orskov 1970; Jann and Westphal 1975). This relationship is believed to arise under certain conditions by the transfer of the acidic capsular heteropolysaccharide to the core-lipid A, resulting in a replacement of the O-antigen side chain in the bacterial LPS.

Information on the glycosylated receptors for lectins in plants would certainly be of value in understanding how these lectins may function *in situ*. Based on immunochemical and genetic studies (Dazzo and Hubbell 1975*c*; Bishop *et al.* 1977; Dazzo and Brill 1979), evidence has been accumulated which suggests that trifoliin exposed on the clover seedling root-surface is bound to a glycoslyated receptor bearing immunological resemblance to saccharide receptors for trifoliin on

R. trifolii. Like trifoliin, antibody to clover seedling roots will specific-ally agglutinate *R. trifolii,* and will immunoprecipitate its trifoliin-binding polysaccharides (Dazzo and Hubbell 1975*c*; Dazzo and Brill 1979). This interaction is specifically inhibited by the hapten, 2-deoxyglucose (Dazzo and Hubbell 1975*c*; Dazzo and Brill 1979). The cross-agglutinating antibody in anticlover root antiserum can be adsorbed by clover seedlings, but not by *M. sativa* or *Aeschynomene americona* seedlings (which are legumes representative of other cross-inoculation groups) (Dazzo and Brill 1979). Antibody to trifoliin-binding polysaccharides of *R. trifolii* cross-reacts with clover root surfaces (Dazzo and Brill 1979). Gene markers for the surface determinants which bind trifoliin and anti-clover root antibody co-transform from *R. trifolii* into *A. vinelandii* with a frequency of 100 per cent (Bishop *et al.* 1977). Monovalent Fab fragments of IgG from anti-clover root antiserum strongly block the agglutination of *R. trifolli* by trifoliin, and inhibit the attachment of these cells to clover root hairs during Phase I attachment (Dazzo and Brill 1979). Based on these data, it is proposed that Phase I attachment is initiated by a selective

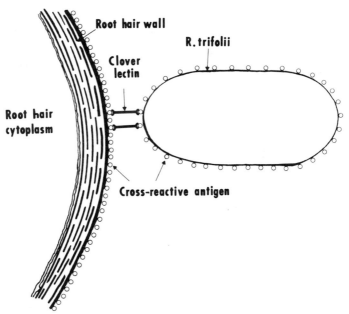

Fig. 9.8. Schematic diagram of the model to explain Phase I attachment. According to this model, antigenically cross-reactive saccharide receptors on *R. trifolii* and clover root hairs are specifically cross-linked with the clover lectin, trifoliin. From Dazzo and Hubbell (1975*c*), courtesy of the American Society for Microbiology.

cross-bridging of similar saccharide receptors for trifoliin on the bacterium and the clover root hair tip (Dazzo and Hubbell 1975*c*; Dazzo and Brill 1979; Dazzo 1980*d*;) (Fig. 9.8). This cross-bridging model is simple in design, but very complex in content. It is designed to describe one of the many delicately controlled events which dictate host specificity.

The theoretical background for the role of common antigens in the formation and maintenance of host–pathogen and host–symbiont interactions has been reviewed (DeVay and Adler 1976). Perhaps a normal, physiological function of trifoliin is to cross-link saccharide structures during tip growth of the root-hair cell wall. The lectin receptor would then serve as the 'address' for trifoliin during its interaction with the root-hair cell wall (Dazzo 1980*d*). Although there is other supporting evidence for cross-reacting antigens which correlates with rhizobial infectivity of *M. sativa* (Kamberger and Nordhoff, personal communication), and evidence for lectin-binding glucans on pea roots which bear chemical resemblances to those on the symbiont of *R. leguminosarum* (Kijne, Ineke and DeVries 1980), other efforts to detect cross-reactive antigens have been unsuccessful (Bohlool, personal communication).

The latter result may be due to the fact that these cross-reactive antigenic determinants are not immunodominant structures. Instead, they are conserved in otherwise immunochemically-distinct polysaccharides of *R. trifolii* strains (Dazzo and Brill 1979; Hrabak, Urbano, and Dazzo 1981). In addition, they can easily be missed since they are present on the cell for only a transient period in normal batch culture (Dazzo *et al.* 1979; Hrabak, Urbano, and Dazzo 1981).

Regulation of recognition in the Rhizobium–*legume symbiosis*

How is the recognition process regulated? Do both symbionts participate in regulating the components of the recognition process? Some answers to these important questions are beginning to emerge. It has been known for many years that fixed nitrogen (e.g. NH_4^+, NO_3^-) is one of the many environmental factors which limits the development and success of the *Rhizobium*–legume symbiosis in nature (for a review, see Fred, Baldwin, and McCoy 1932).

Dazzo and Brill (1978) have found that the immunologically detectable levels of trifoliin and the specific attachment of *R. trifolii* to root hairs decreased in a parallel fashion as the concentrations of either NO_3^- or NH_4^+ were increased in the rooting medium (Fig. 9.9). These experiments support the hypothesis that trifoliin is involved in binding *R. trifolii* to clover root hairs. The levels of these ions which inhibited

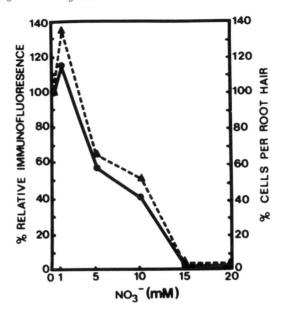

Fig. 9.9. The effect of NO_3^- on adsorption of *R. trifolii* 0403 to root hairs (solid line) and on immunologically detectable trifoliin (dotted line) in the root hair region of clover seedlings. Bacterial adsorption was measured by direct counting and trifoliin was measured by cytofluorimetry using indirect immuno-fluorescence. Values from roots grown in nitrogen-free nutrient solution are taken as 100 per cent, and represent 980 photovolts mm^{-2} and 21 cell root hair^{-1} 200 μm in length. Points along the curve are means from 10 to 15 root hairs or seedling roots, standard deviations vary within 10 per cent of the means. Values are corrected for non-specific rhizobial adsorption, root auto-fluorescence, and nonspecific adsorption of conjugated goat antirabbit λ globulin. From Dazzo and Brill (1978), courtesy of the American Society of Plant Physiologists.

attachment were well below the levels which stunted seedling growth. Interestingly, the levels of trifoliin and rhizobial attachment were increased at a relatively low concentration of NO_3^-. Thus, low levels of NO_3^- enhanced the recognition process, and high levels shut it off. Recent ligand-binding studies using radiolabeled $^{13}NO_3^-$ have detected no direct interactions between NO_3^- and trifoliin, and this anion does not interfere with the direct interaction of trifoliin and its specific antibody or *R. trifolii* (Dazzo *et al.* 1981). There are fewer accessible receptor sites for trifoliin on purified cell walls of roots grown in NO_3^- than in nitrogen-free nutrient medium. Thus, regulation of recognition by NO_3^- is modulated through some intervening process. Umali-Garcia, Hubbell, Gaskins, and Dazzo (1980) have observed also that adherence of *Azospirillum brasilense* to root hairs of the grass *Pennisetum americanum*, (pearl millet) is suppressed when the roots are grown in critical concentrations of NO_3^--containing medium. Scanning electron microscopy showed a distinct degranulation in the surface topography of root hairs grown in NO_3^-. These observations have opened new avenues of investigation which may ultimately provide a solution to the important problem of how fixed nitrogen regulates the developmental events of the *Rhizobium*–legume symbiosis.

Bauer and coworkers (Bhuvaneswari *et al.* 1977) have observed a transient appearance and disappearance of the receptor on *R. japonicum* which specifically binds soyabean lectin. While most strains had their highest percentage of soyabean lectin-binding cells and the greatest number of soyabean-lectin-binding sites per cell in the early and mid-log phases of growth, one strain accumulated the lectin receptor as cells left exponential growth and entered stationary phase. The proportion of galactose residues in the capsular polysaccharide is high at a culture age when the cells bind the galactose-reversible soyabean lectin (Mort and Bauer 1980). A decline in lectin-binding activity accompanying increasing age of cultures is concurrent with a decline in galactosyl residues and a rise in 4-*O*-methylgalactosyl residues on the poly-saccharide. The hydroxyl group at the C-4 position of the galactosyl residues is very important for soyabean lectin-binding, and the observed substitution of hydroxyl at C-4 with *O*-methyl—is likely to prevent or diminish lectin binding (Mort and Bauer 1980). Poly-saccharides which bind soyabean lectin accumulate in the culture fluid of rotory-shaken cultures of *R. japonicum* (Tsien and Schmidt 1980).

Under certain growth conditions, the transient appearance of trifoliin receptors on *R. trifolii* may influence the ability of these bacteria to attach to clover root hairs (Dazzo *et al.* 1979). Cells grown on agar plates of defined media were most susceptible to agglutination by trifoliin when they were harvested at five days of growth. The antigenic determinants which were cross-reactive with clover roots were exposed for only short periods on the bacteria as broth cultures left lag phase and again as they entered stationary phase (Fig. 9.10). Clover roots adsorbed the bacteria in greatest quantities when they harvested from plate cultures incubated for five days and from broth cultures in early stationary phase.

In addition to the capsular polysaccharide, the LPS from *R. trifolii* 0403 could combine with trifoliin when isolated from cells entering stationary phase but not during exponential growth (Hrabak, Urbano, and Dazzo 1981).Thus, the binding of trifoliin to LPS was very dependent upon the growth phase of the cells in batch culture, as predicted (Dazzo *et al.* 1979). Age-dependent differences in the glycasyl composition in the LPS were detected as their alditol acetates using

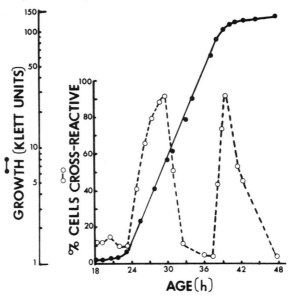

FIG. 9.10. The effect of culture age on the percentage of *R. trifolii* 0403 cells reactive with anti-clover root antiserum. Cells were grown in a chemically defined broth at 30 °C with shaking. Culture growth (closed circles, log scale) was measured with a Klett–Summerson colorimeter at 660 nm and the cross-reactive antigen (open circles, linear scale) by indirect immunofluorescence. From Dazzo *et al.* (1979), courtesy of Springer-Verlag.

combined gas chromatography–mass spectrometry. For instance, more quinovosamine (2-amino-2, 6-dideoxyglucose) was detected in LPS from cells in early stationary phase than in mid-exponential growth. Like 2-deoxyglucose, β-linked glycosides of this amino-substituted dideoxyhexose are effective hapten inhibitors of trifoliin. The fact that neither D-glucose nor 2-deoxygalactose inhibit trifoliin indicates that the molecular environments of C-2 and C-4 on 2-deoxyglucose are important to its interaction with trifoliin. These significant observations re-emphasize the critical importance of culture age in the composition of rhizobial surface polysaccharides, and draws attention to a major underlying cause of conflicting compositional data on rhizobial polysaccharides from one laboratory to another. *Rhizobium* polysaccharides are known to contain *O*-acetyl and pyruvate (carboxyethylidene) noncarbohydrate substitutions which can be important to their antigenic structures (Dudman 1977). It is possible that these substitutions may influence lectin-binding of some rhizobial polysaccharides. There is a recent report of age-dependent increases in pyruvate substitution in polysaccharides of some *R. trifolii* strains (M. Cadmus, personal communication).

It is possible that the effect of culture age on the expression of lectin receptors may influence the competitiveness and ecological behaviour of different rhizobial strains in soil and on the roots of their legume hosts (Dazzo *et al.* 1979). The delay in appearance of the capsule surrounding cells of *R. trifolii* TAl (Dudman 1968; Humphrey and Vincent 1969) may improve their competitiveness for nodulation of clover roots in field soils (Dudman 1968). Inocula prepared from early stationary-phase cells of *R. trifolii* strains NA 30 and 0403 give rise to more clover root-hair infections than equivalent inocula prepared from exponentially growing or late stationary-phase cells (Napoli 1976; Hrabak, Urbano, and Dazzo 1981). The one strain of *R. japonicum* (3Ilb123) found to accumulate the soyabean-lectin binding capsule during the brief transition into exponential and early stationary phases (Bhuvaneswari *et al.* 1977) has been recognized as the most frequently found serogroup in soyabean nodules from many soils of the central United States (Damirgi, Frederick, and Anderson 1967; Ham, Frederick, and Anderson 1971).

The brief accumulation of lectin receptors on cells in the transition between lag and exponential growth may be ecologically important (see Fig. 9.10). This physiological state of arithmetic growth may resemble what happens when quiescent rhizobia in soil encounter the nutrient-enriched enviroment of a growing legume root.

With regard to genetic regulation, exciting evidence is beginning to emerge that genetic elements important to symbiotic recognition are encoded on plasmids of *Rhizobium* (Higashi 1967; Johnson, Beynon, Buchanan-Wollaston, Setchell, Hirsch, and Beringer 1978; Zurkowski and Lorkiewicz 1978, 1979; Nuti, Cannon, Prakash, Lepidi, and Schilperoort 1979; Van der Schaal and Kijne 1979). The genes responsible for the 2-deoxyglucose inhibitable attachment of *R. trifolii* to clover root hairs are encoded on the large nodulation plasmid designated pWZ2 (Zurkowski 1980). The ability of *R. meliloti* to induce polygalacturonase in alfalfa is encoded on plasmid genes (Palomares, Montoya, and Olivares 1979). The report that a fast-growing strain of *R. trifolii* has all the genes necessary to enable the organism to nodulate soyabean effectively (O'Gara and Shanmugam 1978) has been retracted.

Cell wall penetration

9.2 Introduction

The mechanism whereby rhizobia infect the roots of leguminous plants has posed an insurmountable conceptual problem for many years. As Nutman (1956) pointed out, the bacteria must penetrate a solid barrier, the plant cell wall, in such a way that wall integrity is maintained to the extent that host cell viability is not destroyed by plasmoptysis. When the latter occurs, infection is immediately aborted. This requirement for a subtle entry would suggest that delicately regulated cell wall-hydrolyzing enzymes may be operative. Bateman and Miller (1966) pointed out the routine implication of pectic enzymes as a feature of host–pathogen interactions. McCoy (1932) first addressed the question of rhizobial penetration. She investigated two possible mechanisms of cell-wall penetration: (a) mechanical entry through broken walls, or (b) enzymatic hydrolysis. She disproved the first possibility statistically and clearly related the root-hair curling phenomenon to successful infection. McCoy established that the rhizobia were able to produce 'a substance capable of modifying the wall', but she was unable to detect hydrolytic enzymes presumably required for penetration. However, the possibility apparently remained open in her mind since she stated that they were not shown, rather than that they did not exist. If one considers the relatively few and highly localized sites of successful infection, and the very low level of enzyme activity which would be required to accomplish wall penetration without cell destruction, then it is easy to see that the enzyme assays would not detect the low levels required for infection.

Evidence for hydrolytic enzymes in Rhizobium

Recently, Hubbell, Morales, and Umali-Garcia (1978) found pectolytic enzymes in 24 strains of temperate and tropical rhizobia using a qualitative and non-specific pectin agar plate assay. Subsequently, hemicellulase and cellulase activities were found in six strains of temperate and tropical rhizobia (Martinez-Molina, Morales, and Hubbell 1979). Glenn and Dilworth (1979) could not find periplasmic or extracellular activity of any of ten hydrolytic enzymes in two strains of *R. leguminosarum*, including cellulase, β-glucuronidase, and polygalacturonic acid transeliminase. However, in these studies the cells were grown without substrates which could induce the synthesis of these enzymes.

Observations possibly related to rhizobial hydrolytic enzyme production

In vitro *studies*

Anderson (1933) found that the viscosity of broth cultures of certain strains of *R. trifolii, R. leguminosarum, R. japonicum,* and *R. phaseoli* decreased in viscosity between the seventh and fourteenth days of incubation. Martinez-Molina and Morales (unpublished) have confirmed these observations. It is possible that rhizobia accumulate hydrolytic enzymes which gradually depolymerize the polysaccharides they produce.

Rhizosphere studies

The extracellular hydrolytic enzymes of *Rhizobium* may have important effects in the rhizosphere and rhizoplane, prior to cell-wall penetration. Dart and Mercer (1964) examined *Medicago* roots by transmission electron microscopy and found that the uninoculated root-hair surface had a rather uniform granular matrix, in contrast with a randomly oriented network of cellulose fibrils seen on the surface of inoculated root hairs. Munns (1968*b*) proposed that the plant enzymes hydrolysed the non-fibrillar amorphous encrusting matrix (pectin and hemicellulose), thus exposing the microfibrils. We suggest that rhizobial hydrolytic enzymes may be involved in modifying the host wall. Utilization of the products of hydrolysis could then contribute to the nonspecific stimulation of rhizobial growth in the rhizosphere. Many rhizobia can utilize the pentoses, D-xylose and L-arabinose, which are monomeric subunits of xylan and arabinan, the common hemicellulosic constituents of plant cell walls (Graham 1964). Thus, the possession of the enzymes capable of hydrolysing the cell-wall components of legumes may be critical for rhizosphere stimulation of rhizobia, and should be examined as a possible explanation for strain differences in competitive ability, since the activity of the requisite enzymes might vary appreciably among rhizobial strains.

Rhizoplane studies

Phenomena of root-hair curling, deformation, and shepherd's crook formation and their possible relation to the infection process have been recently reviewed (Dart 1977; Broughton 1978). Yao and Vincent (1969) defined 'branched', 'moderately curled' and 'markedly curled' categories as distinct types of root-hair deformation resulting from inoculation. Almost all infection threads were initiated in root hairs exhibiting the 'markedly curled' condition (commonly called the shepherd's crook), and this was associated with the presence of viable cells of only the homologous rhizobia.

Munns (1968*a*) identified an 'acid-sensitive step' as an early stage of infection. At pH 4.4, the rhizobia accumulated in high population in the rhizosphere but root-hair curling and infection did not occur. Raising the pH to 5.4 allowed curling (also penetration?) to proceed; the pH could then be lowered to 4.4 without further hindering of the infection thread formation and nodulation.* Munns later (1968*b*) reported increased pectic enzyme activity in inoculated root solutions, in agreement with Ljunggren and Fåhraeus (1961). The enzyme activity was pH-dependent and Munns proposed that the 'acid-sensitive' step was, in fact, a manifestation of acid inhibition of pectic enzymes. Bateman and Millar (1966) indicate that most pectic enzymes have a pH optimum above 5.0. Pectinase is usually the first enzyme to appear in the pectinase–hemicellulase–cellulase complex observed in plant–phytopathogen associations involving hydrolysis (Cooper 1977). In the absence of more detailed information, the involvement of other polysaccharide-degrading enzymes should not be discounted.

Ljunggren and Fåhraeus (1961) found increased polygalacturonase activity in root solutions containing homologous, but not heterologous, rhizobia. This work has not been consistently repeatable and remains controversial (scc Dart 1977). Ljunggren and Fåhraeus' 'poly-galacturonase' theory should be re-investigated to determine if the enzyme activity is of both plant and microbial origin.

In another study, Munns (1970) determined that success of infection was also dependent on calcium concentration in a manner entirely analogous to the previously established acid sensitivity of infection. The acidity and calcium effects were independent of one another. Broughton (1978) cited studies demonstrating stimulation of rhizobia in the rhizosphere in the presence of levels of calcium and magnesium which were much higher than those required for growth of the bacteria. Blevins, Barnett, and Bottino (1977) reported that the calcium ionophore A23187, which stimulates calcium-dependent reactions, affected critical steps in the infection process or nodule development in soyabeans. Calcium is required in high concentration as an activator of pectolytic enzymes. The calcium-and pH-sensitive stage of infection may in fact be the penetration stage, in which the activity of pectolytic enzymes is critical.

Localization of infection sites

Nutman (1965) has reviewed evidence that legume root infection is not a randomly occurring process. There is clear evidence of discrete zones of root-hair curling and deformation along the root axis in temperate legumes. These zones correspond to loci of infection or areas where a disproportionately high number of root hairs bear infection threads.

It is possible to conceive of excessive hydrolytic enzyme activity which would result in localized root hair lysis. In support of this idea, Napoli (personal communication) has observed that root hairs of *Trifolium* and *Medicago* seedlings, when non-inoculated or inoculated with homologous rhizobia, showed no significant tendency to lyse

---

* Instability of the lectin receptor on *R. trifolii* to pH below 5.0 (Dazzo and Hubbell 1975c) could also contribute to an 'acid-sensitive step'.

(generally < 3.5 per cent). In contrast, inoculation of seedlings with heterologous rhizobia markedly increased (approximately 30 per cent) the incidence of root hair lysis in distinct zones of the root where infections by rhizobia normally occur.

In temperate legumes there are localized sites of infection (penetration) on individual root hairs. Most commonly, infection occurs at the tight fold created by 'marked curling' of the growing root-hair tip or the growing tip of a lateral branch of the root hair. Less commonly, infections may arise laterally where little or no deformation has occurred. Also, infections can arise where one root-hair tip contacts another (Napoli, Dazzo, and Hubbell 1975*a*; Higashi and Abe 1980*b*). Napoli and Hubbell (1975) proposed that infection from within the fold of curled root hair tips results from entrapment of adherent rhizobia within the confines of the curl, thus, '. . . localizing and concentrating the biochemical interactions . . .' required to accomplish infection.

In the case of infections occurring at lateral sites on the root hair, the microfibril-associated Phase II adherence of a large aggregate of rhizobial cells to the cell wall may concentrate required biochemical interactions (Napoli *et al.* 1975*a*; Napoli and Hubbell 1975). These infections may be less frequent because there is minimal physical confinement of the interaction and putative substances required for infection would be more subject to loss by diffusion. Loss by diffusion of enzymes important to penetration would be reduced when adhering cells or aggregates are sandwiched between the root-hair wall and another solid surface, such as another root hair (Napoli *et al.* 1975*a* Higashi and Abe 1980*b*), a glass coverslip (Dazzo, unpublished observation), or soil particles. Under these conditions, the enzymes may then act at a more concentrated level to initiate an infection. Some support for this comes from the findings of Napoli (unpublished observation) that inoculum composed almost entirely of single cells produced infections primarily at the curled root hair tips. Inoculum composed of aggregates of cells produced a greatly increased number of infections at undeformed lateral sites.

Mechanism of cell wall penetration

Current theories

Nutman (1956) proposed a hypothesis of infection via a process of invagination. According to this idea, active wall growth is redirected from the root-hair tip to the site of infection. The direction of growth is inverted in the process, forming an invagination in the wall which appears as an interior tubular structure, the infection thread, with an open pore at the site of initiation. No wall penetration is involved and, theoretically, bacteria within the infection thread are still exterior to the root hair cytoplasm.

Ljunggren and Fåhraeus (1961) proposed a 'polygalacturonase' hypothesis, resulting from their finding that ECPS from rhizobia elicited increased plant pectic enzyme activity only in homologous host plants.* The plant cell wall was thus thought to be softened sufficiently at the site of infection to permit occasional individual cells to penetrate to the plasmalemma, where growth of the host-synthesized infection thread was then initiated. These two hypotheses are discussed in detail by Dart (1977).

Ultrastructure studies

Napoli and Hubbell (1975) studied infected root hairs by transmission electron microscopy to resolve the controversy between the divergent 'invagination' and 'polygalacturonase' hypotheses of legume infection. Thin serial sections were made through infection sites at curled root-hair tips and at lateral, undeformed sites on the root-hair wall. The root-hair wall was continuous with the infection thread wall, although the junction was lightly stained and lacked detail, suggesting some cell-wall modification. The pore of the infection thread was occluded with encapsulated rhizobia. The conclusion of this study was that physical penetration of the wall does not occur and that the infection thread is initiated via a process of invagination analogous to that proposed by Nutman (1956).

Subsequently, Callaham (1979) confirmed the experimental results of Napoli and Hubbell (Figs. 9.11–13), but the greater clarity of his electron micrographs led him to consider the activity of hydrolytic enzymes (see p.294) in his interpretation. His results present a compelling case for the following proposed sequence of events (Fig. 9.14). The rhizobia adhere to the root hair cell wall. The root hair wall is then hydrolysed at the highly localized site of adhesion of the infecting bacteria. Simultaneous with hydrolysis of the existing wall, the root hair synthesizes a thin layer of new cell wall as a low dome over the advancing zone of hydrolysis (Fig. 9.14). This results in an embryonic infection thread which morphologically resembles an invagination of the wall. However, ontogenically, it is a dome of new growth over the infection site. The 'markedly curled' condition of root-hair tips is closely related to infection and requires the presence of live rhizobia (Yao and Vincent 1968). A possible sequence of events whereby this 'markedly curled' condition may be induced is indicated in Fig. 9.15. The remaining root hair deformation categories ('moderately curled' and

---

* Bauer *et al.* (1979) have recently reported that lectin-binding ECPS from homologous rhizobia may condition the root hair to an infectable state in a short period of time.

'branched' as defined by Yao and Vincent 1968) may involve altered tip growth induced by incomplete wall penetration at potential sites of infection. Clearly, more information on root hair growth and physiology is needed to test the validity of these hypotheses.

The dome of new wall was reported to be callose by Kumarasinghe and Nutman (1977). Callaham (1979) examined the dome of wall material surmounting the infection site by histochemical-fluorescence microscopy and found that it contained pectin and cellulose but no callose. The plug of wall material beneath this dome is eventually hydrolysed and no longer recognizable and the dome extends to form the infection thread with an apparent pore (hydrolysed zone of cell wall) or opening to the rhizosphere at the localized point of cell wall

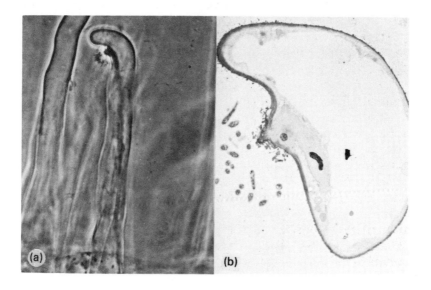

Fig. 9.11. The infection of a *T. repens* root hair by *R. trifolii* from an unenclosed site (figures and legends from Callaham 1979).

Fig. 9.11(a). Phase contrast photomicrograph of the straight infected root hair showing the rhizobial colony (RC) clustered about the dark, pimple-shaped infection site on the concave side of the gently curved tip. The infection thread (IT) is seen passing internally from the infection site to the root hair base. 350 ×.

Fig. 9.11(b). Low magnification electron micrograph of cross-section near the tip of the infected root hair showing the extracellular rhizobial colony (RC) and origin of the infection thread from the inside of the distended wall region. 3000 ×.

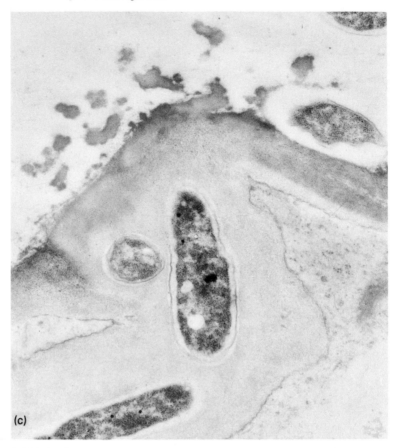

Fig. 9.11(c). Electron micrograph of median cross-section of the infection. The origin of the infection thread is marked by the abrupt termination of the original hair wall layer (HW) in contact with the thread matrix (TM, arrows) which bulges slightly beyond the hair wall. The line of the new inner layer which continues into the thread wall (TW) is marked in this region by small arrows. Extracellular encapsulated *Rhizobium* (ERh) cell is very firmly attached laterally to the hair wall by the capsule (C). The lipopolysaccharide of invading *Rhizobium* cells, (1pp). 25 000 ×.

hydrolysis. This interpretation supports the 'polygalacturonase' hypothesis of Ljunggren and Fåhraeus (1961), involving a direct physical penetration of the root-hair wall mediated by the localized activity of hydrolytic enzymes. Since wall-modifying enzymes of plant origin are involved in wall synthesis and growth (Lamport 1970), the relative contributions of hydrolysing enzymes from both the bacteria and the host plant to the localized wall hydrolysis remain to be determined.

Control of infection

The control of infection is expressed at several levels. Control is exerted in the mutual recognition and Phase I attachment of rhizobia to the cell wall of its homologous host root, as discussed in the first part of this chapter. Once the bacteria specifically adhere to the plant cell wall, penetration becomes a critical step as another control of infection. Penetration may be favoured by modification of root hair growth which forms the cell wall overlap in shepherd's crook formation. There is substantial evidence from time lapse cinephotomicrography (Nutman *et al.* 1973) that the root hair nucleus exerts control of the initiation and continued growth of the infection thread. This presupposes some means

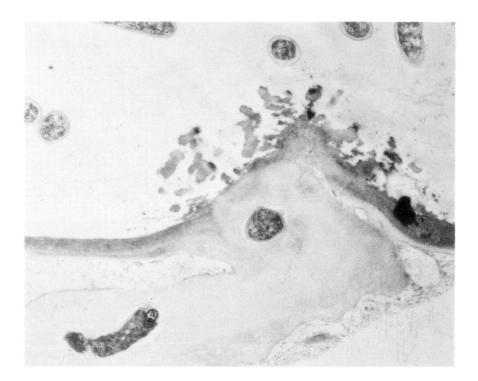

Fig. 9.12. Section sequential to Fig. 9.12 showing the non-encapsulated rhizobia (black Rh) of the colony surrounding the infection site. The thread matrix (TM) is protruding from the gap in the hair wall produced by penetration. The hair wall is degraded in the region of the capsule (C) of the extracellular encapsulated bacterium (not evident in this section). The protruding end of the degraded hair wall (DHW) is surrounded by many blebs of loosened wall material. 12 000 ×.

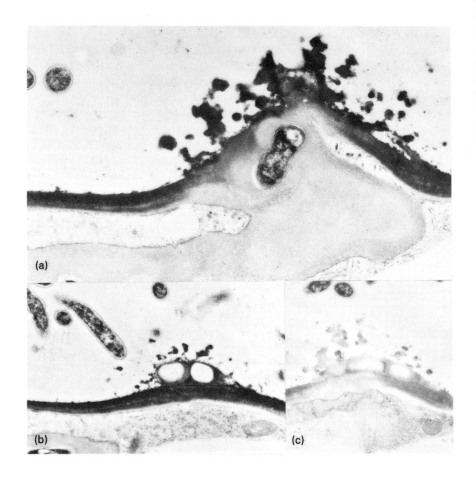

Fɪɢ. 9.13
(a) Electron micrograph showing the degraded wall (large arrow) in the region where the encapsulated bacterium (Fig. 9.11(c)) was attached. *Rhizobium* (Rh) is present just within the opening of the infection site. Thread wall, TW. 13 000 ×.

(b) Section just beyond the infection site showing the rhizobial capsule (C) attached below the plane of section in Fig. 13(a). *Rhizobium* cell, Rh. Infection thread, IT. 7500 ×.

(c) Section through the capsules (C) of rhizobial (not evident) just beyond the infection site of Fig. 9.13(a). The hair wall appears degraded beneath the capsules. Infection thread, IT. *Rhizobium* cell (Rh). 7500 ×.

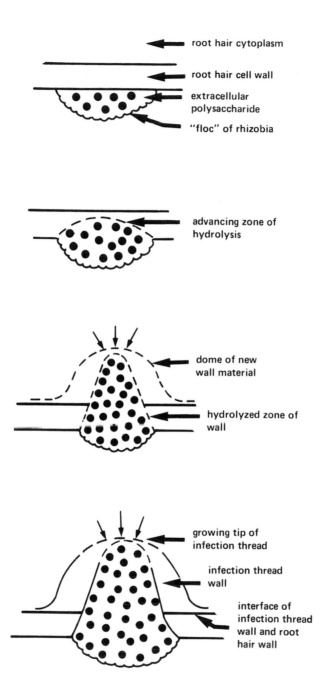

root hair cytoplasm

root hair cell wall

extracellular polysaccharide

"floc" of rhizobia

advancing zone of hydrolysis

dome of new wall material

hydrolyzed zone of wall

growing tip of infection thread

infection thread wall

interface of infection thread wall and root hair wall

FIG. 9.14. Proposed sequence of events in penetration of a legume root-hair cell wall by *Rhizobium*.

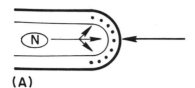

(A)

Dots indicate region of active wall growth at tip of root hair; N=plant cell nucleus; arrows indicate direction of nuclear control of new wall synthesis.

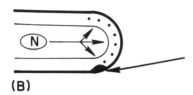

(B)

Point of infection; dark area represents highly localized zone of hydrolysis of cell wall by invading rhizobia; this results in cessation of wall growth at this point but active wall growth and extension continues elsewhere at the growing tip; this unequal growth results in progressively curled form of the root tip (C).

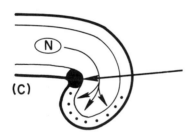

(C)

Rhizobia penetrate sufficiently to establish communication with the host nucleus; cell wall growth redirected from the root hair tip to the point of infection (D). Requirement of rhizobial penetration for nuclear communication is hypothetical.

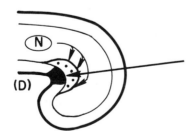

(D)

Dome of newly deposited wall material constituting the tip of the infection thread; curling does not occur when infection is initiated at a non-growing region of the root hair wall.

Fig. 9.15. Proposed sequence of events resulting in marked curling of root hair tips.

of biochemical communication between the infective rhizobia and the host nucleus. The induction or derepression of wall-hydrolysing enzymes may be critical points of control of infection. The existence of additional but currently unsuspected levels of control should remain an open possibility.

Higashi and Abe (1980) have demonstrated the presence of low molecular-weight ($\leq$ 6000 daltons) substance in the periplasmic space of *R. trifolii*, which promotes its infection of clover root hairs. The substance was missing in a spontaneous, non-infective mutant strain of *R. trifolii*. Further studies on the nature of this substance may uncover additional important steps which control infection.

Growth (Peters and Alexander 1966) and chemotaxis (Currier and Strobel 1976) of rhizobia in the rhizosphere may influence the number of infections, but the absence of these events will not absolutely prevent infection in rhizobia–legume combinations which are genetically capable of forming nodules. Likewise, infection thread abortion in root hairs occurs frequently, and is very influential in determining the extent of nodulation at the strain-variety level of homologous rhizobia–legume combinations. However, abortions occur after the barrier to the specificity of infection has already been breached.

Present knowledge of the infection process does not allow a detailed description of the molecular events which explains why non-nodulating or heterologous rhizobia fail to advance root hair penetration to the stage of infection thread development, despite their low attachment to root hairs. Nor is there sufficient information available to explain the infrequent and anomalous nodulation of legumes by rhizobia which fall outside of the normal cross-inoculation groups (for example, Hepper 1978; Trinick, this volume).

Combinations of tropical rhizobia and legumes present a special case with regard to nodulating specificity. Tropical legumes are considered to be the primitive ancestors of the temperate legumes (Norris 1959). In general, the tropical legumes are susceptible to nodulation by a wide range of strains of rhizobia isolated from different tropical legumes. As a corollary to this, there is a basic difference in the early stages of infection in temperate and tropical legumes. In the temperate combinations, nodules arise from infection threads initiated in root hairs. In tropical combinations, infection threads have not been observed in root hairs (Allen and Allen 1940; Napoli *et al.* 1975*b*; Dart 1977; Chandler 1978). The bacteria enter the root at the junction of the root hair and the epidermal and cortical cells. The bacteria spread intercellularly through the middle lamella and enter the cortical cells through a structurally altered host cell wall. The intracellular rhizobia multiply within the cortical cells, which divide repeatedly to form the nodular tissue.

The bacteria also gain access to potential sites of infection in the interior of the root through void spaces in the epidermis created by lateral root emergence and desquamation. Control of infection in these tropical legumes may rely therefore upon other mechanisms, such as control of penetration, establishment of rhizobia–host nucleus communication, etc. Considering the apparent absence of root hair attachment and specificity in tropical combinations, the expanded host range ('promiscuity') of most tropical rhizobia is both logical and expected.

This promiscuity suggests that recognition processes governing specificity in these more primitive legumes are less specific and have broad boundaries. Consistent with this hypothesis is the observation that the lectin, Concanavalin A, from the tropical legume *Canavalia ensiformis* (jackbean), can bind in a biochemically specific manner (inhibited by α-methyl mannoside) with surface determinants produced by many rhizobia, regardless of their ability to nodulate *C. ensiformis* (Dazzo and Hubbell 1975*b*; Chen and Phillips 1976; Kamberger 1979*a*).

Implications of the research

Studies of host recognition and penetration processes in this symbiosis are timely because of the awareness that nitrogen is the nutrient most-commonly limiting to crop productivity, and legumes can offset that major limitation by establishing efficient nitrogen-fixing symbioses. An understanding of the mechanism and control of infection (host recognition and penetration) would not only help to elucidate the developmental events in the legume symbiosis, but also indicate ways in which *Rhizobium* and the host plant may be manipulated genetically to increase the range of agricultural crops which can form nitrogen-fixing symbioses. In addition, a greater understanding of the infection process may lead to development of rapid methods of screening for rhizobia with increased infectiveness and/or legumes with increased susceptibility to rhizobial infection.

There are other benefits to be gained from better understanding of cell-recognition processes between micro-organisms and their plant hosts. For example, the recognition of phytopathogens by plants may reduce the burden of plant pathogenesis on agricultural crops by the induction of disease resistance (Graham, Sequeira, and Huang 1977). An important attribute of many microbial pathogens is their adhesion to the host tissue, which is one of the dominating factors determining successful colonization (Gibbons 1977). Interactions which display selective adherence also serve as useful models to study the biochemistry of cellular recognition, cell surface receptors, fertilization, and compatibility–incompatibility of self and non-self tissues which are topics central to biology.

Acknowledgements

Michigan Agricultural Experiment Station Journal No. 9540. Portions of this research were supported by the Michigan Agricultural Experiment Station, the College of Natural Science, and the College of Osteopathic Medicine, Michigan State University, and by NSF Grants PCM 78–22922 and 80–21906 and the Science and Education Administration, USDA-CGO Grant 78–00099. Support was also provided by the Institute of Food and Agricultural Sciences of the University of Florida, the Department of Plant Pathology of Cornell University, and by NSF Grant DEB 75–14043. The critical comments of numerous colleagues have been invaluable and are gratefully acknowledged.

Addendum

After submitting this manuscript, we have renamed the 50 000 dalton white clover lectin as 'trifoliin A'.

References

Albersheim, P., Ayers, A., Valent, B., Ebel, J., Wolpert, J., and Carlson, R. (1977). *J. Supramol. Structure*. **6,** 559–616.

Allen, O. N. and Allen, E. K. (1940). *Bot. Gaz.*, **102,** 121–42.

Anderson, A. D. (1933). *Res. Bull.* **158,** Iowa Agric. Exp. Sta., Ames, Iowa.

Bal, A. K., Shantharam, S., and Ratnam, S. (1978). *J. Bacteriol.* **133,** 1393–1400.

Bateman, D. F. and Miller, R. L. (1966). *A. Rev. Phytopathol.* **4,** 119–46.

Bauer, W. D. (1977). *Basic Life Sci.* **9,** 283–97.

—— Bhuvaneswari, T. V., Mort, A. J., and Turgeon, G. (1979). *Pl. Physiol., Lancaster* Suppl. **63,** 135.

Bhuvaneswari, T. V. and Bauer, W. D. (1978). *Pl. Physiol., Lancaster* **62,** 71–4.

—— Pueppke, S. G., and Bauer, W. D. (1977). *Pl. Physiol., Lancaster* **60,** 486–91.

Bishop, P. E., Dazzo, F. B., Appelbaum, E. R., Maier, R. J., and Brill, W. J. (1977). *Science, N.Y.* **198,** 938–40.

Blevins, D. G., Barnett, N. M., and Bottino, P. J. (1977). *Physiol. Plant.* **41,** 235–8.

Bohlool, B. B. and Schmidt, E. L. (1974). *Science, N.Y.* **185,** 260–71.

—— —— (1976). *J. Bacteriol.* **125,** 1188–94.

Bowles, C., Lis, H., and Sharon, N. (1979). *Planta* **145,** 143–8.

Broughton, W. J. (1978). *J. appl. Bacteriol.* **45,** 165–94.

Calbert, H. E., Lalonde, M., Bhuvaneswari, T. V., and Bauer, W. D. (1978). *Can. J. Microbiol.* **24,** 785–93.

Callaham, D. A. (1979). M.Sc. thesis, University of Massachusetts, Amherst.

Carlson, R. W., Saunders, R. E., Napoli, C., and Albersheim, P. (1978). *Pl. Physiol., Lancaster* **62,** 912–17.

Chandler, M. R. (1978). *J. exp. Bot.* **29,** 749–55.

Chen, A. T. and Phillips, D. A. (1976). *Physiol. Plant.* **38,** 83–8.

Cooper, R. M. (1977). In *Cell wall biochemistry related to specificity in host-plant pathogen interactions* (ed. B. Solheim and J. Raa) pp. 163–211. Universitetsforlanget, Oslo.

Currier, W. W. and Strobel, G. A. (1976). *Pl. Physiol., Lancaster* **57**, 820–4.

Damirgi, S. M., Frederick, L. R., and Anderson, I. C. (1967). *Argon. J.* **59**, 10–12.

Dart, P. J. (1971). *J. exp. Bot.* **22**, 163–8.

—— (1977). In *A treatise on dinitrogen fixation.* Section III, *Biology* (ed. R. W. F. Hardy and W. S. Silver) pp. 367–472. John Wiley, New York.

—— and Mercer, F. (1964). *Arch. Mikrobiol.* **47**, 334–78.

Dazzo, F. B. (1979). *Am. Soc. Microbiol. News.* **45**, 238–40.

—— (1980a). In *Adsorption of microorganisms to surfaces* (ed. G. Bitton and K. Marshall) pp. 253–316. John Wiley, New York.

—— (1980b). In *The cell surface: mediator of developmental processes* (ed. S. Subtelny and N. Wessels) pp. 277–304. Academic Press, New York.

—— (1980c). In *Advances in legume science* (ed. R. Summerfield and A. H. Bunting) pp. 49–59. British Publications, London.

—— (1980d). In *Nitrogen fixation* II (ed. W. Orme-Johnson and W. E. Newton) pp. 165–87. University Park Press, Maryland.

—— and Brill, W. J. (1977). *Appl. Environ. Microbiol.* **33**, 132–6.

—— —— (1978). *Pl. Physiol., Lancaster* **62**, 18–21.

—— —— (1979). *J. Bacteriol.* **137**, 1362–73.

—— Hrabak, E., Urbano, M., Sherwood, J., and Truchet, G. (1981). In *Current perspectives in nitrogen fixation* (ed. A. H. Gibson and W. E. Newton) Australian Academy of Science, Canberra.

—— and Hrabak, E. (1981). *J. Supramol. Struc. Cell Biochem.* In press.

—— and Hubbell, D. H. (1975a). *Appl. Microbiol.* **30**, 172–7.

—— —— (1975b). *Pl. Soil.* **43**, 713–17.

—— —— (1975c). *Appl. Microbiol.* **30**, 1017–33.

—— Napoli, C. A., and Hubbell, D. H. (1976). *Appl. Environ. Microbiol.* **32**, 168–71.

—— Urbano, M. R. and Brill, W. J. (1979). *Curr. Microbiol.* **2**, 15–20.

—— Yanke, W. E., and Brill, W. J. (1978). *Biochim. Biophys. Acta* **539**, 276–86.

DeVay, J. E. and Alder, H. E. (1976). *A. Rev. Microbiol.* **30**, 147–68.

Dudman, W. F. (1968). *J. Bacteriol.* **95**, 1200–01.

—— (1977). In *A Treatise on dinitrogen fixation*, Section IV (ed. R. Hardy and W. Silver) pp. 487–508. John Wiley, New York.

Fåhreus, G. (1957). *J. Gen. Microbiol.* **16**, 374–81

Fred, E. B., Baldwin, I. L., and McCoy, E. (1932). *Root nodule bacteria and leguminous plants.* University of Wisconsin Press, Madison.

Gibbons, R. J. (1977). In *Microbiology 1977* (ed. D. Schlessinger) pp. 395–406, American Society for Microbiology, Washington, DC.

Glenn, A. R. and Dilworth, M. J. (1979). *J. gen. Microbiol.* **112**, 405–9.

Gmeiner, J. and Schelecht, S. (1979). *Eur. J. Biochem.* **93**, 609–20.

Graham, P. H. (1964). *Antonie van Leenwenhoek* **30**, 68–72.

Graham, T. L., Sequeira, L., and Huang, T. R. (1977). *Appl. Environ. Microbiol.* **34**, 424–32.

Ham, G. E., Frederick, L. R., and Anderson, I. C. (1971). *Agron. J.* **63**, 69–72.

Hepper, C. M. (1978). *Ann. Bot.* **42**, 109–15.

Higashi, S. (1967). *J. gen. Appl. Bacteriol.* **13**, 391–403.

—— and Abe, M. (1980a). *Appl. Environ. Microbiol.* **39**, 297–301.

—— and Abe, M. (1980b). *Appl. Environ. Microbiol.* **40**, 1094–9.

Hrabak, E., Urbano, M., and Dazzo, F. B. (1981). *J. Bacteriol.* In press.

Hubbell, D. H., Morales, V. M., and Umali-Garcia, M. (1978). *Appl. Environ. Microbiol.* **35**, 210–13.

Hughes, T., Leece, J., and Elkan, G. H. (1979). *Appl. Environ. Microbiol.* **37**, 1243–44.
—— and Elkan, G. H. (1979). In *North Am. Rhizobium Conference*, Texas (Abstract).
Humphrey, B. and Vincent, J. M. (1969). *J. gen. Microbiol.* **59**, 411–25.
Jann, B., Jann, K., Schmidt, G., and Orskov, F. (1970). *Eur. J. Biochem.* **15**, 29–39.
Jann, K. and Westphal, O. (1975). In *The antigens* (ed. M. Sela) Vol. III, pp. 1–125. Academic Press, New York.
Johnson, A. W., Beynon, J. L., Buchanan-Wollaston, A. V., Setchell, S. M., Hirsch, P. R., and Beringer, J. E. (1978). *Nature, Lond.* **276**, 634–6.
Kamberger, W. (1979a). *Arch. Microbiol.* **121**, 83–90.
—— (1979b). *FEMS Microbiol. Lett.* **6**, 361–363.
Kato, G., Maruyama, Y., and Nakamura, G. (1979). *Agric. Biol. Chem.* **43**, 1085–92.
—— —— —— (1980). *Agric. Biol. Chem.* **44**, 2843–55.
Kijne, J., Van der Schaal, I. A. M., and De Vries, G. E. (1980). *Plant Sci. Lett.* **18**, 65–74.
Kumarasinghe, R. M. K. and Nutman, P. S. (1977). *Exp. Bot.* **28**, 961–76.
Lamport, D. T. A. (1970). *A. Rev. Plant. Physiol.* **21**, 235–70.
Leps, W., Paau, A., and Brill, W. J. (1980). In *Proc. Am. Soc. Microbiol. Meet.*, Miami, FL K167 (Abstract).
Ljunggren, H. and Fåhraeus, G. (1961). *J. Gen. Microbiol.* **26**, 521–28.
Li, D. and Hubbell, D. H. (1969). *Can. J. Microbiol* **15**, 1133–6.
Maier, R. J. and Brill, W. J. (1978). *J. Bacteriol.* **133**, 1295–9.
Martinez-Molina, E., Morales, V. M., and Hubbell, D. H. (1979). *Appl. Environ. Microbiol.* **38**, 1186–8.
McCoy, E. (1932). *Proc. Roy. Soc. London* **B110**, 514–33.
Munns, D. N. (1968a). *Pl. Soil.* **28**, 129–46.
—— (1968b). *Pl. Soil.* **30**, 117–20.
—— (1970). *Pl. Soil.* **32**, 90–102.
Napoli, C. A. (1976). Ph.D. thesis, University of Florida, Gainesville.
—— Dazzo, F. B., and Hubbell, D. H. (1975a). *Appl. Microbiol.* **30**, 123–31.
—— —— —— (1975b). *Proc. 5th Austr. Leg. Conf.* pp. 15, Brisbane.
—— and Hubbell, D. H. (1975). *Appl. Microbiol.* **30**, 1003–9.
Norris, D. O. (1959). *J. Aust. Inst. Agric. Sci.* **25**, 202.
Nuti, M. P., Cannon, F. C., Prakash, R. K., Lepidi, A. A., and Schilperoort, R. A. (1979). *Proc. Am. Rhizobuim Conf. VIII.* (Abstr.)
Nutman, P. S. (1956). *Biol. Rev. Camb. Phil. Soc.* **31**, 152–93.
—— (1965). In *Ecology of soil-borne plant pathogens* (ed. D. F. Baker and W. C. Snyder) pp. 231–47. University of California Press, Berkeley.
—— Doncaster, C. C., and Dart, P. J. (1973). *Infection of clover by root-nodule bacteria.* British Film Institute, London.
O'Gara, E. and Shanmugam, K. T. (1978). *Proc. Nat. Acad. Sci. U.S.A.* **75**, 2343–7.
Osborne, M. J., Cynkin, M. A., Gilbert, J. M., Muller, L., and Singh, M. (1972). *Meth. Enzymol.* **28**, 583–601.
Palomares, A., Montoya, E., Olivares, J. (1979). *Microbios* **22**, 7–11.
Peters, R. J. and Alexander, M. (1966). *Soil Sci.* **102**, 380–7.
Planqué, K. and Kijne, J. W. (1977). *FEBS Lett.* **73**, 64–6.
Reeke, G. N., Becker, J. W., Cunningham, B. A., Wang, J. L., Yahara, I., and Edelman, G. M. (1975). *Adv. exp. Med. Biol.* **55**, 13–33.
Saunders, R. E., Carlson, R. W., and Albersheim, P. (1978). *Nature, Lond.* **271**, 240–2.
Schmidt, E. L. (1979). *A. Rev. Microbiol.* **33**, 355–76.
Sequeira, L. (1978). *A. Rev. Phytopathol.* **16**, 453–81.
Solheim, B. (1975). In *Specificity in plant disease*, NATO Advanced Study Institute, Sardinia.
—— and Paxton, J. (1980). In *Rockefeller foundation int. conf. alternative plant disease management concepts.* Lake Como, Italy.

Stacey, G., Paau, A., and Brill, W. (1980). *Pl. Physiol., Lancaster* **63,** 609–14.

Tsien, H. C. and Schmidt, E. L. (1977). *Can. J. Microbiol.* **23,** 1274–84.

—— —— (1980). *Appl. Environ. Microbiol.* **39,** 400–4.

Umali-Garcia, M. Hubbell, D., Gaskins, M., and Dazzo, F. B. (1980). *Appl. Environ. Microbiol.* **39,** 219–26.

Urbano, M. R. and Dazzo, F. B. (1980). In *Proc. Am. Soc. Microbiol. Meet.*, Miami, FL N1 (Abstract).

Van der Schaal,. I. and Kijne, J. W. (1979). In *North Am. Rhizobium Conf*, College Station, Texas (Abstract).

Wolpert, J. S. and Albersheim, P. (1976). *Biochem. Biophys. Res. Commun.* **70,** 729–37.

Yao, P. Y. and Vincent, J. M. (1969). *Aust. J. Biol. Sci.* **22,** 413–423.

Zurkowski, W. (1980). *Microbios* **27,** 27–32.

—— and Lorkiewicz, Z. (1978). *Genet. Res. Cambridge.* **32,** 311–14.

—— —— (1979). *Arch. Microbiol.* **123,** 195–201.

10 Development of leguminous root nodules

E. G. M. MEIJER

10.1 Introduction

Leguminous root nodules are highly specialized structures which arise as the result of a sequence of interactions between the host plant and the invading rhizobia. Root-nodule formation begins with infection of the host plant's root tissue by means of a process in which the host's natural defence mechanisms are overcome. It proceeds through a sequence of events which has many of the characteristics of new organ formation in higher plants: (1) dedifferentiation: division of differentiated cells, (2) initiation: organized cell divisions lead to the formation of a meristem, (3) differentiation: the organ becomes recognizable as such, (4) growth and maturation (Libbenga and Bogers 1974).

A unique feature of leguminous root nodules is the tissue with cells containing the nitrogen-fixing bacteroids; vegetative rhizobia which have undergone a series of physiological changes to become virtually nitrogen-fixing organelles.

In this review the successive stages in the formation of root nodules will be described, followed by sections on some biochemical aspects of nodule development and ineffective associations. Bacteroid development and nodule senescence will be discussed by W. D. Sutton in Volume 3 of this series.

10.2 Infection

In the vast majority of the legumes investigated so far, infection is via the root hairs. Many steps are involved including: (1) growth of rhizobia in the rhizosphere, (2) adsorption of the bacteria onto the surface of the cell wall of the root hair, (3) curling and branching of the root hair, (4) infection thread formation, (5) growth of the infection thread into the cortical cells of the plant host's root; (6) release of rhizobia into the cortical cells (e.g. Dart 1977; Broughton 1978; Schmidt 1978).

Recently another mode of infection has been reported for *Arachis hypogaea* (Chandler 1978). In this species direct invasion of the cortical cells of the root occurs with the invaded cells being capable of division. It is possible that this mode of infection is more widespread, as in many species (listed by Dart 1977) no infection threads have been found. The earlier stages of root-hair infection are discussed by Dazzo and Hubbell (Chapter 9) and will therefore not be described here.

Infection via infection threads

Growth of the infection thread

Within the root hair the nucleus is usually found near the site of infection, and it stays close to the tip of the infection thread during its growth towards the base of the root-hair cell (Dart 1977; Fåhraeus and Sahlman 1977). These observations seem to suggest that the nucleus, which is enlarged in infected root-hair cells, directs the growth of the infection thread. Moreover, when the nucleus moves away from the thread its growth stops, but may start again when the nucleus returns (Dart 1977). Involvement of the Golgi bodies in the synthesis of infection thread walls has been suggested by Robertson, Farnden, Warburton, and Bank (1978). Their arguments were based on electron-microscopic observations and the fact that the Golgi bodies may be involved in cell-wall polymer synthesis in a process that is under hormonal control (Shore and MacLahan 1975). Rhizobia can produce auxins (Kefford, Brockwell, and Zwar 1960; Dullaart 1970; Rigaud 1970) and cytokinins (Phillips and Torrey 1972; Puppo and Rigaud 1978), while relatively high concentrations of auxins (Pate 1958; Dullaart 1970) and cytokinins (Puppo, Rigaud, and Barthe 1974; Henson and Wheeler 1976; Syōno and Torrey 1976; Syōno, Newcomb, and Torrey 1976) have been found in root nodules. This could then imply a hormonal control mechanism affecting or regulating cell wall and membrane formation in growing infection threads.

The rhizobia grow and divide within the infection thread and become aligned along the axis of growth, often in a single row. At this stage it appears as if elongation and growth of the thread result from bacterial division. The rhizobia produce and become embedded in a mucopoly-saccharide-containing matrix, termed 'zoogleal matrix' by Dart (1977). As the infection thread grows towards the root cortex, it crosses already existing cell walls, becoming attached to and continuous with them (Dart 1977). The thread-wall material has the same staining properties as the plant cell wall (Robertson, Lyttleton, Bullivant, and Grayston

1978). Within the cortex, extensive branching of the thread occurs, resulting in the infection of many cells by the same infection thread (Dart 1977). A close association of the infection thread with the nucleus of the cell through which it passes (Fig. 10.1 (a)) has been reported by several workers (Arora 1956; Mosse 1964; Libbenga and Harkes 1973; Newcomb 1976; Roughley, Dart, and Day 1976; Dart 1977). The nuclei in the outer and middle root cortex cells of *Pisum sativum* which are penetrated by the infection thread are enlarged and have the ability to incorporate ^{3}H-thymidine (Libbenga and Harkes 1973) (Fig. 10.1 (b)) indicating the occurrence of DNA synthesis. The close association of the infection thread with the host cell's nucleus would, of course, facilitate molecular exchange between the invading bacteria and the nucleus if this were necessary (Meijer and Broughton 1982).

Release from the infection thread

Usually rhizobia are released shortly after the penetration of the cortical cells by the infection thread. Prior to release disintegration of the thread wall takes place, often at the tip, but also along the sides (Goodchild and Bergersen 1966; Newcomb 1976; Bassett, Goodman, and Novacky 1977a; Roughley *et al.* 1976; Robertson *et al.* 1978). This breakdown of thread wall material may be due to *Rhizobium*-induced cellulase activity (Verma and Zogbi 1978; Verma, Kazazian, Zogbi, and Bal 1978). Vesicles derived from the thread membrane are formed at the sites where disintegration of the thread wall has occurred. From these vesicles, rhizobia embedded in the zoogleal matrix and surrounded by a membrane, termed peribacteroid membrane by Robertson, Taylor, Craig, and Hopcroft (1974), are budded off by a process which seems to be identical with endocytosis (Goodchild and Bergersen 1966; Bassett *et al.* 1977a; Dart 1977; Robertson *et al.* 1978). Initially the peribacteroid membrane is formed from the infection thread membrane rather than by new synthesis (Bassett *et al.* 1977a; Robertson *et al.* 1978). During the course of nodule development it retains the biochemical and ultrastructural characteristics of the host plasma membrane (Tu 1975; Verma *et al.* 1978).

Direct inter-cellular infection

Infection threads have not been observed in root hairs or nodules of several, mainly tropical or subtropical species (Dart 1977). Nodules of *Arachis hypogaea* (Allen and Allen 1940; Chandler 1978), a number of *Aeschynomene* species (Suessenguth and Beyerle 1936; Arora 1954; Napoli, Dazzo, and Hubbell 1975), and several *Stylosanthes* species (Rango Rao 1977; M. R. Chandler, personal communication), arise in the junctions of lateral roots. In addition stem nodules occur on

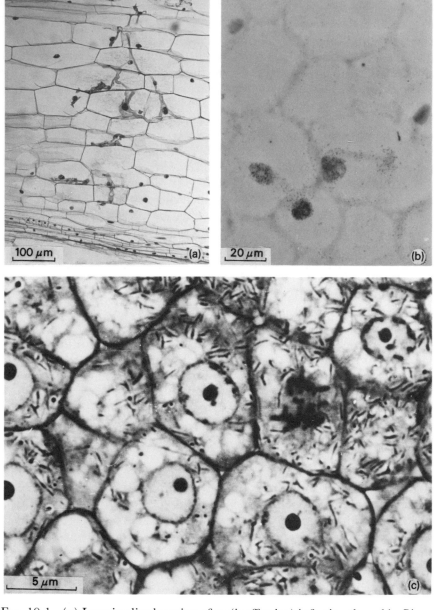

Fig. 10.1. (a) Longitudinal section of an (ineffective) infection thread in *Pisum sativum*. Note the close association with the nuclei. (From Libbenga and Harkes 1973.) (b) Microautoradiograph of an infection thread in *P. sativum* cultured in ^{3}H-thymidine. Note the heavily labelled nuclei associated with the infection thread, which can be recognized by the labelled rhizobia. (From Libbenga and Harkes 1973.) (c) Young nodule of *Arachis hypogaea* showing rod shaped rhizobia and a mitotic figure. Note the large nuclei and prominent nucleoli. (M. R. Chandler, unpublished.)

Aeschynomene indica (Arora 1954; Yatazawa and Yoshida 1979), *A. paniculata* (Suessenguth and Beyerle 1936), *Sesbania sesban* and *Inga laurina* (M. Yatazawa, personal communication). The latter two species have, however, not been investigated for the presence of infection threads. The stem nodules in *A. indica*, *S. sesban*, and *I. laurina* have been examined for nitrogenase activity, with positive results (Yatazawa and Yoshida 1979; M. Yatazawa, personal communication). The implication of the above is that it is likely that a different mode of infection exist.

Of the aforementioned species only the infection process in *Arachis hypogaea* has been studied in detail (Chandler 1978). Root hairs in this species develop only where lateral roots emerge. They are thick-walled, long, occasionally septate, and may have enlarged basal cells. Curling and branching of the root hairs can be induced by rhizobia, but they are not infected. Infections are only initiated where the enlarged basal cells occur. The bacteria enter the space between the root hair wall and the adjoining epidermal and cortical cells. The large basal cells are the first to be invaded, but infections in these cells do not develop further. The cells in the infected area separate at the middle lamella, and intercellular zoogleal strands of rhizobia are formed in the resulting spaces. Increased numbers of mitochondria and endoplasmic reticulum in cells adjacent to the intercellular zones of infection suggest a *Rhizobium*-induced increase in metabolic activity. The cortical cells in the invaded area often have enlarged nuclei with a prominent nucleolus. Nodules develop only from infected cortical cells of emerging lateral roots. As with infection thread walls, invasion of the cells is preceded by structural changes in the cell wall, perhaps produced by similar enzymatic mechanisms. The cell wall forms protrusions into the cell and its cellulose structure appears to be broken down. Membrane-bound rhizobia are released into the host cell, often in close association with the nucleus. A phase of rapid multiplication of the bacteria follows, at the end of which each bacterium becomes enclosed in its own membrane envelope. The rhizobia are distributed entirely by repeated divisions of infected cells (Fig. 10.1(c)) and develop into bacteroids when these cells have ceased to divide. In the early stages of nodule development in *Arachis hypogaea*, only infected cells divide so that no uninfected cells are found in the bacteroid zones of peanut nodules.

A similar infection process of direct-intercellular infection followed by repeated divisions of invaded cells has been observed in *Stylosanthes* species (M. R. Chandler, personal communication). The mode of infection is controlled by the plant host genome, as some strains of *Rhizobium* can produce nodules via infection threads in, e.g. *Glycine max* or *Vigna unguiculata*, and through the direct intercellular infection process on others (Dart 1977).

10.3 Nodule initiation and development

The external shape of root nodules varies considerably among the legumes. In a review on the systematic value of root nodule shapes, Corby (1980) recognized six nodule types. Within the Fabaceae the distribution coincides remarkably with the tribal classification, whereas within the Mimosaceae and Caesalpinaceae a large branched type of nodule, called astragaloid by Corby, predominates. Dart (1977) distinguished three types of nodules: apical or cylindrical nodules which are often branched (as found on *Medicago, Pisum, Trifolium* and *Vicia* spp), spherical type nodules (e.g. *Desmodium, Glycine, Phaseolus* and *Vigna* spp) and collar-shaped nodules which are found on *Lupinus* spp. Nodules of species in which direct intercellular infection occurs are always spherical. The apical type nodule is mainly found on temperate legumes whereas spherical type nodules are typical of legumes of tropical or sub-tropical origin.

Nodule shape is clearly a characteristic of the plant host, as fast- and slow-growing rhizobia can produce the same type of nodule on one species (e.g. *V. unguiculata)* and the same strain of rhizobia can produce different types of nodule on different host species.

Detailed reviews on morphological, cytological, and ultrastuctural aspects of nodule development have been published by Libbenga and Bogers (1974), Dart (1977), and Goodchild (1977) and therefore only the important characteristics of apical and spherical types of nodule will be described here, with emphasis on the earlier stages in nodule formation.

Apical-type nodules

As most detailed work on the morphogenesis and ultrastructure of the development of this type of nodule has been done on members of the Viceae (*Pisum, Vicia*) and the Trifolieae (*Trifolium, Medicago*), the following description is largely based on studies of the developing nodules of *Pisum sativum,* by Libbenga and Harkes (1973), Newcomb (1976), and Newcomb, Sippel, and Peterson (1979).

When the infection thread has penetrated three to six radial layers of the cortex, mitotic activity in a small group of cortical cells directly ahead of the thread is initiated. The first divisions of these cells are always anticlinal. Newcomb *et al.* (1979) observed anticlinal divisions of cortical cells when the infection thread was still within the root hair cell. A meristematic area is thus formed and is penetrated by the infection thread. The meristematic cells are invaded by rhizobia, cease to divide and develop into nitrogen-fixing bacteroid-containing cells. Kodama (1978) however observed mitoses in infected cells of *Sophora flavescens* (Sophorae) and *Cytisus scoparius* (Genistae). The nuclei and nucleoli in infected cells are generally enlarged, as are the nuclei in adjacent cells. New mitotic activity is initiated in neighbouring cortical cells. The infection thread continues to branch and more cells are invaded by the rhizobia. Additional cortical layers contribute to the nodule as development proceeds. The thread eventually extends backward towards the epidermis and an apical meristem develops. The apical meristem and the pericycle layers of the nodule are not penetrated by the infection thread.

Provascular strands are formed from the nodule meristem at an early stage. They develop into a discontinuous vascular system connected to the root vascular system and growing from the apical meristem. In pea nodules transfer cells have been observed in the pericycle and xylem parenchyma before the onset of nitrogen fixation (Newcomb and Peterson 1979). Transfer cells were also found in root nodules of a number of other species all belonging to the Viceae and Trifoleae with the exception of *Lupinus*. They are probably involved in the secretion of amides and amino acids (Pate, Gunning, and Briarty 1969). Eventually the mature nodule develops with its characteristic regions: a meristem, a differentiating zone where release and establishment of bacteria in the host cells occurs, a central bacteroid region, a vascular system and, at later stages, a degenerate zone (e.g. Newcomb 1976; Truchet 1978). Recently Fyson and Sprent (1980) reported the occurrence of effective nodules on the first stem node (epicotyl) of *Vicia faba*. These nodules were similar in morphology and anatomy to normal root nodules. Infection took place via the stem hairs, nodule development was initiated in the cortical cells of the stem and proceeded as with root nodules. It seems therefore that the conditions in the epicotyl- and root tissue are similar enough for nodule development to be possible in both.

Spherical type nodules

Relatively few accounts of the morphogenesis of this type of nodule have been published, with *Glycine max* being the best studied species. Goodchild and Bergersen (1966) and Bassett *et al.* (1977*a*) dealt with the ultrastructural aspect of the infection process and bacteroid development. Newcomb *et al.* (1979) investigated the early development of *G.max* nodules.

In the earliest stage observed, anticlinal divisions occur in all layers of the cortex of soyabean roots, while the infection thread is still confined to the root hair or has just penetrated the basal cell wall. Subsequent divisions in the outermost layers are mainly periclinal. The cells in the outer layers are characterized by enlarged nuclei and nucleoli, small vacuoles, and densely-staining cytoplasm. The extensive meristematic activity in these cells results in the formation of a spherical mass of cytoplasmically-rich cells near the base of the infected root hair. Later, the infection thread branches and penetrates further into the cortex. Eventually nodules develop with an inner spherical mass of dividing cytoplasmically-rich cells, covered by a layer of cells that vary in size and cytology. Many of these inner cells have enlarged nuclei and nucleoli. Some of them become infected and develop into a central bacteroid-containing zone. Other mitotically active cells develop into vascular strands. More mature nodules contain a central bacteroid zone, with uninfected cells, and meristematic regions at the edge of this zone. Mitotic activity ceases in the mature nodule.

Nodules of soyabean and other Phaseoleae have vascular strands which fuse at the apex and contain no transfer cells (Pate *et al.* 1969; Sprent 1980). A possible relationship between nodule vascular anatomy and nodule export products has recently been discussed by Sprent (1980).

10.4 Biochemical aspects of nodule development

In addition to the morphological and cytological changes, a series of biochemical changes occur in the plant cells and the rhizobia. The bacteria develop into bacteroids and eventually fill most of the cytoplasm often leaving one rather small central vacuole (Fig. 10.2(a)). Number of bacteroids per membrane envelope, one to several, and their shape and size depend on the host species (Dart 1977). The differentiation into bacteroids, which is often reversible (Gresshoff, Skotnicki, Eadie, and Rolfe. 1977; Sutton, Jepsen, and Shaw 1977; Tsien, Cain, and Schmidt 1977), must therefore result from interactions with the host cells, although the exact nature of these interactions is unknown. The nodule environment induces the synthesis of bacteroid-specific proteins, however (Shaw and Sutton 1979). In the mature bacteroid, interactions with the cell environment result in the induction of nitrogenase synthesis and activity, although under certain conditions many rhizobia can fix atmospheric nitrogen in the free-living form (e.g. Kurz and LaRue 1975; McComb, Elliot, and Dilworth 1975; Pagan, Child, Scowcroft, and Gibson 1975).

A number of biochemical changes in the plant cells result from the presence of rhizobia. The enlarged nuclei observed in the vicinity of the invading bacteria (Fig.10.2(a)) seem to suggest the occurrence of DNA synthesis and polyploidy. The meristematic activity is perhaps induced and maintained by increased levels of phytohormones, possibly *Rhizobium* produced (Libbenga and Bogers 1974; Syōno *et al.* 1976). Nodule-specific proteins ('nodulins'—Legocki and Verma 1979, 1980) are elaborated and a massive synthesis of leghaemoglobin takes place. This section of the review will concentrate on biochemical changes taking place in the plant cell during nodule development. Biochemical

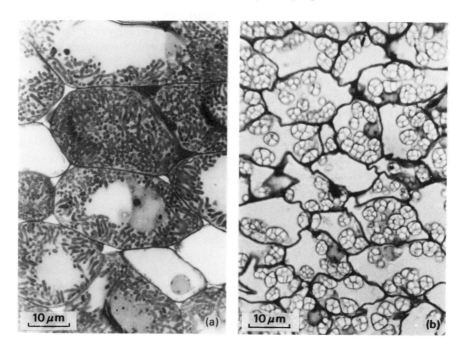

Fig. 10.2. (a) Cells in the bacteriod region of an effective *Medicago sativa* nodule. Infected cells contain enlarged nuclei. (From Truchet *et al*. 1980.) (b) Cells in the central region of a Leu⁻ induced nodule of *M. sativa* supplied with urea. The cells are devoid of bacteria, have small nuclei and contain numerous starch granules. (From Truchet *et al*. 1980.)

aspects of bacteroid development, nitrogen fixation and assimilation will be discussed in Volume 3 of this series.

DNA synthesis and ploidy levels in nodule cells

Controversy has arisen as to the ploidy levels of root nodule cells. Observations by Wipf and Cooper (1940) and Torrey and Barrios (1969) seem to suggest that the mitotically active cells in the root cortex of *Pisum sativum*, from which nodules arise, are disomatic. Mitchell (1965) reported the occurrence of mainly disomatic cells in the meristem and higher ploidy levels in the bacteroid zone of pea nodules. Kodama (1970, 1977) investigated 26 species (from 21 genera and 12 tribes) which possess apical-type nodules, and obtained results that were in close agreement with those of Mitchell. Cortical cells of *Astragalus sinicus* roots apparently become disomatic when they are approached by the infection thread (Kodama 1970). An extensive study of nuclear variations in nodule cells of *P. sativum* and *Medicago sativa* using cytological, histo-autoradiographical, and microspectro-photometrical methods has been published by Truchet (1978). He found DNA levels per nucleus of 2c and 4c in the apical meristem as well as in the original meristematic cells, 4c to 32c in the differentiating zone and 16c to 32c in the bacteroid zone. A gradient of DNA content per nucleus in the differentiating zone was observed. DNA synthesis (as measured by ^{3}H-thymidine incorporation) took place in the meristem and the differentiation zone. Cell divisions occurred in the meristem only. The conclusions that can be drawn from this work are that meristematic cells in apical type nodules are monosomatic, in disagreement with other workers, and that the immediate presence of rhizobia induces DNA-synthesis without cell division.

Six of the seven legumes with spherical nodules examined by Kodama (1970, 1977) showed predominantly monosomatic mitoses in the nodule meristems. More detailed examination of *Vigna unguiculata* nodules demonstrated the presence of mainly monosomatic cells in the peripheral meristems, and disomatic nuclei in the cells of the bacteroid zone (Kodama 1970). It seems likely, therefore, that rhizobia can induce endoreduplication in apical-type as well as in spherical-type nodules.

Phytohormones and nodule development

As mentioned in the Introduction, similarities exist between the development of leguminous root nodules and normal plant organs. Since plant organ development is thought to be controlled, at least to some extent, by hormone balance it seems only logical to expect phytohormone involvement in nodule formation.

Relatively large concentrations of indole acetic acid (IAA) have been found in root nodules of *Pisum arvense* (Pate 1958) and *Lupinus luteus* (Dullaart 1970). In addition, rhizobia have the ability to convert L-tryptophan into IAA (Kefford *et al.* 1960; Dullaart 1970; Rigaud 1970). Increased amounts of gibberellic acid have been detected in root nodules of *Phaseolus vulgaris, Pisum sativum* (Radley 1961), and *Lupinus luteus* (Dullaart and Duba 1970).

High levels of zeatin and its riboside were present in pea nodules (Syōno and Torrey 1976; Syōno *et al.* 1976) and nodules of *Vicia faba* (Henson and Wheeler 1976). Puppo, Rigaud, and Barthe (1974) reported the presence of isopentenyladenine (2iP) in nodules of *Phaseolus vulgaris. Rhizobium leguminosarum* can produce 2iP in culture (phillips and Torrey 1972) and *R. phaseoli* 2iP and some zeatin in the presence of bean roots (Puppo and Rigaud 1978). In pea nodules, cytokinin activity is highest in young nodules and is positively correlated with mitotic indices, while in older nodules most cytokinin is present in the apical meristem (Syōno *et al.* 1976).

Involvement of hormones in nodule initiation has been suggested by several workers (e.g. Libbenga, Van Iren, Bogers, and Schraag-Lamers 1973; Phillips and Torrey 1972; Libbenga and Bogers 1974; Syōno *et al.* 1976). Some evidence for this has come from *in vitro* experiments with pea root explants. When these are cultured on a defined medium in the presence of an auxin and a cytokinin the cortical cells undergo two rounds of DNA synthesis prior to mitosis, thereby becoming disomatic (Libbenga and Torrey 1973; Phillips and Torrey 1972). Moreover, in this system three meristematic areas are formed opposite the radii of the central cylinder (Libbenga *et al.* 1973) which is where nodule initiation in pea root usually occurs (Libbenga and Harkes 1973). Both polyploid mitoses in cultured segments of pea root cortex and nodulation, but not infection thread formation are inhibited by abscisic acid (Phillips 1971).

These similarities in response in *in vivo* and in *in vitro* systems combined with the data on hormone content in nodules and rhizobial hormone production are indeed striking. It could be therefore that phytohormones play a role in initiating and/or maintaining meristematic activity during nodule development, but it is still not clear whether (hormone induced?) endoreduplication is a prerequisite for nodule initiation.

Leghaemoglobin and 'nodulins'

The major nodule-specific protein is leghaemoglobin (lHb), a myoglobin-like protein which may regulate oxygen tensions in the root nodules (Wittenberg, Appleby, and Wittenberg 1972). In mature soyabean nodules it may account for up to 20 per cent of total protein synthesis (Sidloi-Lumroso and Schulman 1977). lHb seems to be a true product of symbiosis in that the protein component is coded for by plant genes (Dilworth 1969; Broughton and Dilworth 1971; Cutting and Schulman 1971; Sidloi-Lumbroso, Kleiman, and Schulman 1978; Baulcombe and Verma 1978) while the haem group is thought to be synthesized by the bacteroids (Cutting and Schulman 1972; Godfrey, Coventry, and Dilworth 1975; Nadler and Avissar 1977). Hybridization of lHb cDNA to soyabean DNA showed that the haploid soyabean genome contains about 40 copies of the lHb genes (Baulcombe and Verma 1978). In *G.max* (Bergersen and Goodchild 1973; Verma, Bal, Guérin, and Wanamaker 1979), *Lupinus angustifolius* (Robertson *et al.* 1975), *Pisum sativum* and *V. unguiculata* (Bisseling, Moen, Van Den Bos, and Van Kammen (1980) lHb synthesis begins before nitrogenase activity is detectable.

In soyabean there are two major electrophoretically distinguishable forms of lHb: lHb-S (slow moving) and lHb-F (fast-moving) coded for by separate genes (Appleby, Nicola, Hurrell, and Leach 1975). The biosynthesis of lHb-F predominates in young soyabean nodules, and that of lHb-S in mature nodules (Fuchsman and Appleby 1979; Verma *et al.* 1979). By means of iso-electric focusing, four major and four minor components can be distinguished (Fuchsman and Appleby 1979). Some of these components differ in amino acid sequences but others appear to be derived from related forms by means of post-translational processing during nodule development (Whittaker, Moss, and Appleby 1979). Synthesis of lHb seems to be mainly under transcriptional control, however, as lHb mRNA, which represents 18–20 per cent of the soyabean nodule poly (A) + polysomal RNA, is absent in uninfected tissue (Auger, Baulcombe, and Verma 1979). Moreover the order of appearance of lHb-F and lHb-S is associated with the appearance of their respective mRNAs (Verma *et al.* 1979). A more detailed account of lHb structure, synthesis and function will be given in Volume 3.

Nodule specific proteins other than lHb (nodulins) have recently been identified (Legocki and Verma 1979, 1980). Legocki and Verma (1980) detected 18–20 nodule specific polypeptides in the *in vitro* translation products of soyabean nodule polysomes, using immuno-logical techniques, which may represent 7–10 per cent of the total protein synthesized in the nodule cell cytoplasm. Nodule specific mRNAs are, like lHb mRNA, not detected in uninfected root tissue so that 'nodulin' biosynthesis is probably under transcriptional control (Verma, Auger, Brisson, Goodchild, Legocki, and Haugland 1980*a*). It seems likely that at least some of these nodule-specific proteins are involved in the establishment of a successful symbiotic association.

10.5 Ineffective associations

In ineffective associations, root nodule formation is initiated, but little or no atmospheric nitrogen is fixed. Such nodules are often small and white, suggesting absence of leghaemoglobin, and the plants usually small and chlorotic. As we have seen, nodule formation involves a complex series of interactions between legume host- and rhizobial genome. Therefore genetic variation in either symbiotic partner can lead to different nodulation responses, ranging from non-infectivity to normal root nodules. Additionally, environmental conditions such as temperature, mineral deficiency and water-logging can result in ineffective nodulation (Jordan 1974).

As the subject of ineffectiveness has been discussed in detail by Jordan (1974), only some of the better-studied systems will be described here.

Host conditioned ineffectiveness

In *G. max*, four genes have been reported as conditioning an ineffective response. The gene rj_1, when homozygous, inhibits nodulation at an early stage, as evidence of nodule development is barely visible (Williams and Lynch 1954). Homozygous rj_1rj_1 plants are able to form nodules, however, with certain strains of *R. japonicum*. These strains invariably produce rhizobitoxine-induced foliar chlorosis, suggesting involvement of this *Rhizobium*-produced glycine derivative in the nodulation process (Devine and Breithaupt 1980). Roots of Rj_2 plants (a dominant factor reported in some cultivars) develop either small white cortical proliferations or rudimentary nodules when inoculated with certain strains or *R. japonicum* (Caldwell 1966; Caldwell, Hinson, and Johnson 1966). Other dominant strain-specific genes (Rj_3 and Rj_4) with similar phenotypic effects have been reported by Vest (1970) and Vest and Caldwell (1972). No detailed observations of the development of these four types of ineffective soyabean nodules have been published, however. Another strain-specific ineffective association exists in soyabeans. On cultivar Clark-63 ineffective nodules are produced by *R. japonicum* strain 8-0, which produces normal nodules on other host varieties. In these nodules early development proceeded in the usual way, but after release into the host cells, the rhizobia degenerate rapidly (Bassett, Goodman, and Novacky 1977*b*).

Four single recessive genes (*i*, *ie*, *n*, *d*) have been reported to be responsible for ineffective nodulation in *T. pratense* (Bergersen and Nutman 1957; Nutman 1969). In nodules on *i*, *i*, plants inoculated with *R. trifolii* strain A, release of bacteria from the infection thread is restricted. The few bacteroids that develop degenerate rapidly. Nodules on *ie ie* plants are often normal in external appearance. The infection thread invades the cortical cells, but the bacteria usually remain enclosed in vesicles. These vesicles presumably consist of cell wall material derived from the infection thread. Tumor-like structures within the infected regions of the nodules develop due to abnormal division on invaded and uninvaded cells. The *ie ie* response is elicited by several *R. trifolii* strains. Both *ie ie* and i_1 i_1 plants can be restored to normal nodulation by the activity of single recessive (and independently segregating) suppressor genes. Nodules on *n n* and *d d* plants are small, contain few infection threads, and infected cells. The few bacteria that are released senesce rapidly. All four types of ineffective nodules were characterized by the presence of abundant starch in nodule cells and the absence of lHb.

Viands, Vance, Heichel, and Barnes (1979) described an ineffective recessive mutation in *Medicago sativa* which has some of the characteristics of the i_1 *T. pratense* mutation: abundance of starch granules in the nodule cells, absence of lHb, and rapid degeneration of the few bacteroids that form. In addition, the nodules were abnormally large and irregular in shape.

In *Pisum sativum* a strain-specific ineffective system controlled by a single recessive gene has been reported (Holl 1975; Degenhardt, La Rue, and Paul 1976). The inoculated roots of *P. sativum* cultivar Afghanistan developed white swellings which contained enlarged cells devoid of bacteroids. Apparently nodule formation was initiated, but ceased due to abortion of the infection thread. Lie (1978) showed that the ineffective response in cultivar Afghanistan also depends on the rhizobial strain used to inoculate the plants.

Rhizobium-conditioned ineffectiveness

A major problem in the search for mutations affecting the legume–*Rhizobium* symbiosis is that the desired mutants cannot directly be selected for *in vitro*, but have to be tested on the host plant. Many rhizobial strains with an abnormal symbiotic phenotype, therefore, were isolated originally as antibiotic resistant, anti-metabolite resistant or as auxotrophic mutants (Pankhurst, Schwinghamer, and Bergersen 1972; Truchet and Dénarié 1973*a*, *b*; MacKenzie and Jordan 1974; Pankhurst 1974; Dénarié, Truchet, and Bergeron 1976; Pankhurst 1977; Schwinghamer 1977; Pain 1979).

In nodules formed by a riboflavin requiring auxotroph of *R. trifolii* on *T. pratense*, a large proportion of rhizobia in the nodule cells failed to develop into bacteroids and often degenerated rapidly (Pankhurst *et al.* 1972). In older nodules, release from the infection thread was impaired, possibly due to increased amounts of polysaccharide material in which the bacteria were embedded within the thread. When riboflavin was added to the plant culture medium, most rhizobia were transformed into normal bacteroids and effective nodules developed. Pankhurst (1974) examined ineffective *T. pratense* nodules formed by a number of anti-metabolite and antibiotic-resistant mutants of two *R. trifolii* strains. Several mutants, all with an apparently modified cell envelope, produced tumour-like structures, usually at the junctions of the main and lateral roots. These growths contained no distinct meristem or vascular system and were devoid of infection threads and bacteria. In ineffective nodules formed by other mutants various defects were observed, such as failure to release from the infection thread, failure of bacteria to develop into bacteroids, incomplete development of rhizobia into bacteroids, or bacteroids with abnormal morphology. In some cases these abnormalities may be associated with alterations in the bacterial cell wall (Pankhurst 1974).

Infection of *M. sativa* with a viomycin- or a tuberactinomycin-resistant mutant of *R. meliloti* was initially followed by the normal pattern of nodule formation (MacKenzie and Jordan 1974). After release from the infection thread, however, the bacteria disintegrated rapidly. In nodules formed by R20, a naturally-occurring ineffective strain of *R. meliloti*, bacteroids developed, but degenerated prematurely. All three types of ineffective *M. sativa* nodules were characterized by a build-up of endoplasmic reticulum and an accumulation of starch granules in the nodule cells. Other mutants of *R. meliloti* were studied by Truchet and Dénarié (1973*a*, *b*) and Truchet, Michel, and Dénarié (1980). Multiplication of bacteria inside the host cytoplasm and conversion into bacteroids were inhibited in nodules produced by a uracil-requiring mutant. The latter abnormality was also observed in nodules formed by an adenine-requiring *R. meliloti* mutant (Truchet and Dénarié, 1973*a*). A leucine-requiring mutant (Leu⁻) was reported by Truchet and Dénarié (1973*b*) and a detailed account of its symbiotic properties was recently published (Truchet *et al.* 1980). In the small and spherical nodules produced by the Leu⁻ mutant, the bacteria were not released from the infection threads. Meristematic activity was limited and most cells in the central nodule tissue were monosomatic. Addition of urea (as a nitrogen source) to the plant growth medium resulted in increased mitotic activity and the formation of vascular tissue, but the rhizobia were not released into the host cells and the central cells remained monosomatic (Fig. 10.2(b)). Effective nodules were formed, however, when this system was supplied with L-leucine or one of its precursors (*a*-ketoisovalerate, *a*-ketoisocaproate). These nodules had all the characteristics of normal apical-type nodules including a central zone with infected polyploid cells.

   R. leguminosarum strain 1019 formed small, white nodules on *P. sativum* in which the bacteria released from the infection thread were not surrounded by a peribacteroid membrane (Newcomb, Syōno, and Torrey 1977). Mitotic activity and cytokinin content declined at an early stage and infected cells contained reduced numbers of polysomes and mitochondria and only small amounts of endoplasmic reticulum. Apparently impaired biosynthetic capacity led to defective membrane synthesis and thus prevented bacteroid formation.

   A breakdown of the peribacteroid membrane takes place in ineffective soyabean nodules produced by *R. japonicum* strain 61-A-24. Additionally the nodule cells contain an unknown amino acid-like compound (Werner, Mörschel, Stripf, and Winchenbach 1980). No lHb-F was detected in 61-A-24 induced nodules, while low concentrations of lHb-S and lHb-F were present in nodules formed by ineffective *R. japonicum* strain SM5 (Verma, Goodchild, Brisson, Thomas, Haugland, Sullivan, and Lacroix 1980*b*). Low concentrations of lHb-mRNA were detected in these ineffective nodules by means of hybridization of nodule mRNA with lHb-cDNA (Verma *et al.* 1980*b*). Reduced synthesis of some 'nodulins' and absence of some others was observed in these nodules

(Legocki and Verma 1980), probably due to reduced levels of nodule-specific mRNA sequences (Verma *et al.* 1980*a*).

Other ineffective associations

Ineffective nodules often arise as the result of non-homologous legume–*Rhizobium* associations (e.g. Pankhurst, Craig, and Jones 1979; Trinick 1980). For example, infection of *Lotus pedunculatus* with strain NZP2213 (isolated from effective nodules of *L. tenuis*) induced the formation of small ineffective nodules derived from cortex cells. They contained vascular tissue and a central zone of dividing cells with abundant starch granules. The nodule cells were not infected and rhizobia were found only between the outer two or three cell layers of the nodule. Nodule development appeared to be blocked at an early stage (Pankhurst *et al.* 1979). A negative correlation between the ability to nodulate *Lotus* effectively and flavolans sensitivity of rhizobia was found by Pankhurst and Jones (1979).

Many rhizobial strains can induce tumour-like structures on legume hosts that they do not normally nodulate (MacGregor and Alexander 1971; Hamdi 1973; Hepper 1978). This type of nodule was termed pseudonodule by MacGregor and Alexander (1971). They seem to originate from the epidermal cell layers of the root rather than from cortex cells. Furthermore, they consist of disorganized cells, have no meristematic centre and are devoid of bacteria.

10.6 Concluding remarks

The establishment of a successful symbiosis depends on the differential expression of both legume host- and rhizobial genes at successive stages of nodule development. Regulation of this process could be through exchange(s) of specific chemical signals between the symbiotic partners and/or through induced changes in the physiological environment. Bruenning and Wullstein (1972) attempted to demonstrate the existence of such signals. They observed a transfer to radioactivity from rhizobia, which had been cultured in the presence of tritiated leucine and proline, to the polyploid nuclei of developing root nodules of *Trifolium repens*. It is not known, however, in what form this transfer occurred. Truchet *et al.* (1980) also proposed involvement of chemical signals in nodule form-ation in their account of the ineffective *M. sativa*–*R. meliloti* (Leu⁻) association. Their results suggest that rhizobia can trigger initiation of nodule development from a distance by means of a compound(s) which is capable of crossing both cell walls and plasmalemma, such as a phytohormone. They termed this factor 'nodule organogenesis induc-ing principle' (NOIP). As it seems that the physical presence of the membrane-bound rhizobia within the nodule cells is required for the differentiation of the central nodule tissue (including endoreduplication) they also suggested the existence of a 'central tissue differentiation inducing principle' (CTDIP), which can only pass across the peribacteroid membrane. It will have to be seen, however, whether these inducing factors directly derepress host genes or act via induced changes in the physiological environment of the root cells. Chemical messengers could, of course, also be involved in the conversion of free-living rhizobia into bacteroids.

A number of authors have compared the process of root nodule formation with crown-gall tumour induction by *Agrobacterium tumefaciens* (e.g. Libbenga and Bogers 1974; Dénarié and Truchet 1979; Meijer and Broughton 1980; Schell and Van Montagu 1980). Fast-growing rhizobia and *A. tumefaciens* are closely-related species (Lippincott and Lippincott 1975) and both are able to induce similar morphogenetic changes in the moss *Phylaisella selwynii* (Spiess, Lippincott, and Lippincott 1977). Crown-gall tumour formation involves binding of the bacteria to the host cells (determined by plasmid and chromosomal genes, Whatley, Margot, Schell, Lippincott, and Lippincott 1978), followed by injection of part of a plasmid (Ti plasmid) which then becomes stably integrated into the plant nucleus (Willmitzer, De Beuckeleer, Lemmers, Van Montagu, and Schell 1980; Yadav, Postle, Saïki, Thomashow, and Chilton 1980). At least some of the genes involved in rhizobial infection of legumes are located on plasmids (see Chapter 5) and a similar mechanism of transfer of genetic information could be envisaged (e.g. Meijer and Broughton 1982). There are important differences between these two systems however. *A. tumefaciens* always remains extracellular and its presence is required for only a limited period for transformation to occur. Also root nodules senesce, whereas crown-gall tissue can be cultured *in vitro* for long periods. Furthermore, plant tumours tend to show uncontrolled growth, whereas cell division is confined to distinct meristematic regions of root nodules and a functional organ-like structure develops after rhizobial infection.

An important approach in the study of the interactions between *Rhizobium* and legume host in nodule development is genetic analysis of microsymbiont and macrosymbiont. Vincent (1980) has listed the currently recognizable steps in nodule formation: nodule initiation (Noi), bacterial release (Bar), and bacteroid development (Bad). Plant- and rhizobial mutations affecting each step are known. It should be possible to identify certain nodule-specific gene products (*Rhizobium*- and plant-produced) and associate them with each particular step in nodule development. It will then be of great interest to investigate the mechanisms by which the synthesis of these products is controlled, and if and how they affect nodule formation.

Acknowledgements

I would like to thank E. Schölzel (MPI) for typing the manuscript and M. R. Chandler (Rothamsted Experimental Station) for providing the photograph used in the Fig. 10.1.

References

Allen, O. N. and Allen, E. K. (1940). *Bot. Gaz.* **102,** 121.
Appleby, C. A., Nicola, N. A., Hurrell, J. G. R., and Leach, S. J. (1975). *Biochemistry* **14,** 4444.
Arora, N. (1954). *Phytomorphology* **4,** 211.
—— (1956). *Phytomorphology* **6,** 367.
Auger, S., Baulcombe, D., and Verma, D. P. S. (1979). *Biochim. biophys. Acta* **563,** 496.
Bassett, B., Goodman, R. N., and Novacky, A. (1977*a*). *Can. J. Microbiol.* **23,** 573.
—— —— —— (1977*b*). *Can. J. Microbiol.* **23,** 873.
Baulcombe, D. and Verma, D. P. S. (1978). *Nucl. Acids Res.* **5,** 4141.
Bergersen, F. J. and Goodchild, D. J. (1973). *Aust. J. Biol. Sci.* **26,** 741.
—— and Nutman, P. S. (1957). *Heredity* **11,** 175.
Bisseling, T., Moen, A. A., Van Den Bos, R. C., and Van Kammen, A. (1980). *J. gen. Microbiol.* **118,** 377.
Broughton, W. J. (1978). *J. appl. Bacteriol.* **45,** 165.
—— and Dilworth, M. J. (1971). *Biochem. J.* **125,** 1075.
Bruenning, M. L. and Wullstein, L. H. (1972). *Physiol. Plant* **27,** 244.
Caldwell, B. E. (1966). *Crop Sci.* **6,** 427.
—— Hinson, K., and Johnson, H. W. (1966). *Crop Sci.* **6,** 495.
Chandler, M. R. (1978). *J. exp. Bot.* **29,** 749.
Corby, H. D. L. (1980). In *Advances in legume systematics* (ed. R. M. Polhill and P. H. Raven). Royal Botanical Press, Kew (In the press).
Cutting, J. A. and Schulman, H. M. (1971). *Biochim. biophys. Acta* **229,** 58.
—— —— (1972). *Biochim. biophys. Acta* **184,** 432.
Dart, P. (1977). In *A treatise on dinitrogen fixation.* Section III: *Biology* (ed. R. W. F. Hardy and W. S. Silver) p. 367. John Wiley, New York.
Degenhardt, T. L., La Rue, T. A., and Paul, E. A. (1976). *Can. J. Bot.* **54,** 1633.
Dénarié, J. and Truchet, G. (1979). *Physiol. Vég.* **17,** 643.
—— —— and Bergeron, B. (1976). In *Symbiotic nitrogen fixation in plants* (ed. P. S. Nutman) p. 47. Cambridge University Press.
Devine, T. E. and Breithaupt, B. H. (1980). *Crop Sci.* **20,** 394.
Dilworth, M. J. (1969). *Biochim. biophys. Acta* **184,** 432.
Dullaart, J. (1970). *Acta Bot. Neerl.* **19,** 573.
—— and Duba, L. I. (1970). *Acta Bot. Neerl.* **19,** 877.
Fåhraeus, G. and Sahlman, K. (1977). *Ann. Acad. Reg. Sci. Upsaliensis* **20,** 103.
Fuchsman, W. H. and Appleby, C. A. (1979). *Biochim. biophys. Acta* **579,** 314.
Fyson, A. and Sprent, J. I. (1980). *J. exp. Bot.* **31,** 1101.
Godfrey, C. A., Coventry, D. R., and Dilworth, M. J. (1975). In *Nitrogen fixation by free-living organisms* (ed. W. D. P. Stewart) p. 311. Cambridge University Press.
Goodchild, D. J. (1977). *Int. Rev. Cytol. Suppl.* **6,** 235.
—— and Bergersen, F. J. (1966). *J. Bacteriol.* **92,** 204.
Gresshoff, P. M., Skotnicki, M. L., Eadie, J. F., and Rolfe, B. G. (1977). *Plant Sci. Lett.* **10,** 299.
Hamdi, Y. A. (1973). *Zbl. Bakt. Abt. II* **128,** 532.
Henson, I. A. and Wheeler, G. T. (1976). *New Phytol.* **76,** 433.

Hepper, C. M. (1978). *Ann. Bot.* **42,** 109.
Holl, F. B. (1975). *Euphytica* **24,** 767.
Jordan, D. C. (1974). *Proc. Indian Acad. Sci.* **40,** 713.
Kefford, N. P., Brockwell, J., and Zwar, J. A. (1960). *Aust. J. Biol. Sci.* **13,** 456.
Kodama, A. (1970). *J. Sci. Hiroshima Univ. Series* B, Div. 2, **13,** 223.
—— (1977). *Bull. Hiroshima Agr. Coll.* **5,** 389.
—— (1978). *Jap. J. Genet,* **53,** 381.
Kurz, W. G. W. and La Rue, T. A. (1975). *Nature, Lond.* **256,** 407.
Legocki, R. P. and Verma, D. P. S. (1979). *Science, N.Y.* **205,** 190.
—— —— (1980). *Cell* **20,** 153.
Libbenga, K. R. and Bogers, R. L. (1974). In *The biology of nitrogen fixation* (ed. A. Quispel) p. 430. North Holland, Amsterdam.
—— and Harkes, P. A. A. (1973). *Planta* **114,** 17.
—— and Torrey, J. G. (1973). *Am. J. Bot.* **60,** 293.
—— Van Iren, F., Bogers, R. J., and Schraag-Lamers, M. F. (1973). *Planta* **114,** 29.
Lie, T. A. (1978). *Ann. appl. Biol.* **88,** 462.
Lippincott, J. A. and Lippincott, B. B. (1975). *A. Rev. Microbiol.* **29,** 377.
MacGregor, A. N. and Alexander, M. (1971). *J. Bacteriol.* **105,** 728.
MacKenzie, C. R. and Jordan, D. C. (1974). *Can. J. Microbiol.* **20,** 755.
McComb, J. A., Elliot, J., and Dilworth, M. J. (1975). *Nature, Lond.* **256,** 409.
Meijer, E. G. M. and Broughton, W. J. (1982). In *Molecular biology of plant tumors.* (ed. G. Kahl and J. Schell), Academic Press, New York (In the press).
Mitchell, J. P. (1965). *Ann. Bot.* **29,** 371.
Mosse, B. (1964). *J. gen. Microbiol.* **36,** 49.
Nadler, K. D. and Avissar, Y. L. (1977). *Plant Physiol.* **60,** 433.
Napoli, C. A., Dazzo, F., and Hubbell, D. L. (1975). In *Proceedings of the fifth Australian legume nodulation conference* (ed. J. Vincent), Brisbane (Abstract in *Rhizobium Newsl.* **20,** suppl., 35).
Newcomb, W. (1976). *Can. J. Bot.* **54,** 2163.
—— and Peterson, R. L. (1979). *Can. J. Bot.* **57,** 2583.
—— Sippel, D., and Peterson, R. L. (1979). *Can. J. Bot.* **57,** 2603.
—— Syōno, K., and Torrey, J. G. (1977). *Can. J. Bot.* **55,** 1891.
Nutman, P. S. (1969). *Proc. R. Soc. Lond.* B **172,** 417.
Pagan, J. D., Child, J. J., Scowcroft, W. R., and Gibson, A. H. (1975). *Nature, Lond.* **256,** 406.
Pain, A. N. (1979). *J. appl. Bacteriol.* **47,** 53.
Pankhurst, C. E. (1974). *J. gen. Microbiol.* **82,** 405.
—— (1977). *Can. J. Microbiol.* **23,** 1026.
—— and Jones, W. T. (1979). *J. exp. Bot.* **30,** 1095.
—— Craig, A. S., and Jones, W. T. (1979). *J. exp. Bot.* **30,** 1085.
—— Schwinghamer, E. A., and Bergersen, F. J. (1972). *J. gen. Microbiol.* **70,** 161.
Pate, J. S. (1958). *Aust. J. biol. Sci.* **11,** 516.
—— Gunning, B. E. S., and Briarty, L. G. (1969). *Planta* **85,** 11.
Phillips, D. A. (1971). *Planta* **100,** 181.
—— and Torrey, J. G. (1972). *Plant Physiol.* **49,** 11.
—— —— (1973). *Develop. Biol.* **31,** 336.
Puppo, A. and Rigaud, J. (1978). *Physiol. Plant* **42,** 202.
—— —— and Barthe, P. (1974). *C. R. Acad. Sci. Paris, Série D* **279,** 2039.
Radley, M. (1961). *Nature, Lond.* **191,** 684.
Rango Rao, V. (1977). *J. exp. Bot.* **28,** 241.
Rigaud, J. (1970). *Physiol. Plant* **23,** 171.

Robertson, J. G., Farnden, K. J. F., Warburton, M. P., and Bank, J. M. (1975). *Aust. J. Plant Physiol.* **2**, 265.

—— Lyttleton, P., Bullivant, S., and Grayston, G. F. (1978). *J. Cell Sci.* **30**, 129.

—— Taylor, M. P., Craig, A. S., and Hopcroft, D. H. (1974). In *Mechanisms of regulation of plant growth* (ed. R. L. Bieleski, A. R. Ferguson, and M. M. Cresswell). p. 31. Royal Society of New Zealand, Wellington.

Roughley, R. J., Dart, P. J., and Day, J. M. (1976). *J. exp. Bot.* **27**, 431.

Schell, J. and Van Montagu, M. (1980). In *The molecular basis of microbial pathogenicity* (ed. H. Smith, J. J. Skehel, and M. J. Turner) p. 225, Verlag Chemie GmbH, Weinheim.

Schmidt, E. L. (1978). In *Interactions between non-pathogenic soil micro-organisms and plants* (ed. Y. R. Dommergues and S. V. Krupa) p. 269, Elsevier, Amsterdam.

Schwinghamer, E. A. (1977). In *A treatise on dinitrogen fixation.* Section III: *Biology* (ed. R. W. F. Hardy and W. S. Silver) p. 577. John Wiley, New York.

Shaw, B. D. and Sutton, W. D. (1979). *Biochim. biophys. Acta* **563**, 216.

Shore, G. and MacLahan, G. A. (1975). *J. Cell Biol.* **64**, 557.

Sidloi-Lumbroso, R. and Schulman, H. M. (1977). *Biochim. biophys. Acta* **476**, 295.

—— Kleiman, L., and Schulman, H. M. (1978). *Nature, Lond.* **273**, 558.

Spiess, L. D., Lippincott, B. B., and Lippincott, J. A. (1977). *Bot. Gaz.* **138**, 35.

Sprent, J. (1980). *Plant, Cell Environ.* **3**, 35.

Suessenguth, K. and Beyerle, R. (1936). *Hedwigia* **75**, 234.

Sutton, W. D., Jepsen, N. M., and Shaw, B. D. (1977). *Plant Physiol.* **59**, 741.

Syōno, K. and Torrey, J. G. (1976). *Plant Physiol*, **57**, 602.

—— Newcomb, W., and Torrey, J. G. (1976). *Can. J. Bot.* **54**, 2155.

Torrey, J. G. and Barrios, S. (1969). *Caryologia* **22**, 47.

Trinick, M. J. (1980). *J. appl. Bacteriol.* **49**, 39.

Truchet, G. (1978). *Ann. Sci. Nat., Bot. Biol. Vég.* **19**, 3.

—— and Dénarié, J. (1973a). *C. R. Acad, Sci. Paris, Série D* **277**, 841.

—— —— (1973b). *C. R. Acad. Sci. Paris, Série D* **277**, 925.

—— Michel, M., and Dénarié, J. (1980). *Differentiation* **16**, 163.

Tsien, H. C., Cain, P. S., and Schmidt, E. L. (1977). *Appl. Environ. Microbiol.* **34**, 854.

Tu, J. C. (1975). *J. Bacteriol.* **122**, 691.

Verma, D. P. S. and Zogbi, V. (1978). *Plant Sci. Lett.* **13**, 137.

—— Bal, S., Guérin, C., and Wanamaker, L. (1979). *Biochemistry* **18**, 476.

—— Kazazian, V., Zogbi, V., and Bal, A. K. (1978). *J. Cell Biol.* **78**, 919.

—— Auger, S., Brisson, B., Goodchild, R., Legocki, R., and Haugland, R. (1980a). *Eur. J. Cell Biol.* **22**, 52 (Abstract).

—— Goodchild, B., Brisson, N., Thomas, D. Y., Haugland, R., Sullivan, D., and Lacroix, L. (1980b). *Pl. Physiol.* **6** (Suppl.), 124 (Abstract).

Vest, G. (1970). *Crop Sci.* **10**, 34.

—— and Caldwell, B. E. (1972). *Crop Sci.* **12**, 692.

Viands, D. R., Vance, C. P., Heichel, G. H., and Barnes, D. K. (1979). *Crop Sci.* **19**, 905.

Vincent, J. (1980). In *Nitrogen fixation.* Vol. II (ed. W. E. Newton and W. H. Orme-Johnson) p. 103. University Park Press, Baltimore.

Werner, D. W., Mörschel, E., Stripf, R., and Winchenbach, B. (1980). *Planta* **147**, 320.

Whatley, M. H., Margot, J. B., Schell, J., Lippincott, B. B., and Lippincott, J. A. (1978). *J. Gen. Microbiol.* **107**, 395.

Whittaker, R. G., Moss, B. A., and Appleby, C. A. (1979). *Biochem. Biophys. Res. Commun.* **89**, 552.

Williams, L. F. and Lynch, D. L. (1954). *Agron. J.* **46**, 28.

Willmitzer, L., De Beuckeleer, M., Lemmers, M., Van Montagu, M., and Schell, J.

(1980). *Nature, Lond.* **287,** 359.

Wipf, L. and Cooper, D. C. (1940). *Am. J. Bot.* **27,** 821.

Wittenberg, J. B., Appleby, C. A., and Wittenberg, B. A. (1972). *J. biol. Chem.* **247,** 527.

Yadav, N. S., Postle, K., Saïki, R. K., Thomashow, M. F., and Chilton, M. D. (1980). *Nature, Lond.* **287,** 458.

Yatazawa, M. and Yoshida, S. (1979). *Physiol. Plant* **45,** 293.

Index